ACVP & ASVCP Concurrent Annual Meeting 2016

Annual Meeting was combined between the American College of Veterinary Pathologists (ACVP) and American Society for Veterinary Clinical Pathology (ASVCP)

New Orleans, Louisiana, USA
3 - 7 December 2016

ISBN: 978-1-5108-5328-7

Printed from e-media with permission by:

Curran Associates, Inc.
57 Morehouse Lane
Red Hook, NY 12571

Some format issues inherent in the e-media version may also appear in this print version.

Printed by Curran Associates, Inc. (2018)

For permission requests, please contact American College of Veterinary Pathologists (ACVP) at the address below.

American College of Veterinary Pathologists (ACVP)
2424 American Lane
Madison, WI 53704
USA

Phone: (608) 443-2466
Fax: (608) 443-2478

info@acvp.org

Additional copies of this publication are available from:

Curran Associates, Inc.
57 Morehouse Lane
Red Hook, NY 12571 USA
Phone: 845-758-0400
Fax: 845-758-2633
Email: curran@proceedings.com
Web: www.proceedings.com

TABLE OF CONTENTS

CLINICAL PATHOLOGY FOCUSED SCIENTIFIC SESSION I

CLINICAL PATHOLOGY FOCUSED SESSION II

CLINICAL PATHOLOGY POSTERS

DIAGNOSTIC PATHOLOGY FOCUSED SCIENTIFIC SESSION I

DIAGNOSTIC PATHOLOGY FOCUSED SCIENTIFIC SESSION II

DIAGNOSTIC PATHOLOGY POSTERS

EDUCATION FOCUSES SCIENTIFIC SESSION

EDUCATION FOCUSED POSTERS

EXPERIMENTAL DISEASE POSTERS

INDUSTRIAL AND TOXICOLOGIC PATHOLOGY FOCUSED SCIENTIFIC SESSION I

INDUSTRIAL AND TOXICOLOGIC PATHOLOGY FOCUSED SCIENTIFIC SESSION II

INDUSTRIAL AND TOXICOLOGIC PATHOLOGY POSTERS

NATURAL DISEASE FOCUSED SCIENTIFIC SESSION I

NATURAL DISEASE FOCUSED SCIENTIFIC SESSION II

NATURAL DISEASE POSTERS

VETERINARY STUDENT POSTERS

ACVP LATE-BREAKING POSTER SESSION

Clinical Pathology Focused Scientific Session I

December 4, 2016
1:30 PM – 1:45 PM

FELINE MATURE ERYTHROCYTE MIRNOME; INSIGHTS INTO THE RED BLOOD CELL BIOLOGY

Pierre L Deshuillers

Background: Human and mouse erythrocytes contain abundant micro RNAs (miRs). However, miRs' role in the mature red blood cell is unknown. In the bone marrow, miRs are essential for differentiation, maturation and fine-tuning of the erythroid response to stressors such as oxidative stress. MiRs dysregulation makes miRs promising disease biomarkers and therapeutic targets, resulting in a rapidly expanding field of research. So far, feline miR profiles have not been well characterized.

Objectives: The goal of this study was to document and sequence miRs in feline mature erythrocytes as a baseline for further studies of erythrocyte biology.

Methods: A method for mature erythrocyte purification involving centrifugation, leukoreduction and immunodepletion was developed and used on 8 whole blood specimens from clinically healthy cats. MiRs were subsequently isolated and sequenced by high-throughput RNA-Seq technique. Sequencing data was analyzed with MiRDeep2 to assemble the mature erythrocyte miRNome.

Results: Mature erythrocytes were successfully isolated and the subsequent miR extraction yielded good quality specimens for sequencing. Sequencing and bioinformatics analysis revealed an abundant miR population in the erythrocytes (230-260 sequences). The most abundant sequences are involved in the regulation of erythropoiesis (miR-486, miR-191) and antioxidant defenses (miR-451, miR-144) in people and mice.

Conclusion: Further investigations on the roles that miRs play in erythropoiesis and cell-to-cell communications in feline mature erythrocytes are currently underway.

December 4, 2016
2:45 PM – 3:00 PM

BIOMARKER DEVELOPMENT FOR HYPERTROPHIC OSTEODYSTROPHY IN DOGS

Noa Safra, Peta L Hitchens, Emanual Maverakis, Anupam Mitra, Courtney Korff, Eric Johnson, Amir Kol, Michael J Bannsch, Niels C Pedersen, Danika L Bannasch

Background: Hypertrophic osteodystrophy (HOD) is a developmental inflammatory disease in dogs with episodic recurrences and an unexplained etiology.

Objective: Identify biomarkers that may have diagnostic, prognostic or therapeutic implications HOD.

Methods: A prospective case control study was performed in 26 affected and 102 unaffected dogs. A diagnosis of HOD was based on signalment, clinical signs of systemic inflammation, characteristic radiographs, and response to treatment. Dogs in remission were adult dogs no longer experiencing episodes of HOD and puppies free of symptoms after the completion of treatment. Candidate biomarkers representing pathways of innate and adaptive immunity were measured in serum of affected dogs and compared with unaffected controls. Wilcoxon rank-sum test was used to compare mean ranks and to evaluate correlations with clinical signs.

Results: Significantly higher serum levels were observed for innate immunity markers IL-6, TNF, GM-CSF, IL-1beta, IL-18, CXCL10, IL-10 in HOD dogs compared to unaffected dogs. Moreover, the key adaptive immunity cytokines, IL-2 and 1L-7, were not significantly different. Unexpectedly, dogs in remission were not different from dogs during active disease. HOD clinical signs associated with tested markers were pyrexia (IL-6), diarrhea (IL-8), vomiting (IL-1beta), pustules (IL-8), and relapsing episodes (IL-6, GM-CSF).

Conclusions: Results from this study suggest that HOD etiology involves aberrant innate immunity. Furthermore, dogs in remission have increased serum levels of markers, providing a possible explanation for HOD predisposition. Candidate biomarkers should be further investigated for their diagnostic and prognostic use and possibly as targets for specific therapy.

December 4, 2016
3:30 PM – 3:45 PM

GREY EOSINOPHILS IN A MINIATURE SCHNAUZER WITH A POORLY DIFFERENTIATED MAST CELL TUMOR

Katherine L Irvine, Lauren C Smith, Rose E Raskin, Kristen R Friedrichs

Background: A 10-year-old female spayed miniature schnauzer was presented for investigation of an intra-nasal mass diagnosed by histopathologic examination as an undifferentiated round cell neoplasm with an infiltrate of segmented leukocytes, interpreted as neutrophilic inflammation. During subsequent monitoring over several months, the peripheral leukocyte count was repeatedly normal to slightly increased with low numbers of toxic neutrophils. The most recent analysis revealed a significant leukocytosis of 66,100 cells/μL, of which 39,700 cells/μL had variably mature, lobed nuclei and grey cytoplasm with clear vacuoles, resembling grey eosinophils.

Objective: Our objective was to identify grey eosinophils definitively in a miniature schnauzer and highlight their misidentification as toxic neutrophils. We also sought to identify the neoplasm as a poorly differentiated mast cell tumor with an infiltrate of grey eosinophils.

Methods: Peripheral blood smears from the patient and a canine control were stained for alkaline phosphatase (AP), peroxidase, and esterase activities, and with Luxol fast blue (LFB). Histopathologic sections of the nasal mass were stained with LFB and immunohistochemically for tryptase.

Results: On blood smears, the cytoplasm of the suspected grey eosinophils stained for AP and the granules stained with LFB, confirming an eosinophilic lineage. Peroxidase staining was weak and esterase staining absent. On histopathologic sections from the nasal mass, the segmented leukocytes contained LFB-staining granules, indicating an eosinophilic infiltrate. Neoplastic cells expressed tryptase, confirming their mast cell lineage.

Conclusions: Grey eosinophils may be under-recognized and interpreted incorrectly as toxic neutrophils. This report expands the canine breeds in which these eosinophils have been identified.

December 4, 2016
3:45 PM – 4:00 PM

EFFECT OF TIME AND STORAGE ON TOXIC CHANGE IN CANINE NEUTROPHILS

Liza Bau-Gaudreault, Christian Bédard, Carolyn N. Grimes

Background: The presence of toxic change in neutrophils is a frequently used clinical indicator of inflammation in dogs.

Objective: To evaluate the effect of time and storage conditions on toxic change in canine neutrophils.

Methods: 150uL of blood (in EDTA) from eight dogs with no toxic change on fresh blood smears (T0) was stored at room temperature (ROOM), in a box with a frozen icepack (BOX), and at 4C (FRIDGE). For each storage condition, blood smears were prepared 2 (T2), 4 (T4), 8 (T8), and 24 (T24) hours post blood-draw. Smears were randomized and each smear was evaluated for the presence of toxic change.

Results: A statistically significant effect of time and storage conditions on the presence of toxic change was observed. When compared with T0, the number of neutrophils containing toxic change was significantly higher at T8 and T24 for ROOM ($p<0.0001$) and BOX ($p<0.0001$) samples and at T24 for FRIDGE samples ($p<0.0001$). Additionally, smears were falsely classified as having 1+ toxic change in 0/8 (T2), 1/8 (T4), 3/8 (T8), and 8/8 (T24) for ROOM samples; 0/8 (T2 and T4), 2/8 (T8), and 5/7 (T24) smears for BOX samples; and 0/8 (T2, T4, and T8), and 2/8 (T24) for FRIDGE samples.

Conclusion: Smears may be falsely classified as having toxic change as early as 4 hours post blood-draw in samples stored at room temperature, 8 hours when stored on ice, and 24 hours when stored at 4C. These effects should be considered when interpreting toxic change in canine neutrophils.

December 4, 2016
4:00 PM – 4:15 PM

COMPARISON OF IMMUNOCYTOCHEMICAL STAINING FOR CD3, CD20, CYTOKERATIN AND VIMENTIN IN UNSTAINED AND PREVIOUSLY WRIGHT-STAINED CYTOLOGIC SMEARS OF CANINE TUMORS.

Daniela Hernandez, Heather Priest, Tracy Stokol

Background: Immunocytochemical (ICC) staining of cytologic specimens can yield a definitive diagnosis of tumor type but is limited by availability of adequately cellular and diagnostic smears. We reasoned that ICC staining of previously Wright's stained smears (prestained) might solve this issue. **Objective:** To compare ICC staining on unstained and prestained smears from certain canine neoplasms. **Methods:** Cytologically diagnosed cases of lymphoma (5 T, 5 B), carcinoma (5), and sarcoma (5) with at least 4 available smears were prospectively included. After acetone fixation and antigen retrieval in EDTA, smears were stained with antibodies against CD3 and CD20 (lymphoma), cytokeratin (carcinoma), and vimentin (sarcoma) using standard protocols. Outcome measures were categorical grades of percentage positive cells (0 to 4) and staining intensity (SI, 0-3). Median grades were compared with a Wilcoxon-sign test. **Results:** Median grades for percentage positive cells and SI were not significantly different between unstained and prestained smears. The percentage positive cells was equal for cytokeratin and vimentin in unstained and prestained smears but lower in prestained smears (3/5 each) for CD20 and CD3. For all markers except cytokeratin, SI was lower in prestained smears (2/5 CD20, 1/5 CD3, 4/5 vimentin). In one T cell lymphoma, CD3 immunostaining was absent in the prestained smear. **Conclusions**: ICC staining of unstained and prestained smears is comparable for CD20, cytokeratin and vimentin, with mild decreases in percentage positive cells and SI in prestained smears. In contrast, CD3 immunostaining of prestained smears showed inconsistent results that can result in a false negative diagnosis.

December 4, 2016
4:15 PM – 4:30 PM

FELINE LYMPHOCYTOSIS: IMMUNOPHENOTYPIC CHARACTERIZATION AND CLINICAL OUTCOME

Emily D. Rout, Janna A. Yoshimoto, Julia D. Labadie, Anne C. Avery, Paul R. Avery

Background: Lymphocytosis is a more frequent hematologic abnormality in cats than dogs, but few studies have characterized the lymphocyte population.

Objective: Evaluate the immunophenotype, clonality and clinical outcome in cats with lymphocytosis.

Methods: We examined the flow cytometry immunophenotype of 309 cats >1 year of age with a persistent lymphocytosis (>6,000 cells/uL). PCR for antigen receptor rearrangement was performed on a subset of cases. Clinical follow-up was obtained from 95 cats.

Results: T lymphocytes were most frequently expanded, with CD4+T-lymphocytosis being the most common immunophenotype (36% of cases), and fewer CD5+/CD4-/CD8- (7%) and CD8+ (5%) cases; 82% of T-lymphocytosis cases exhibited a clonal T cell population, indicating these are cases of T cell lymphoproliferative disease; 16% of cases had a monomorphic expansion of B cells, but only 7% of these were clonal; 11% of cases had aberrant immunophenotypes, including very low CD5 expression only, lineage negative cells, or CD4+/CD8+cells. Unlike dogs, cats often had multiple lymphocyte subsets expanded, with 24% of cases having elevations in 2 or more populations of B cells, CD4+T cells or CD8+T cells. The median survival was longest in cats with multiple subset expansions (986 days), followed by CD4+T-lymphocytosis (752 days), CD5+/CD4-/CD8-T-lymphocytosis (382 days), and aberrant expression (22 days); B-lymphocytosis median survival was not achieved.

Conclusions: This study corroborates previous findings that CD4+T cell is the most common form of feline chronic lymphocytic leukemia and has an indolent clinical course, and suggests cats with multiple subset expansions or B-lymphocytosis may have reactive disease.

Clinical Pathology Focused Scientific Session II

December 6, 2016
8:00 AM – 8:15 AM

COMPARISON OF THE SEDIVUE DX(TM) ANALYZER WITH MANUAL MICROSCOPY FOR DETECTION OF RED BLOOD CELLS AND WHITE BLOOD CELLS IN URINE SEDIMENTS.

Annalisa M. Hernandez, Graham E.A. Bilbrough, Dennis B. DeNicola, Celine Myrick, Alexandra N. Myers, Mary B. Nabity

Background: Microscopic evaluation of urine sediments is often not performed in veterinary clinics due to lack of time or confidence. The IDEXX SediVue Dx(TM) Urine Sediment Analyzer (SDx) is a recently introduced instrument that performs automated urine sediment analysis.

Objective: To determine sensitivity and specificity of the SDx for detecting clinically significant numbers of white blood cells (WBCs) and red blood cells (RBCs) using manual microscopy as the reference method.

Methods: 392 urine samples (81% canine, 19% feline) were evaluated for WBCs and RBCs using the SDx (1.0.0) and manual microscopy. For SDx analysis, 170uL of well-mixed, uncentrifuged urine was pipetted into a disposable cartridge. Following brief centrifugation, a built-in camera-microscope system captured 70 high magnification images, and image recognition software was employed to identify and quantify formed elements. For manual microscopy, urine was centrifuged using a KOVA system to obtain the sediment, and each element was quantified by averaging the number of cells in 10 high power fields (HPF, 40x). For clinical significance, a threshold of >5 cells/HPF was used for both the SDx and manual microscopy.

Results: On manual microscopy, 27% and 35% of samples contained >5 WBCs/HPF or >5 RBCs/HPF, respectively. The SDx showed moderate sensitivity and high specificity for detecting ≥5 WBCs or RBCs/HPF (sensitivity 74% and 85%; specificity 94% and 95%, respectively) compared with manual microscopy.

Conclusion: The IDEXX SediVue Dx(TM) Urine Sediment Analyzer accurately detects clinically significant numbers of WBCs and RBCs in urine sediments. Further evaluation of other formed elements is underway.

December 6, 2016
8:15 AM – 8:30 AM

VALIDATION OF TWO POINT-OF-CARE METERS FOR MEASURING TRIGLYCERIDE CONCENTRATION IN CHICKEN WHOLE BLOOD AND PLASMA

Katherine L Irvine, Christoph Mans, Kristen R Friedrichs

Background: Disorders of the avian reproductive tract are easy to diagnose, yet monitoring their resolution is problematic. Sex hormones are the best reflection of reproductive status but sample volumes and lack of reference intervals limit their utility clinically. Blood triglyceride concentration rises markedly during sustained estrogen secretion and may be used to monitor avian reproductive status. Portable meters for measuring human blood triglycerides use minimal sample volumes but none have been validated in birds.

Objective: The objectives were 1) to assess the precision and accuracy of two portable meters for measuring blood triglycerides in chickens, and 2) to perform method comparison using a reference analyzer.

Methods: Pooled whole blood and plasma from chickens was used for repeatability, recovery, and dilution experiments. For method comparison, triglyceride concentration was measured in whole blood and plasma from 42 chickens using both meters and in plasma by a reference machine; these were compared using correlation, Deming regression, and Bland-Altman analysis.

Results: Within-run precision was 2.51-11.51% and between-run precision was 3.09-12.23%. The meters performed well in dilution studies, whereas one meter showed limited recovery. Correlations between the meters and reference analyzer were good to excellent (0.86-0.98). Bias was similar between whole blood and plasma for each meter, yet markedly different between meters. The total observed error was consequently within our total allowable error of 20% for one meter but not the other.

Conclusions: The meters have acceptable performance for measuring triglyceride concentration in chicken whole blood and plasma but one meter requires meter-specific reference intervals.

December 6, 2016
8:30 AM – 8:45 AM

VERIFICATION OF THE ELEMENT POC BLOOD GAS INSTRUMENT FOR USE WITH CAMELID VENOUS BLOOD, AND DETERMINATION OF REFERENCE INTERVALS

Lisa Viesselmann, Ricardo Videla, Bente Flatland

Background: Heska Element POC (EPOC) is a blood gas instrument intended for use with canine, feline, and equine whole blood; no verification for use with camelid specimens has been reported.

Objectives: Investigate EPOC analytical performance using camelid specimens and commercial quality control material (QCM) and establish EPOC camelid reference intervals (RI).

Methods: Camelid blood (n=113) was measured using the EPOC (for pH, pCO2, pO2, HCO3, base excess, SO2, sodium, potassium, ionized calcium, chloride, TCO2, anion gap, hematocrit, hemoglobin, glucose, lactate, and creatinine); plasma was measured using a Roche Cobas c501 and Integra 400 Plus (for sodium, potassium, chloride, TCO2, anion gap, glucose, and creatinine). Method comparison data were analyzed using Pearson's correlation, Passing-Bablok regression, and Bland-Altman plots. EPOC precision was evaluated using QCM and camelid blood.

Results: For all measurands except anion gap and chloride, correlation of EPOC vs. each Cobas instrument was $r>0.8$. Mild constant bias was present for chloride, glucose, TCO2, anion gap, and creatinine, and mild proportional bias for chloride, glucose, and anion gap. Except pO2 and pCO2, EPOC precision (QCM and blood) ranged from CV <1% to 6.3%. EPOC observed total error (QCM data) was below ASVCP-recommended allowable total error. Separate blood gas RI were needed for arterial vs. venous/mixed specimens, and RI were partitioned by species for pO2, SO2, sodium, potassium, calcium, chloride, lactate, and creatinine.

Conclusions: The EPOC shows good performance with camelid blood. Lack of complete agreement with automated chemistry analyzers highlights the importance of interpreting patient data using instrument-specific RI.

December 6, 2016
8:45 AM – 9:00 AM

THE EFFECT OF THREE RESUSCITATIVE FLUID THERAPIES ON HEMOSTASIS, MEASURED BY ROTATIONAL THROMBOELASTOMETRY, IN DOGS

Sunita Seshia, Matthew C Gaunt, Beverly A Kidney, Marion L Jackson

Background: There is evidence that administration of intravenous fluids impairs hemostasis. Rotational thromboelastometry (ROTEM) may provide more sensitive measures of fluid effects on hemostasis than traditional coagulation tests.

Objective: To determine if resuscitative fluid therapy affects hemostasis, as measured by ROTEM.

Methods: A randomized crossover design of six healthy dogs administered intravenous fluid treatments of colloid, crystalloid, and hypertonic saline at therapeutic doses. Primary ROTEM values reported for INTEM and EXTEM included coagulation time (CT), clot formation time (CFT), maximum clot firmness (MCF), and alpha-angle. Reported FIBTEM values included CT and MCF. Secondary hemostatic measurements included PT, APTT, and fibrinogen. Samples were collected at baseline, 1, 4, and 8 hours post-treatment. Univariate ANOVA on transformed data evaluated differences between groups and within group effects. When significant differences were noted ($P = 0.003$), Tukey test was performed.

Results: A statistically significant interaction between individual dog and treatment was noted with INTEM and EXTEM variables CT, CFT, MCF, alpha-angle; FIBTEM variables CT and MCF; and PT, APTT, and fibrinogen. INTEM MCF showed significant decrease from baseline for all treatments. EXTEM MCF was significantly lower in crystalloid and hypertonic saline treated dogs. PT was significantly lower in colloid treated dogs. All treatments resulted in a significant increase in PT at 1 hour. Fibrinogen concentration was significantly different between all treatments.

Conclusions: Clinically relevant doses of resuscitative fluids result in decreased clot firmness as measured by TEM, and affect hemostatic parameters in healthy dogs. There is also a significant individual dog response to treatment.

December 6, 2016
9:00 AM – 9:15 AM

RETROSPECTIVE EVALUATION OF SERUM IRON, RDW, AND NUCLEATED RED BLOOD CELLS IN DOGS WITH ACUTE TRAUMA: 129 CASES (2009-2015)

Eric J. Fish, Sonya Hansen, Elizabeth A. Spangler, Philippe R. Gaillard, Shirley Fan, Lenore M. Bacek

Background: Canine patients with acute trauma present with a broad spectrum of systemic injuries and high risk for mortality. Hematologic variables such as low serum iron, the presence of nucleated red blood cells (nRBCs) and increased red cell distribution width (RDW) have been associated with poor prognosis in human and veterinary patients with critical illness, but have not been evaluated in dogs with acute trauma.

Objective: The goal of this study was to compare serum iron, nRBCs, and RDW between surviving and non-surviving dogs with acute trauma.

Methods: CBC, serum biochemistry, and patient data were retrospectively reviewed from 129 dogs with acute trauma (109 survivors and 20 non-survivors) presenting within 24 hours of injury. Univariate, multiple logistic regression, diagnostic performance, and ROC curve statistical analyses were performed.

Results: The presence of nRBCs in circulation was significantly associated with non-survival (p=0.030). Mean serum iron was significantly lower in non-survivors relative to surviving dogs (p=0.026). RDW was not statistically different. Normal or high serum iron (>76 ug/dL) had 100% specificity and 100% positive predictive value (PPV) for survival (95% confidence interval 87-100%). A multiple logistic regression model including age, type of injury, nRBCS, and serum iron had an ROC-AUC of 0.81 for discriminating survivors from non-survivors.

Conclusions: These results suggest that the presence of nRBCs and low serum iron are associated with mortality in patients with acute trauma. Absence of hypoferremia was highly associated with a favorable prognosis in this patient population. These parameters may warrant inclusion in trauma scoring systems.

December 6, 2016
9:15 AM – 9:30 AM

ANALYTICAL VALIDATION OF A NEW POINT-OF-CARE ASSAY FOR QUANTIFICATION OF SERUM AMYLOID A IN HORSES

Diana Schwartz, Nicola Pusterla, Mary M Christopher

Background: Serum amyloid A (SAA) is a major acute phase protein in horses that dramatically increases in response to inflammatory stimuli. A new point-of-care test for SAA (Stablelab, Epona Biotech Ltd., Ireland) is available, but studies evaluating its analytical accuracy are lacking.

Methods: Analytical validation of the Stablelab test was done in accordance with ASVCP guidelines using residual equine serum/plasma and whole blood samples from the Clinical Pathology Laboratory at UC Davis. A turbidimetric immunoassay (TIA, University of Miami) was used as the reference method for comparison. We also determined linearity, imprecision (intra- and inter-assay), and effect of sample type (serum/plasma vs whole blood) and hematocrit.

Results: The Stablelab test was linear between 0-1500 µg/ml (r=0.990). There was moderate agreement between the Stablelab test and TIA within the linear range (r=0.897), with slight constant bias (mean difference = 84.6±53.6 µg/ml). Intra-assay CVs were 13%, 18%, and 15% at high (782 µg/ml), intermediate (116 µg/ml), and low (64 µg/ml) concentrations. Inter-assay (inter-batch) CVs were 45%, 14%, and 15% at high (1372 µg/ml), intermediate (140 µg/ml), and low (56 µg/ml) concentrations. SAA concentrations in whole blood and serum/plasma were positively correlated (r=0.908), with proportional negative bias at >1000 µg/ml; results were not affected by hematocrit (r=0.006).

Conclusion: The Stablelab test has acceptable accuracy and repeatability in horses with SAA concentrations of up to 1000-1500 µg/ml in serum/plasma and whole blood. Lower reproducibility and accuracy at high SAA concentrations may affect serial measurements, for which use of the same batch is recommended.

December 6, 2016
9:30 AM – 9:45 AM

APPLICATION AND UTILITY OF A NEAR-BEDSIDE MULTIPLEX IMMUNOCYTOCHEMISTRY PROTOCOL FOR SIMULTANEOUS VIMENTIN AND CYTOKERATIN STAINING

Kelly S. Santangelo, Sarah Leavell, Russell R. Moore, Amy L. MacNeill

Background: While cytopathology is efficient and non-invasive, diagnostic conundrums can exist due to the morphologic overlap of certain cell types. Ancillary diagnostics, particularly immunocytochemistry (ICC), can help overcome the limitations of conventional cytology and allow timely definitive diagnoses.

Objective: Our primary aim was to evaluate the clinical utility of a validated simultaneous vimentin and cytokeratin multiplex ICC protocol using prospectively collected cavity fluid samples from clinical cases.

Methods: A fluorescent multiplex ICC protocol for vimentin (SP20), cytokeratin (AE1/AE3 + 8/18) and nuclear staining (4',6-diamidino-2-phenylindole, DAPI) was validated for use in canine and feline samples. Recommended quality control practices, including positive and negative sample controls and corresponding assay controls were used. Slides were visualized using a fluorescent microscope and analyzed with associated software. Negative antibody isotype controls allowed calibration to background fluorescence. Cytospin preparations or direct smears of clinical cases were made on positively charged slides, evaluated in real time using the ICC protocol, and then compared to subsequent histologic diagnoses using kappa statistics.

Results: Thirty-five canine and feline samples were assessed, which included an array of primarily epithelial, mesenchymal, and mesothelial lesions. Cohen's kappa coefficient was excellent for the novel immunoassay compared to histologic findings and better than cytologic evaluation, only. For some cases, multiplex ICC was the only antemortem test performed throughout the clinical work-up to provide definitive evidence of neoplasia.

Conclusions: This vimentin and cytokeratin multiplex ICC protocol is a quick and accurate diagnostic tool that adds value to traditional fluid cytopathology evaluation.

December 6, 2016
10:30 AM – 10:45 AM

COMPARISON OF CYTOLOGY AND QUANTITATIVE BIOCHEMICAL ANALYSIS OF TRIGLYCERIDE CONCENTRATION IN LIVER SAMPLES FROM DAIRY COWS

Erica Behling-Kelly, Michael M. Fry, Benjamin Yao, Daryl V. Nydam, Franco L. Yeal-Lapes, Jessica A. McArt, Sabine Mann

Background: Hepatic lipidosis (HL) in dairy cattle is of major clinical and economic importance. The accuracy of cytology for diagnosing bovine HL is understudied.

Objective: Compare cytologic evaluation for visible lipid vacuolation and triglyceride concentration (mg/mg wet weight) ([TG]) in paired dairy cow liver samples.

Methods: Liver biopsies from Holstein cows enrolled in two separate dietary intervention trials were apportioned for solvent extraction of lipids (Folch method) and impression smears for cytologic evaluation. [TG] of lipid extracts was measured via the Hantzsch condensation method and comparison to a Triolein standard curve. HL was evaluated cytologically using an ordinal scale (0-4) based on estimated percentage of hepatocytes with discrete cytoplasmic vacuolation and estimated mean percentage of cytoplasm occupied by vacuoles on Wright-Giemsa stained imprints.

Results: Data from 51 paired samples showed strong, significant correlation ($r = 0.89$, $P < 0.0001$) between cytology score and [TG]. Using a diagnostic threshold of [TG] ≥ 4% for clinically significant HL, diagnostic sensitivity and specificity of cytology score ≥ 2 were 100% (28 of 28) and 96% (22 of 23), respectively. Using a threshold of [TG] ≥ 15% for severe HL, sensitivity and specificity of a cytology score ≥ 3 were 100% (9 of 9) and 93% (39 of 42), respectively.

Conclusion: Results of this study support the hypothesis that cytology is an accurate means of predicting hepatic [TG] and thus diagnosing HL in dairy cows. Cytology has the potential to be a valuable part of the diagnostic repertoire of practicing dairy veterinarians.

December 6, 2016
10:45 AM – 11:30 AM
Dynamics of Peripheral and Mucosal CD4+ T Cells in Simian Immunodeficiency Virus Infection and AIDS
Ronald Veazey

December 6, 2016
11:30 AM – 11:45 AM
DIETARY EXOGENOUS THYROTOXICOSIS IN DOGS
Jennifer Jones, Olgica Ceric, David Rotstein, Renate Reimschuessel, Sarah Nemser, Jackie Queen, Lee Anne Palmer, Lauren Carey, Enrique Yanes Santos, Robert Wilson

FDA Center for Veterinary Medicine received two complaints of pet illness, involving three dogs, from two different households. The cases included two spayed female Chihuahuas, ages 4 and 6 years, and a 10-year-old spayed female Jack Russell Terrier. The dogs exhibited clinical signs of excessive panting, polyuria, and polydipsia. Clinical pathology findings in all dogs included low thyroid stimulating hormone (TSH) and elevated total and free 3,5,3'5'-tetraiodothyronine (T4) that were 1.8 to 3.6 times greater than the upper limit of the reference range. Scintigraphy from one dog in each household was negative for ectopic thyroid gland. While the household cases were temporally and spatially separate, all three dogs had consumed the same type of pet treat. The treat was discontinued in both households, and the clinical signs improved. A normal total T4 was subsequently documented in one case. Treats were analyzed for iodine species 3-monoidodotyrosine (MIT), iodide, 3,5-diiodotyroisine (DIT), 3,5 –

diiodothyronine, 3,5,3'-triiodothyroinine (T3), and T4 using HPLC with inductively coupled plasma mass spectrometry detection (LC-ICP-MS). One of five different product samples had a total of 158 µg iodine/g and was found to have 67.0 µg/g MIT, 119 µg/g DIT, T3<MQL of 12 µg/g, and 52.0 µg/g T4. Microscopic examination of HE-stained treats along with thyroglobulin and calcitonin immunohistochemistry did not indicate the presence of intact thyroid gland follicles, C-cells, or colloid. The suspected source of thyroid gland tissue was from the meat (gullet). Exogenous thyrotoxicosis has been sporadically reported in the literature in both dogs and people.

December 6, 2016
11:45 AM – 12:00 PM

CLINICOPATHOLOGICAL FINDINGS OF FANCONI SYNDROME POSITIVE DOGS FOLLOWING JERKY PET TREAT INGESTION

Jennifer Jones, Renate Reimschuessel, Olgica Ceric, Jake Guag, David Rotstein, Lauren Carey, Lee Anne Palmer, Rachel Han, Urs Giger

Background: Approximately 6-7% of over 5,000 jerky pet treat (JPT) reports FDA received since 2007 indicate canine Fanconi syndrome (FS) – a rare renal tubular defect.

Methods: Since 2012, Vet-LIRN investigated nearly 500 FDA reports for various illness types in dogs after eating JPT. Case investigations help FDA elucidate potential relationships between ingesting JPT and developing FS. Vet-LIRN collects medical records, confirms JPT exposure, rules out known causes of FS, and tests urine samples from exposed dogs with various clinical signs.

Results: As of March 16, 2016, Vet-LIRN confirmed FS in 226 dogs with the urine Fanconi panel at PennGen's Metabolic Genetics laboratory at the University of Pennsylvania. FS-positive dogs had generalized severe amino aciduria, lactic aciduria, ketonuria, and glucosuria. For dogs testing PennGen FS positive, over 55% were female, and over 80% weighed less than 30 pounds. The top pure-bred and mixed breeds testing FS positive were Maltese, Poodle, Shih Tzu, Chihuahua, and Yorkshire Terrier. FS-positive dogs typically had decreased appetite, polydipsia, polyuria, and vomiting, but 7.5% of FS-positive dogs were clinically asymptomatic. Owner interviews confirmed JPT exposure for asymptomatic dogs. FS-positive dogs most commonly had glucosuria, proteinuria, hematuria, hypokalemia, minimally concentrated urine, and azotemia (BUN and/or creatinine).

Conclusions: The FS is often reversible for many afflicted dogs after withdrawing JPT and appropriate treatment. Vet-LIRN continues to investigate any potential causal relationship between JPT ingestion and FS development in order to help FDA and veterinarians better identify and understand the clinical picture of dogs with potential JPT induced FS.

Clinical Pathology Posters

C-01: QUALIFICATION OF A NOVEL MASS SPECTROMETRY ASSAY FOR THE DETERMINATION OF THE SOLUBLE TRANSFERRIN RECEPTOR (STFR) IN SERUM OF CYNOMOLGUS MONKEYS

Michael Winter, Tom Dunkley, Annette Koerner, Juergen Funk

Introduction

Uptake of iron-transferrin complexes into cells is mediated via the cell surface expressed transferrin receptor-1 (CD71). Proteolytic cleavage of the extracellular receptor domain generates the formation of a truncated soluble transferrin receptor (sTfR). Increased serum sTfR levels correlate therefore with an expanded erythropoietic cell mass and/or upregulated cellular expression levels. The determination sTfR has become a standard test for the differentiation of functional iron deficiency from iron depleting anemia in humans.

Materials and methods

Serum samples taken over 9-months from male and female naive cynomolgus monkeys (n=16) were used for the qualification of a modified human SISCAPA (Stable Isotope Standards and Capture by Anti-Peptide Antibodies) sTfR mass spectrometry methodology (MSM) assay. CBC was measured on an ADVIA 120 and biomarkers of iron status were determined by commercial test kits.

Results and discussion

The sTfR MSM assay performed well with no detectable interferences and a low inter/intra assay CV of

Conclusion

The determination of sTfR in serum of cynomolgus monkeys provides a translatable diagnostic test for the functional characterization of anemia or iron metabolism in this species. This biomarker is diagnostically relevant in complex pathologies with coexisting inflammation or systemic toxicity, where ferritin concentrations do not correlate with iron status.

C-02: SORBITOL DEHYDROGENASE DILUTION EXPERIMENT IN CANINE, SWINE AND NON-HUMAN PRIMATE SERUM

Laura C Cregar, Adam D Aulbach

BACKGROUND

Sorbitol dehydrogenase (SDH) is used as a liver-specific indicator of hepatocellular injury. Anecdotally, sample dilution has yielded aberrant results when measuring SDH activity in some species.

OBJECTIVE

To determine the effect of sample dilution on SDH activity in pigs, dogs and non-human primates (NHP).

METHODS

Serum SDH activity was measured in 10 dogs, pigs and NHPs (SDH reagent - Sekisui Diagnostics; Olympus AU2700 analyzer). Neat serum samples were diluted 1:2, 1:5, 1:10, 1:20 and 1:50 with Nerl water or 0.9% saline. Measured values were compared to back-calculated values of prior dilution levels and % CV calculated.

RESULTS

Baseline SDH activity was within the dynamic range of the assay (0.5 – 50 U/L) for all animals. SDH activity was linear within the dynamic range of the assay in all canines when diluted with water or saline. Consistent over-recovery of SDH activity was present in swine serum diluted 1:2 and 1:5 with water, and 1:2 with saline, but became linear at greater dilutions (≥1:10 with water or ≥1:5 with saline). Over-recovery of SDH activity was also present in NHP serum diluted 1:2 with water or saline, but became linear at greater dilutions (≥1:5 with water or saline).

CONCLUSIONS

This study describes a phenomenon of a non-linear relationship between SDH activity and sample dilution in NHP and pig serum that suggests the presence of a negative interferent or other artifact that can be overcome by further sample dilution. Additional studies are needed to confirm and characterize the cause of these aberrant results.

C-03: THE RENIN-ANGIOTENSIN-ALDOSTERONE SYSTEM IN GREYHOUNDS COMPARED TO NON-GREYHOUND DOGS

M Judith Radin, Jennifer Martinez, M Cristina Iazbik, Toni Hoepf, C Guillermo Couto, Barrak M Pressler, Christine Kellogg, Jane Robertson

Background: The renin-angiotensin-aldosterone system (RAAS) plays a key role in control of blood pressure and electrolyte homeostasis. Greyhounds have higher blood pressures and higher serum sodium and chloride compared to other dog breeds and mixed breeds. The objective was to determine whether Greyhounds have alterations in their RAAS which contribute to the hemodynamic and clinicopathologic differences observed in the breed.

Methods: Clinically healthy Greyhounds and non-Greyhound blood donor dogs were sampled (n=20/group). True renal status was unknown but presumed normal.

Results: Greyhounds had significantly higher serum creatinine (p14ug/dL while only 1 non-Greyhound had an SDMA >14ug/dL. SDMA was correlated with serum creatinine (r_s=0.688, p_s=0.295, p=0.068). Greyhounds had significantly higher systolic blood pressure compared to non-Greyhounds. Plasma renin activity and angiotensin converting enzyme (ACE) activity were similar between groups (p=0.598 and p=0.772, respectively). Plasma aldosterone was significantly lower in Greyhounds versus non-Greyhounds (p=0.002), and the ratio of aldosterone to ACE activity was lower in Greyhounds (p=0.049).

Conclusion: Activation of the RAAS did not appear increased in Greyhounds. Lower aldosterone was appropriate given the higher serum sodium and might suggest that tubular effects of angiotensin II predominant over aldosterone effects.

C-04: ANALYTICAL PERFORMANCE OF ELECTROLYTES ON IDEXX CATALYST ONE (TM) CHEMISTRY ANALYZER FOR CANINE AND FELINE BLOOD SAMPLES

Brian A. Meyer, John Christian, Nicole Rosen, Graham E. Bilbrough, Denise Taddeo, Dennis B. DeNicola

Background: The IDEXX Catalyst One™ (C1) is an automated, dry slide technology, in-clinic chemistry analyzer developed for veterinary application. Verification of satisfactory performance by method comparison is necessary for clinical confidence in results. **Objective:** The purpose of this study was to perform a method comparison of electrolytes between C1 and an established reference analyzer, the Ortho Clinical Diagnostics Vitros 5,1 FS, using canine and feline blood samples. **Methods:** Samples were evaluated for sodium (n=110), potassium (n=110), and chloride (n=109). Method comparison analysis was performed using linear regression, correlation coefficients (*r*), bias, and observed total error (TE_{obs}). Electrolyte results ranging across a wide dynamic range (Na:115-181, K: 2.9-5.7, Cl:82- 142 mmol/L) were compared between the analyzers. Imprecision (CV) was assessed using low and high concentration control fluids (Vitros) by performing a Clinical and Laboratory Standards Institute-compliant 5-day precision study. **Results:** The C1 had CV values ≤2.43%, correlation coefficient values ≥0.95, bias values ≤8.66 %, and TE_{obs} between 0.44 and 11.82 for all electrolytes. When classifying electrolyte results bucketed into values below, within and above RI for the particular analyzers, there was 85%, 79% and 92% agreement for Na, K and Cl results, respectively. **Conclusion:** The $TE_{obs} > TE_{allowable}$ for some analytes suggests undesirable results per ASVCP guidelines; however, different methodologies are involved that may explain these results. The C1 had good precision, strong correlation and agreement with the Vitros for all electrolytes.

C-05: IS THE MEASUREMENT OF GLUCOSE CONCENTRATION IN EFFUSIONS A RELIABLE MARKER FOR THE DIAGNOSIS OF SEPTIC EFFUSIONS?
Abigail Guija de Arespacochaga

Background: Low glucose concentrations in effusions are considered diagnostic for septic effusions.

Methods: In 169 effusions (76 canine and 93 feline) total protein (TP), total cell count (TCC), glucose concentration (GLU) were measured and completed by cytologic evaluation. According to TP, TCC and cytologic and/or positive culture findings effusions were classified as protein poor (n=28) and protein rich (n=40) transudates, septic (n=14), non septic (n=33), FIP associated (n=13 cats) exudates and vessel/organ rupture (n=41). Additionally exudates were grouped into TCC 10-50x10^3/µL and >50x10^3/µL. Sensitivity, specificity and predictive values for a glucose concentration cutoff of 50 mg/dL were calculated.

Results: Glucose concentration was statistically significantly lower in septic exudates of both species compared to all other groups. However, between septic exudates with TCC <50x10^3/µl and non septic exudates with TCC >50x10^3/µl, no statistically significant difference could be detected. For a GLU cutoff 50x10^3/µl exudates were neoplastic.

Conclusion: Measurement of glucose concentration in effusions should always be linked to cytological examination; highly cellular effusion often associated with neoplasia can have GLU< 50mg/dl.

C-06: SPURIOUS PLATELET DOT PLOTS IN DOGS: A RETROSPECTIVE STUDY
Laetitia Piane, Marie-Lou Mayaud, Nathalie Bourges-Abella, Catherine Trumel, Anne Geffré

Introduction:

Abnormal platelet dot plots and erroneously high platelet counts have often been reported in humans and less frequently in veterinary practice.

Objectives:

The aims of this study were 1/ to evaluate the frequency of abnormal platelet dot plots in the dog, 2/ to identify diseases in conjunction with abnormal platelet dot plots, 3/ to determine if disease-specific patterns could be identified.

Materials and methods:

In this retrospective study, all canine complete blood cells counts obtained with the Sysmex XT-2000iV between 2011 and 2015 were reviewed. Cases of abnormal platelet dot plots were classified into 6 sub-groups (A to F) according to shape, density and limits.

Results:

61/4713 abnormal patterns (1.3%) have been found.

The following abnormalities have been observed: 20 lipemias, 10 babesiosis, 8 thrombocytopenias and giant platelets, 6 aged specimens collected one to three day before analysis, 5 red blood cells fragmentation, 4 leukemias and one trypanosomiasis.

For the largest group of diseases, 13/20 lipemias were classified as B, 10/10 babesiosis as C, 8/8 thrombopenia and giant platelets as A, 4/6 old blood as B, 4/5 red blood cells fragmentation as D and 3/5 white blood cells fragmentation as C.

Discussion/Conclusions:

In our study, similar diseases as in humans were identified to be responsible for abnormal platelet dot plot in dogs. The recognition of specific patterns could help clinical pathologists to establish a diagnosis and manufacturers to set flags.

C-07: CHARACTERIZATION OF FELINE COAGULOPATHIES USING THE ACL-TOP CTS ANALYZER: A RETROSPECTIVE ANALYSIS OF 103 CATS

Angela D Gwynn, Kim Little, Rae Richardson, Davis Seelig, Kevin Williams, Leslie C Sharkey

Background: Prothrombin time (PT) and activated partial thromboplastin time (APTT) are not strongly predictive of bleeding in human or veterinary medicine. However, subclinical hemostatic abnormalities are thought to be common in cats. Despite this, there is scant work in cats characterizing prolongations in PT or APTT and limited understanding as to the significance of subclinically prolonged PT and APTT values and the overall clinical utility of coagulation testing.

Objective: To characterize the clinical, clinicopathologic, and coagulation profiles (PT, APTT, fibrinogen) of ill cats using the turbidometric ACL-TOP CTS analyzer.

Methods: Retrospectively, coagulation data including PT, APTT, fibrinogen, waveform raw data (first and second derivative curves/deltas), and quantitative Clauss fibrinogen (QFA) for 103 cats were reviewed.

Results: Cats withprolonged PT and/or APTT (n=55) most commonly had hepatobiliary/pancreatic, gastrointestinal, or lymphoid disease. Similarly, cats with normal coagulation profiles (n=48) had respiratory, hepatobiliary/pancreatic, and gastrointestinal disease. Irrespective of coagulation values, all cats with hepatobiliary disease had similar GGT. However, ALT and total bilirubin values were higher and ALP was lower in cats with abnormal coagulation values. Additionally, cats with abnormal coagulation profiles had higher QFA, first derivative, and delta values, but lower second derivative values than cats with normal coagulation profiles.

Conclusions: In cats with hepatobiliary disease, those with prolonged coagulation values hadhigher ALT activity than those with normal coagulation values. Future studies examining the utility of feline coagulation testing, including characterizing novel waveform data and examining its role beyond global coagulation function are warranted.

C-08: EVALUATION OF DIFFERENT METHODS, STORAGE TEMPERATURE AND PROCESSING TIME OF PLATELET COUNT OF DOGS

Julieta Volpato, Antonise M Jaguezeski, Cláudio R.S. Mattoso, Adson Costa, Carla D.L. Cancelier, Julia B Morais, Mariângela Lovatel, Nádia C Weinert, Paulo B Todeschini, Maysa G.N. Xavier, Letícia A Yonezawa, Mere E Saito

Background: Platelet count has great importance in clinical analysis and contributes to diagnosis of several diseases, especially in patients with haemostatic alterations. However, variations between methods may become relevant for clinical diagnosis.

Objective: This study aimed to compare different methods of platelet count, and evaluate the viability of platelets in blood samples over 24 hours and the influence of storage temperature.

Methods: Blood samples were collected in EDTA tubes from 40 mixed breed healthy dogs by jugular venipuncture, and separated into two aliquots: one stored at room temperature and another refrigerated at 4ºC. Platelet counts were performed manually by using blood smear, hemocytometer, and automated count using Labtest SDH-3 vet; immediately after collection, at 30 minutes, 1, 2, 4, 6 and 24 hours after collection for both aliquots. Also, platelet morphology and aggregates were observed in blood smears.

Results: Manual platelet counts remained the same at all times after collection when refrigerated and at room temperature, were lower only at 24h after collection. Automated platelet counts were lower at 6 and 24h in samples kept at room temperature and at 24h in refrigerated samples. Also, automated platelet counts were lower than manual counts at 24h after collection at both storage temperatures. Also, there was mild platelet aggregation (7.5%) at 6h which increased (32.5%) at 24h, at both storage temperatures.

Conclusion: Platelet counts are reliable with manual and automated methods for 4 hours in all storage conditions, are stable longer when refrigerated, and are more reliable for longer with manual methods.

C-09: AN IMPROVED PURIFIED CANINE PLATELET-RICH PLASMA PROTOCOL.
Marina Mitie Monobe, Rodrigo C Silva, Camilo Bulla

Background: Platelet-rich plasma (PRP) is widely used in veterinary medical healing therapies and platelet research. Many studies have inconclusive results due to variable protocols and different degrees of blood contamination in resulting samples.

Objective: To improve the purity and activation rate of the ultra-purified PRP protocol.

Methods: Peripheral blood samples (acid citrate dextrose tubes) were drawn weekly for three weeks from six Walker hound dogs. PRP was obtained after each of six centrifugation steps at 300g for 5 min, and pooled. Ultra-purified PRP (OP) was obtained by adding 3mL of PRP over a 5mL Optiprep 1.063g/mL density barrier, and centrifuging at 350g for 15 min. Platelet recovery and WBC and RBC contamination of fractions were assessed using a hemocytometer. Platelet activation was assessed by flow cytometry (FC) using P-selectin and Annexin V markers in PRP and OP samples.

Results: This method showed reliability and reproducibility when applied to different dogs on different days, but individual ranges were observed. Mean platelet recovery was 65.44% in PRP and 71.93% in OP. PRP had high platelet purity but OP reduced RBC and WBC contamination to $^{9} \pm 0.91\text{x } 10^{9}$ for PRP and $1.03\text{x}10^{9} \pm 0.86\text{x}10^{9}$ for OP. Analyzed fractions demonstrated no platelet activation on FC.

Conclusion: This study described an ultra-purified PRP protocol with high platelet purity and no detectable platelet activation. This protocol could improve applicability in healing therapies and platelet research, but further studies focusing on clinical intervention should be conducted.

C-10: CANINE PLATELET RESPONSES TO DIFFERENT AGONISTS IN LIGHT TRANSMITTANCE AGGREGOMETRY
Sandra C. Bulla, Peres R. Badial, Kari Lunsford, Stephen Pruett, Camilo Bulla

Background: Functional platelets fully aggregate as a result of orchestrated platelet receptor-ligand interactions, involving several molecules. Light transmittance aggregometry (LTA) is the gold standard in evaluating platelet function, but many aspects of aggregation testing lack standardization. Also, there is little consistent information on how canine platelets respond to different agonists using LTA.

Objective: To evaluate the dose-dependent response of canine platelets to different agonists.

Methods: Platelet-rich plasma was obtained by centrifugation from citrated blood of healthy dogs. Using LTA, platelet aggregation was evaluated in pooled platelet-rich plasma in response to different concentrations of ADP, collagen, TXA2 receptor agonist (U46619), arachidonic acid, gamma-thrombin, and protease-activated receptors (PAR) -1 and -4 agonists. Antagonists of ADP receptors, and PAR-1 and -4 were also tested.

Results: ADP induced irreversible and dose-dependent aggregation starting at 10μM, and P2Y1 and P2Y12 receptors antagonists successfully inhibited platelet aggregation. Collagen caused a permanent shape change at 5μg/mL and aggregation of at least 90% at higher concentrations (above 20μg/ml). Although U46619 induced shape change at 100μM, aggregation was only achieved with 4mM. Arachidonic acid induced shape change at 2mM and aggregation at 8mM. Gamma-thrombin promoted shape change at 2.5nM and maximal aggregation at 15nM. PAR agonists fail to cause aggregation and PAR antagonists were not able to block gamma-thrombin aggregation.

Conclusion: Although canine platelets adequately responded to most agonists tested, the concentrations needed were much higher than what is reported for humans. PAR agonists or antagonists did not function on canine platelets in the tested concentrations.

C-11: INFLUENCE OF HIGH- AND LOW-INTENSITY EXERCISE TESTS ON HEMATOLOGICAL PARAMETERS IN AUSTRALIAN CATTLE DOGS

Julieta Volpato, Juliane S. Xavier, Ana C. Dalmina, Carla D.L. Cancelier, Adson Costa, Maysa G.N. Xavier, Estefani L.C. Pereira, Mere E. Saito, Leticia A. Yonezawa

Background: In Southern Brazil, Australian Cattle dogs are very commonly used as herding dogs in animal livestock, which requires great physical effort for working. However, little is known about hematological changes associated with exercise in this breed.

Objective: This study was undertaken to evaluate changes in hematological parameters of Australian Cattle dogs submitted to high- and low-intensity exercise tests on a treadmill.

Methods: Seven Australian Cattle dogs underwent the high-intensity test, which consisted of 5 min walking at 3.5 km/h, 5 min trotting at 7.0 km/h and 5 min walking at 3.5 km/h. Seven days later, the dogs underwent the low-intensity test, which consisted of 40 min walking at 3.5 km/h. Both exercise tests were performed on a motorized treadmill set at a 0% slope. Blood samples were obtained from a jugular vein before testing, immediately after testing, and at 15 and 30 min after testing. The following hematological parameters were determined using an automatic hematology analyzer: red blood cells, hemoglobin and white blood cells. Packed cell volume was determined using the microhematocrit method, and both mean corpuscular volume and mean corpuscular hemoglobin concentration were calculated. White blood cell differential counts were performed manually. Also, plasma protein was determined by refractometry.

Results: There was no significant difference between times or between tests for all hematological parameters.

Conclusion: In conclusion, Australian Cattle dogs seem to be excellent athletes and have excellent aerobic capacity and fitness, and/or the high- and low-intensity exercise tests did not have workload enough to induce hematological changes.

C-12: HEMATOPOIETIC NEOPLASM: CLINICAL, HEMATOLOGICAL, BIOCHEMICAL AND CYTOLOGICAL EVALUATION OF 79 DOGS FROM BRAZIL.

Marina Mitie Monobe, Rodrigo Costa da Silva, Regina Kiomi Takahira

Background: Studies associating prevalence, clinical and laboratory alterations of hematopoietic neoplasms are scarce, old and haven't included pathological progression.

Objective: Determine the prevalence and characterize the clinical, hematological, biochemical and cytological bone marrow alterations of different types of hematopoietic neoplasms.

Methods: 79 patients diagnosed with hematopoietic neoplasms by bone marrow evaluation were included in this study. Periodic laboratorial evaluation from the first clinical day until death included CBC, hepatic and renal biochemical profile of enzymes and urinalysis.

Results: The frequency of hematopoietic neoplasms was 46.84% ALL, 25.32% CLL, 29.73% lymphoma V, 8.86% AML, 2.53% MDS and 2.53% MM. Clinical signs included hepatosplenomegaly (33.33%), anorexia (31.94%), gastroenteritis (29.17%), lymphadenomegaly (20.83%) and hemostatic abnormalities (18.06%); while laboratory alterations included anemia (91.14%), thrombocytopenia (70.89%), lymphocytosis (70.89%), increased activity of hepatic enzymes (42.24%), azotemia (34.15%) and hyperglobulinemia (22.5%). ALL cases had a significantly higher occurrence and development of severe anemia and thrombocytopenia, pancytopenia, percentage of peripheral cellular atypia, medullar dysplasia and peripheral and medullar blast count, and shorter survival time (~30 days); while CLL had a higher occurrence of cyclic lymphocytic leukocytosis, discrete laboratory alterations and survival time >750 days. Evaluation of bone marrow in all AML cases showed the highest hypercellularity associated with erythroid hypoplasia. A significant association between increased survival time and early diagnosis of lymphoma V by bone marrow evaluation was also observed.

Conclusion: This study provides a better understanding of the clinical and laboratory progression of hematopoietic neoplasms, most common bone marrow alterations and its prevalence in dogs from Brazil.

C-13: EVALUATION OF A NAGEOTTE COUNTING CHAMBER METHOD AND FLOW CYTOMETRIC TECHNIQUE FOR QUANTIFICATION OF RESIDUAL LEUKOCYTES IN LEUKOREDUCED CANINE PACKED RED BLOOD CELL UNITS.

Austin K Viall, Dana LeVine

Background: Current blood banking standards recommend leukoreduced blood components contain ^{6} residual white blood cells (rWBC) per unit, corresponding to

Methods: Double leukoreduced pRBC units spiked with canine leukocytes were used to create six serial dilution rWBC standards (

Results: Flow cytometric method exhibited excellent linearity (R^2=0.99) with a slight positive proportional bias (slope=1.128 ± 0.01). For rWBC concentrations ≥ 1.5 WBC/µL, the average accuracy was 79% and average coefficient of variation (CV) was 11%. While the Nageotte chamber method had good linearity (R^2=0.93), a significant negative proportional bias was found (slope=0.552 ± 0.036). For concentrations ≥ 1.5 WBC/µL, average accuracy was 8% and average CV was 15%.

Conclusion: The flow cytometric method appears superior to the Nageotte chamber technique for quantifying rWBC in canine pRBC units.

C-15: HEMATOLOGY REFERENCE INTERVALS OF CATTLE WITH THE XT-2000iV ®ANALYZER

Nicolas Herman, Catherine Trumel, François Schelcher, Nathalie Bourges, Anne Geffre

Background: Hematology reference intervals have not been determined in cattle with the XT-2000iV® (Sysmex) analyzer.

Objective: The aim of this study was to establish cattle hematology reference intervals according to ASVCP's recommendations.

Methods: 4-mL K3-EDTA blood samples were obtained from 156 clinically healthy female dairy and beef cows, of various breed, 2-10 years of age. A maximum of 5 cows at one of 3 stages: early lactation (0-90 days in milk), mid-lactation (90-180 days in milk), or dry period, were taken on each farm. Blood samples were analyzed on the XT-2000iV®. Distributions of results were tested for normality and reference intervals (2.5-97.5 limits) and their 90% confidence intervals were determined by the nonparametric method with the Reference Value Advisor freeware.

Results: Only 122 specimens were included: 49% (60/122) dairy cows and 51% (62/122) beef cows; 31.1% (38/122) in early lactation, 35% (43/122) in mid-lactation and 34% (41/122) in the dry period. The reference intervals were [4.9-7.5]x10E12/L and [4.8-7.8]x10E12/L for erythrocytes measured by cytometry and impedance respectively, [82-129]g/L for hemoglobin, [0.24-0.39]L/L for hematocrit, [41.0-61.1]fL for mean corpuscular volume, [14.2-19.6]pg for mean corpuscular hemoglobin, [321-360]g/L for mean corpuscular hemoglobin concentration, [17.7-25.5]% for red blood cell distribution width, [0.5-6.9]x10E9/L for reticulocytes, [4.4-10.8]x10E9/L for leukocytes, [0.6-5.0]x10E9/L for neutrophils, [1.7-5.1]x10E9/L for lymphocytes, [0.2-1.2]x10E9/L for monocytes, [0.1-2.1]x10E9/L for eosinophils.

Conclusion: Reference intervals obtained were close to previous data obtained with the Advia120. They have been determined according to international recommendations, thus can be used in clinical cases and in interlaboratory comparisons.

C-16: OPTIMIZATION OF LYMPHOCYTE MARKERS FOR IMMUNOCYTOCHEMISTRY OF PREVIOUSLY STAINED SLIDES FROM DOGS AND CATS BY TWO COMMERCIAL DIAGNOSTIC LABORATORIES

Julie Vickers, Rose Raskin, Jennifer Ward, Angus Toland, Andrew Torrance

Background: Romanowsky-stained materials are often used to diagnose lymphoid neoplasia. However, there has been minimal use of immunocytochemistry for lymphocyte markers on previously stained slides.

Objective: The intent was to optimize a protocol for manual immunocytochemistry on pre-stained cytologic material for use in two commercial diagnostic laboratories involving detection of CD3e and CD20 antigens in phenotypically confirmed canine and feline cases.

Methods: Clinical specimens (blood, tissue aspirates, effusions) submitted for cytologic evaluation were stained by a methanolic Romanowsky method and examined by board certified pathologists. Additional slides from fluid and extra unstained aspirate smears were used for comparison with pre-stained materials. Antigen retrieval involved citrate buffer at 95C for 25 minutes in a decloaking chamber (Biocare). Following peroxidase and casein protein blocking, CD3e (Dako) and CD20 (ThermoFisher) rabbit polyclonal antibodies were applied. Negative controls for specimens lacked primary antibody. Concurrent positive controls involved samples of known reactivity. Signal amplification consisted of polymer antibody (Biocare/Dako) with horseradish peroxidase. The chromogen (3,3'diaminobenzidine) was used with hematoxylin counterstain. Signals were compared between unstained and pre-stained slides; reactions graded by intensity and percent stained cells.

Results: Optimized protocols for ICC staining of pre-stained material (minimum 3.5 hrs) were developed for phenotyping lymphoid neoplasia. Antibody dilutions ranged 1:200-500 (CD3e) and 1:100-200 (CD20) with 30 min incubation. Unstained and pre-stained slides had similar membrane/cytoplasm graded reactions. CD3e and CD20 antibodies with negative control were interrogated on a single cellular slide.

Conclusions: Effective manual immunocytochemistry of pre-stained cytologic specimens was achieved in a commercial diagnostic laboratory setting.

C-17: REPEAT PATIENT TESTING (RPT) SHOWS PROMISE AS A QUALITY CONTROL METHOD FOR HEMATOLOGY TESTING

Bente Flatland, Kathleen P Freeman

Background: RPT-QC has potential for in-clinic laboratories (e.g., without access to or that cannot afford quality control material [QCM]).

Objectives: Determine whether RPT-QC provides P_{ed} and P_{fr} comparable to QC using QCM and allows similar TE_a.

Methods: Control limits (for RBC, HGB, HCT, MCV, WBC, and PLT) for an Advia 120™ (n = 23) and Scil Vet ABC™ (n = 22) were calculated using normal canine data from the routine UTCVM caseload. Samples were measured at accession and again within 24 hr after overnight storage and rewarming. Control limits were validated using 24 additional canine samples tested similarly. Achievable TE_a, P_{ed}, and P_{fr} were investigated using Westgard EZRules3™ and compared to those achievable with QCM.

Results: 1_{2s}, $1_{2.5s}$, and 1_{3s} rules were chosen most commonly; optimal rule choice varied by measurand and instrument. Achievable TE_a with P_{ed} >85% and P_{fr} < 5% were: RBC 10%; HCT 10-15%; MCV 12%; and WBC 15-35%. No acceptable solution was found for HGB (too little data variation) and PLT (too much variation).

Conclusions: RPT-QC (N = 1 or 2) provides TE_a, P_{ed}, and P_{fr} comparable to QCM for RBC, HCT, MCV, and WBC. Advantages include species-specific matrix, low cost, and no deterioration over time (a fresh sample is used each day). A potential disadvantage is daily access to healthy patient samples. A challenge is determining control limits (rule selection); consultation with a quality expert is advised. Further study is needed to confirm performance of RPT-QC and investigate whether RPT-QC using abnormal samples is feasible.

C-18: PRIMARY INTRAORBITAL, EXTRAOCULAR B CELL LYMPHOMA WITH INTRACYTOPLASMIC GRANULES IN A DOG

Jeremie Korchia, Dodd Sledge, Shin A Park, Tanit Kasantikul, Simon M Petersen-Jones, Anthony Pease, Cheryl L Swenson

Primary intraorbital, extraocular lymphoma is rarely reported in dogs, but when it occurs it is most commonly conjunctiva-associated. This case report describes an intraorbital, extraocular B-cell lymphoma with an atypical clinical presentation and cytologic appearance. An 11-year-old spayed female Miniature Pinscher presented with swelling under the left upper eyelid and mild exophthalmos. There were no other clinical signs or impaired vision. Funduscopy revealed indentation of the dorsal wall of the globe. A lobular, homogenous, hypoechoic mass extended dorsal and posterior to the globe on ocular ultrasound. A fine needle aspirate of the mass yielded individualized, pleomorphic, monotypic round cells that had few small vacuoles, few pale light pink granules, a moderate nuclear to cytoplasmic ratio, and pleomorphic nuclei ranging from ovoid to multilobulated. On histopathologic examination of an orbital exenteration specimen, a highly cellular proliferation of intermediate-sized neoplastic lymphocytes effaced orbital connective tissue and muscle superior and posterior to the globe, but neoplastic cells did not extend to the conjunctival epithelium. Thirty percent of overtly neoplastic cells had perimembranous immunoreactivity for CD20, and none were labeled for CD3. On clinical staging including thoracic radiographs and abdominal ultrasound, there was no evidence of systemic involvement and the dog has had no related clinical signs at 2 months post exenteration. The location of the mass, lack of association with the conjunctival epithelium, and the cytologically apparent intracytoplasmic pink granules are unusual findings. This case demonstrates that primary B cell lymphoma should be considered as a differential for cases with a similar presentation.

C-19: PROGNOSTICATION OF CANINE T CELL LYMPHOMA

Nariman Deravi, Dorothee Bienzle, Olaf Berke, Anthony Mutsaers, Michelle Oblak, Stefan Keller, Veronica Parsons

Background: Lymphoma in dogs is frequently treated with chemotherapy and grouped into B- or T-cell types identifying CD79a or CD21 (B) and CD3 (T) markers. T-cell lymphoma is considered to have a worse prognosis than B-cell lymphoma, however, there are different types of T-cell lymphoma with variable biological behaviors. Few studies have examined whether prognosis differs between subtypes of T-cell lymphoma.

Objective: The objective of this study was to determine prognosis of patients with T-cell lymphoma characterized by cytopathology, histopathology, immunohistochemistry and flow cytometry.

Methods: Lymph node aspirates/biopsies were collected from dogs with T-cell lymphoma. Flow cytometric analysis included detection of CD3, CD4, CD8, CD21, CD22, CD45, and MHCII antigens. Formalin-fixed, paraffin embedded sections were assessed for architecture and cytomorphology, and immunohistochemistry was performed for expression and location of CD3, CD79a, and granzyme. On frozen sections, CD4, CD8, and CD21 expression was determined. Tumors were classified following World Health Organization guidelines, and results compared to flow cytometry and cytopathology results, response to chemotherapy and survival.

Results: Using flow cytometry, four cases were $CD3^{+}/CD8^{+}$, three were $CD3^{+}/CD4^{+}$, and four were $CD3^{+}/CD4^{-}/CD8^{-}$. Histomorphologically, tumors had diffuse architecture with peripheral displacement and compression of remnant follicles. Tumors expressed CD3 diffusely; four were also granzyme positive, and CD79a expression was restricted to remnant follicles. In frozen sections, CD4, CD8 and CD21 antigens were expressed in accordance with other immunophenotypic results.

Conclusions: Based on preliminary analysis, patients with $CD8^{+}$ T-cell lymphoma had a different response to chemotherapy than those with $CD4^{+}$ T-cell lymphoma.

C-20: MYELOFIBROSIS ASSOCIATED WITH ERYTHROID HYPOPLASIA SECONDARY TO CHRONIC FLUCONAZOLE ADMINISTRATION IN A DOG

Valerie M. Wong, Daniel S. Foy

A 7-year-old, female, spayed Boxer dog was presented to the Animal Health Institute at Midwestern University (MWU-AHI) for wellness examination (day 0). The patient was previously diagnosed with coccidioidomycosis and had been receiving fluconazole therapy (5.4 mg/kg, PO, q24h) for >5 years. Physical examination was unremarkable.

Hematology showed a moderate, normocytic, normochromic, non-regenerative anemia (Hct: 27%, reference interval: 36-60%). The IgG titer for *Coccidioides* spp.was positive at 1:8. Further diagnostics were declined by the owner, and fluconazole therapy was discontinued due to suspicion that Coccidioidomycosis was in remission. On day 35, the patient had a marked, macrocytic, hypochromic anemia with poor regeneration (Hct: 17%; reticulocyte count: 82,200/µL). Abdominal ultrasonography was unremarkable. On day 42, the nonregenerative anemia persisted (Hct: 17%; reticulocyte count: 67,700/µL); histopathology of the bone marrow biopsy showed marked erythroid hypoplasia and myelofibrosis. The granulocytic and megakaryocytic cell lines were unremarkable. The packed cell volume (PCV) improved progressively without additional treatment (22% on day 55; 26% on day 63). On day 82, hematocrit and reticulocyte count were 35% and 51,700/µL, respectively. The hematocrit was within the reference interval at 47% on day 211. IgG titer for *Coccidioides* spp. remained positive at 1:4. This is the first reported case of myelofibrosis associated with erythroid hypoplasia secondary to chronic fluconazole administration in a dog. A retrospective study at the MWU-AHI found similar hematocrit values between dogs receiving chronic fluconazole therapy and healthy controls (P=0.085). We conclude that this case represents an idiosyncratic reaction to chronic fluconazole treatment.

C-21: CONCURRENT GRANULAR LYMPHOMA AND MAST CELL NEOPLASIA WITH MULTIPLE ORGAN INVOLVEMENT IN A DOG

Francisco O Conrado, Elizabeth A Kieran, Julia A Conway, Michele D James, Mary K Leissinger

A 15-year-old spayed female Jack Russell terrier presented to the UF Veterinary Hospital for lethargy, fluctuating appetite, and weight loss. In aspirates of the enlarged spleen and liver, large granular lymphocytes predominated, accompanied by a moderately cellular and atypical erythrophagocytic mast cell infiltrate consistent with concurrent granular lymphoma and mast cell neoplasia. L-Asparaginase therapy was instituted, but the patient deteriorated rapidly and was euthanized two weeks after diagnosis. Histopathologic evaluation of the left axillary lymph node and spleen at necropsy supported the presence of mast cell neoplasia and lymphoma in these tissues. Immunohistochemistry and polymerase chain reaction for antigen receptor rearrangements (PARR) were performed. CD3 and CD20 markers were inconclusive and PARR analysis on both the paraffin block and cytologic preparation failed to reveal a clonal rearrangement, raising suspicion for an NK cell origin. The mast cell neoplasm did not contain internal tandem duplications (ITDs) involving exon 11 or 8 by C-*kit* mutation molecular analysis. The coexistence of two or more primary tumors at the same anatomic site is relatively rare, and microenvironment alteration by one neoplasm has been a proposed precipitating event to development of a second. The term collision tumor has been applied to clinical entities where two distinct tumors occur at a single anatomic location, though it was unclear in the case herein if either neoplasm represented primary or metastatic disease. Regardless, these cases present both a challenge to clinicians in case management and to pathologists in diagnosis, and selection and interpretation of appropriate ancillary tests.

C-22: CEREBROSPINAL FLUID ANALYSIS WITH THE PROCYTE DX HEMATOLOGY ANALYZER AND COMPARISON WITH HEMOCYTOMETER-DERIVED CELL COUNTS

Holly M. Brown, Jessica Wilson-Hess, Fred L. Metzger, Dennis B. DeNicola

Background: Cerebrospinal fluid (CSF) is routinely collected in private practice while the fluid analysis is most commonly performed at reference laboratories. Given the low total nucleated cell count (TNCC) in CSF and the low protein content that is problematic for cellular preservation, timely CSF analysis is beneficial in minimizing cell degradation and providing accurate results. Manual cell counts can be performed in-clinic via hemocytometer and microscopy; however, such analysis requires availability of personnel to process samples promptly and specialized training to obtain reliable results. A simple and accurate in-clinic automated process for timely CSF TNCC is needed to obtain useful diagnostic information.

Objective: The aim of this study was to compare manual hemocytometer TNCC to the in-clinic IDEXX ProCyte Dx™ Hematology Analyzer (PDx) TNCC in CSF samples from dogs and cats.

Methods: CSF samples were obtained from client-owned dogs (n=34) and cats (n=2) as part of their diagnostic work-up for neurologic disorders. Automated TNCC were determined by the PDx and compared to that determined by manual hemocytometer. Method comparisons were made with Spearman's correlation and Bland-Altman bias analyses.

Results: Manual TNCC ranged from 0 to 858 cells/microliter; the majority (26/36, 72%) were ≤5. The Spearman's correlation coefficient (rho:0.75) supports good correlation between the two counting methods. In cases with TNCC

Conclusion: Automated analysis of CSF on the PDx provides useful and reliable TNCC. As such, it is an effective tool allowing practices to derive important and timely diagnostic information from their in-clinic laboratory.

C-23: PRELIMINARY HEMATOLOGY STUDY OF WILD SPOTTED EAGLE RAYS (AETOBATUS NARINARI) OFF OF SARASOTA, FLORIDA

Jill E. Arnold, Brent R. Whitaker

Background: Reports of hematology values for elasmobranchs are sparse, especially for stingray species. There are no published blood values for the Near-Threatened (IUCN) spotted eagle ray, *Aetobatus narinari*. Our interest in this species was prompted by the 2010 Deepwater Horizon oil spill that contaminated waters in the Gulf of Mexico, where the rays could be exposed to components of the oil accumulated in their primary food source, bivalve mollusks. The aim of this study was to obtain blood specimens for CBC and plasma chemistry profiles from at least 40 free-ranging rays, in partnership with Mote Marine Laboratory's program for monitoring the summer population in coastal waters near Sarasota, Florida from 2010 - 2013.

Methods: A total of 42 specimens were obtained from sub-adult to adult rays for CBCs by established methods for elasmobranchs. Unfortunately, the frozen plasma specimens were shipped incorrectly and rendered unusable for chemistry profiles.

Results: Results for cell counts, packed cell volume, and leukocyte differentials were comparable to reports for other stingray species. The most notable finding was the presence and prevalence of immature cells, both erythrocyte and leukocyte. Immature granulocytes were commonly observed as small round cells, approximately 8-10 µm, with a large eccentric nucleus and light blue cytoplasm filled with eosinophilic granules. Polychromatophilic erythrocytes were common and occasional mitotic cells were observed on all of the blood smears (n=40).

Conclusion: Future studies comparing complete blood counts and cell morphology from aquarium-maintained eagle rays would be beneficial to determine the significance of this finding.

C-24: HEMATOLOGIC EVALUATION IN GUANS (PENELOPE OBSCURA) PHYSICALLY RESTRAINED AND ANESTHETIZED WITH ISOFLURANE

Adson Costa, Julieta Volpato, Cláudio R.S. Mattoso, Bruno Lunardeli, Aury N Moraes, Letícia A Yonezawa, Mere E Saito

Background: Hematology is a great tool used for early diagnosis in many species including birds. However, the capture stress as well as sedation can cause changes in hematologic analyses.

Objective: This study aimed to evaluate the differences found in hematologic parameters between guans physically restrained and anesthetized with isoflurane.

Methods: The study evaluated 32 free-living guans living in the Brazilian Institute of Environment and Renewable Natural Resources Park in Painel, SC, which were captured using a trapdoor. Blood samples were obtained from the ulnar vein after physical restraint of the birds and collected in sodium heparin tubes. The birds were anesthetized with isoflurane and after 30 minutes, a new blood sample was collected from the jugular vein of the anesthetized birds also in sodium heparin tubes. Blood cell count, packed cell volume, hemoglobin, total protein and fibrinogen evaluation were performed on the heparinized blood from the physically and chemically restrained birds, and mean corpuscular volume and mean corpuscular hemoglobin concentration were calculated. Data were analyzed using the paired t-test and the differences were analyzed with the Tukey test.

Results: There was a decrease in the erythrocyte count, hemoglobin and packed cell volume measurement and an increase in mean corpuscular hemoglobin concentration in chemically restrained birds. The total protein values, fibrinogen and thrombocytes were also decreased with chemical restraint.

Conclusion: It was concluded that isoflurane anesthesia influences hematological parameters and should be taken into account when interpreting the hematology from chemically restrained guans.

C-25: FIRST DESCRIPTION OF OSLERUS (FILAROIDES) OSLERI LARVAE FOUND IN PERIPHERAL BLOOD OF A DOG FROM BRAZIL.

Marina Mitie Monobe, Bruna dos Santos, Regina K Takahira

Background: Nematodes affecting the bronchopulmonary system are uncommon and rarely cause pulmonary disease in dogs. However, the increased number of cases in non-endemic regions shows the importance of including *Oslerus (Filaroides) osleri* as a differential diagnosis in small animal internal practice.

Objective: The aim of this study was to report the first occurrence of *O. osleri* filariae found in peripheral blood from a dog from Sao Paulo state, Brazil.

Case Report: A young male mixed-breed street dog presented to primary clinical care for lethargy, oral papillomas, anorexia, and hematochezia. At initial clinical evaluation, the dog was diagnosed with and treated for oral papillomatosis by autogenous vaccine (Infervac®); lesions resolved a month later. Lymphadenomegaly was also seen on initial presentation. No evident radiographic abnormalities were observed. Blood was collected and evaluated weekly for a month. Laboratory alterations included hyperproteinemia evaluated by refractometer (9.8 g/dL; reference range 6.0-8.0 g/dL), lymphocytic leukocytosis (23,800 WBC/µL; reference range 1,000-4,800 cells/µL), eosinophilia (2,400 cells/µL; reference range 100-1,250 cells/µL), discrete basophilia (500 cells/µL; reference range 0-100 cells/µL) and microfilaria (2-3 filariae/slide); laboratory alterations persisted for up to two weeks. The morphology of the first stage larvae found in peripheral blood and feces as evaluated by light microscopy were consistent with *Oslerus osleri*.

Conclusion: This report indicates that osleri infection should be considered as a differential diagnosis in young dogs presenting with cough and respiratory distress. This finding reveals the importance of more studies focusing on the epidemiology of the disease, which is not well known in Brazil.

C-26: HEMATOLOGIC DISTURBANCES CAUSED BY NATURALLY ACQUIRED INFECTION WITH RANGELIA VITALII (PIROPLASMIDA) OF DOGS IN BRAZIL

Stella F Valle, João Fábio Soares, Luciana Sonne, Caroline P Andrade

Background: Rangeliosis is a disease of domestic and wild canines caused by the protozoan *Rangelia vitalii*, which is transmitted by the tick *Amblyomma aureolatum*. Different stages of *R. vitalii* infection are found in endothelial cells prior to blood cells, and the main laboratory findings of infection are thrombocytopenia, hemolytic anemia, and reactive leukogram.

Objectives: The aim of this retrospective study was to investigate hematologic disturbances and platelet indices in dogs with naturally acquired *R. vitalii* infection.

Methods: CBCs and platelet indices were obtained from 8 dogs that presented to a veterinary teaching hospital with nonspecific clinical signs and a diagnosis of *R. vitalii* infection from 2014 to 2016. EDTA blood samples were examined in a hematology analyzer (ProCyte Dx, Idexx Laboratories), followed by light microscopy and conventional PCR.

Results: CBCs revealed mild to severe anemia (HCT: 9 to 34%), thrombocytopenia (platelets: 0 to 181.000/mL), and, in one dog, leukocytosis (WBC: 65.190/mL). Due to marked thrombocytopenia, platelet indices were only measured in two dogs (MPV: 10.8 fL; PDW: 15 fL; PCT: 0.06 and 0.19%). Although the anemia of rangeliosis is associated with hemolysis, a reticulocyte response was absent to slight in two dogs. *R. vitalii* was visualized within erythrocytes in blood smears from two animals. Two dogs with follow-up experienced normalization of platelet indices within 20 days of antiprotozoal therapy.

Conclusion: The pathologic mechanisms of naturally acquired rangeliosis are unclear. Our findings suggest that presence of thrombocytopenia could be found in a specific stage of infection associated mainly with endothelial damage.

C-27: CROTALUS DURISSUS SNAKE BITE: PRE AND POST TREATMENT THROMBOELASTOMETRIC CHANGES – CASE REPORT

MAURICIO O WILMSEN, MARINA M MONOBE, BRUNA SANTOS, JOAO A RIBEIRO, GABRIELA A OLIVEIRA, REGINA K TAKAHIRA

A 3-year-old male dog was admitted to the Veterinary Hospital with pale mucous membranes, nasal discharge, myasthenia, bilateral ptosis, nystagmus and lateral recumbency. One hour before admission, it had been bitten by a *Crotalus durissus* snake. The dog received five vials of antiophidic antivenom (Master soro plus – Vencoforma®), which is able to neutralize 150 mg of crotalic venom according to the package insert. Citrated blood samples were collected for thromboelastometry (ROTEM® – Delta), WBCT (whole blood clotting time), PT (prothrombin time), aPTT (activated partial thromboplastin time) and fibrinogen concentration (FIB) before and 24 hours after antivenom treatment. All clotting times were prolonged above the reportable range (WBCT >30 minutes; TP and aPTT >3 minutes). Platelet count was 119,000/uL. The dog also demonstrated an increased clot formation time (CT) with in-TEM and ex-TEM activators (542 seconds and 646 seconds, respectively). Clot formation time (CFT), maximum clot firmness (MCF) and α angle could not be recorded, since the clot size didn't reach 10 mm. A hyperfibrinolytic state was confirmed by the 100% index of maximun lysis (ML) and fibrinogen concentration below the test detection limit. All variables returned to reference range 24 hours after antivenom administration: PT (8.1 seconds), aPTT (7.3 seconds), CT (35 seconds and 77 seconds), CFT (126 seconds and 133 seconds), MCF (59 mm and 42 mm) and ML (1% and 65%) with in-TEM and ex-TEM activators, respectively. The results showed a consumption of clotting factors associated with a hyperfibrinolytic state that was completely reverted by antivenom therapy.

C-28: FIRST REPORT OF THE EMERGING BASIDIOMYCETE PATHOGEN SCHIZOPHYLLUM COMMUNE IN THE URINARY TRACT OF A CAT

Andrea P Santos, Ewan D Wolff, Allison Kendall, Lynn F Guptill, Larry G Adams, John A Christian, Craig A Thompson

An eight-year-old spayed female Domestic Shorthair with International Renal Interest Society Stage III chronic kidney disease was presented for flushing of bilateral subcutaneous ureteral bypass (SUB) devices. Physical examination was unremarkable. The cat was azotemic (creatinine 4.9 mg/dL; reference interval: 0.9-2.3 mg/dl). Urine cytology revealed moderate to marked neutrophilic and macrophagic inflammation. Numerous round, clear 4-5 micron sized unidentified organisms were noted. They were occasionally phagocytized. Results of bacterial and fungal cultures and *Histoplasma* sp. urine antigen assay were negative, but an assay for serum beta-D glucan was positive. A pan-fungal PCR followed by DNA sequencing revealed the pathogen to be *Schizophyllum commune*. Treatment included fluconazole, combination fluconazole and terbinafine, and later clotrimazole SUB infusions. Treatment failed to resolve the infection, however, no clinical signs related to the infection were observed for 14 months. *Schizophyllum commune* are mushroom-forming fungi (genus *Schizophyllum*, order Agaricales, phylum Basidiomycota) commonly found in rooting trees worldwide. Human and animal infections with *S. commune* are rare. Molecular diagnosis of *S. commune* became available in 2001; since then, *S. commune* has been recognized in human medicine as an emerging cause of sinusitis. In immunocompromised people, respiratory and ocular infections, as well as brain abscesses have been reported. In dogs, *S. commune* infection was first confirmed from a granulomatous neck mass, and later in a dog with fungal osteomyelitis. To the authors' knowledge, this is the first report of *S. commune* infection in the urinary tract of a vertebrate. Furthermore, this organism has not been reported in cats.

C-29: CLINICAL PATHOLOGY AND FEEDING HISTORY DATA FOR NECROPSIED DOGS WITH FANCONI SYNDROME THAT DIED AFTER EATING JERKY PET TREATS

Jennifer Jones, Mary Nabity, Olgica Ceric, David Rotstein, Lee Anne Palmer, Lauren Carey, Cathy Brown, Rachel Cianciolo, Renate Reimschuessel

Background: FDA has received over 5200 reports of canine illness after jerky pet treat (JPT) consumption, with over 1100 reported canine deaths. Approximately 7% of reports indicated Fanconi-like syndrome (FS). Most dogs with FS reported to FDA recover with supportive care and JPT withdrawal.

Methods: Vet-LIRN collected medical records and feeding histories for a subset of these reported deaths. FDA collaborated with noted renal pathologists to evaluate clinical data and feeding histories from 61 necropsied dogs with various pre-mortem illnesses and reported JPT consumption.

Results: Eleven of 61 dogs fit our case definition of clinical FS with pre-mortem normoglycemic glucosuria and/or a positive urine FS screening test. Over 60% of the 11

FS dogs were female and weighed less than 30 pounds. Two were Shih Tzus, and 9 were other pure or mixed breeds. Most had either elevated creatinine or blood urea nitrogen. Seven dogs had two or more blood electrolyte abnormalities, while one dog had normal electrolytes. Over 70% of the 11 dogs had proteinuria. Eight of 11 dogs' feeding histories contained limited information in the FDA report and medical records, which may provide an incomplete picture of the diet. Three of 11 dogs had expanded feeding histories with owner interviews, and the reported amounts of JPT fed to them were ranging from 7 to 24 treats per week, fed over a week to many years.

Conclusions: Vet-LIRN continues gathering data to help detect a root cause of the reported illnesses after JPT exposure.

C-30: IMPACT OF URINE FORMED ELEMENT SETTLING AND SAMPLE ASPIRATION LOCATION ON MICROSCOPIC URINALYSIS

Jeremy Hammond, Carelton Ericson, Celine Myrick, Bailey Clock, Dennis B DeNicola

Background: Urine microscopy can be performed with neat and concentrated samples. For both methods, thorough mixing of the sample is recommended; however, evaluation of sample aspiration timing, and location of aspiration within the sample tube is deficient in the literature.

Objective: To assess the effect of sample settling and within specimen tube sample aspiration location on the numerical recovery of formed elements.

Methods: A full factorial design of experiment including mixing duration, aspiration time, and location of aspiration within the specimen tube was performed. For each condition, counts of formed elements were performed with the recently introduced IDEXX SediVue Dx™ Urine Sediment Analyzer, which uses only 165 microliters of neat urine. Theoretical calculations based on physics principles were employed to develop a model to predict settling rate for various formed elements in urine.

Results: Increased settling time and higher aspiration location within the specimen tubes yielded the lowest crystal counts. For example, waiting only 15 seconds after mixing a marked struvite crystalluria sample and sampling from the top, middle and bottom of the specimen tube yielded 3.6, 23.1 and 151.5 crystals/hpf, respectively. The mathematical model predicted non-crystal elements to settle at 1 micron/second. Predicted crystals settling were at rates between 10-60 microns/second, depending on the size and density of the crystal.

Conclusions: Sample aspiration immediately upon mixing is essential to ensure consistent and accurate urine formed element counts for all methods of urine microscopic analysis.

C-31: MONITORING FELINE HEMATURIA AT HOME: ASSESSMENT OF A POTENTIAL SCREENING TEST

Elodie Khenifar, Jérôme del Castillo, Carolyn Gara-Boivin

Background: Detecting hematuria is a challenge for cat owners. A granular chromogenic screening test for the litterbox is being developed to detect hematuria.

Objective: To determine the hematuria lower limit of detection in vitro, and to verify its robustness in altered urine composition.

Methods: Firstly, the product was tested in quadruplicate by pouring 50 µL of pooled male feline urine fortified with feline red blood cells (RBC) at concentrations ranging between 0 and 25000 rbc/µL. Secondly, 12 healthy male and female cats were used in a Greco-latin square design, in which they received 4 distinct fluid therapies that caused transient polyuria with differential effects on urinary specific gravity and pH (3 cats/sequence). Urine was collected individually and fortified ex vivo with RBC identically to the first experiment. Chromogenic intensity was scored with an ordinal scale, 0 (none) to 3+ (strong), and analyzed with polytomous multivariate logistic regression.

Results: In the 1st experiment, the test's lower limit of detection on pooled urine was 160rbc/µL. In the 2nd experiment, the intravenous fluids successfully changed the urinary SG, pH and proteins. The limit of detection significantly increased with increasing pH (320rbc/µL with pH≥8.5) and proteinuria (640rbc/µL with Prot≥15 mg/dL) but continued to detect microhematuria. No significant differences were observed in the detection efficiency with variations in SG, sex or neutering ($p>0.05$).

Conclusions: Based on these results, this test showed potential for hematuria detection. A clinical study, currently in progress, is verifying the performance of the screening test in the household.

C-32: SERUM FIBROBLAST GROWTH FACTOR 23 IN NORMAL DOGS AND DOGS WITH CHRONIC KIDNEY DISEASE

Keren E Dittmer, Kalyani C Perera, Peter A Elder

Background: Chronic kidney disease (CKD) is a significant cause of morbidity and mortality in dogs. In human medicine, fibroblast growth factor 23 (FGF23) is considered to be one of the key mediators of CKD. FGF23 secretion by osteocytes is increased in response to hyperphosphatemia; in humans and cats increases in serum FGF23 concentration parallel the stage of kidney disease. The aim of this study was to determine if FGF23 concentrations were increased in dogs with CKD.

Methods: A retrospective case series study was designed where serum samples submitted to a commercial laboratory from dogs with a history of polyuria/polydipsia were collected over a 15-month period. The human FGF23 ELISA kit was first validated for use in dogs, and then serum FGF23, and phosphorus, urea and creatinine concentrations were measured. The Mann-Whitney test was used to test for a difference between non-azotemic and CKD groups; a one-way ANOVA with Tukey pairwise comparisons was used to test for differences between the International Renal Interest Society stages.

Results: The median serum FGF23 concentration of dogs with CKD was 5194.6 pg/mL, which was significantly greater (P

Conclusions and clinical relevance: Serum FGF23 concentration is increased in dogs with CKD. This phosphatonin pathway may be a useful target for the development of future treatments to control plasma phosphorus concentration in chronic kidney disease.

C-33: QUALITATIVE EVALUATION OF CALCIUM OXALATE DIHYDRATE AND STRUVITE CRYSTAL NUMBERS WITH AN IN-CLINIC AUTOMATED URINE SEDIMENT MICROSCOPY SYSTEM

Arthur R Alleman, Graham Bilbrough, Jeremy Hammond, Donald J McCrann, Christine Myrick, Dennis B DeNicola

Background: Manual microscopic characterization of urine crystals is difficult in many practice situations. The IDEXX SediVue DxTM Urine Sediment Analyzer (SDx) automates this process with gentle centrifugation and microscopic analysis of neat urine. Seventy high-resolution microscopic images are processed with veterinary-specific convolution neural network image evaluation software.

Objective: To compare SDx results to manual review of SediVue images for the presence or absence of significant calcium oxalate (CaOxDi) and struvite crystalluria.

Methods: A mixture of 300 canine and feline urine samples from a database of 1056 manually cataloged urine samples previously processed by the SDx were selected to assure adequate representation of common urine formed elements. Two experienced microscopists, blinded to image selection and SDx performance (SW1.0.0), characterized all 70 images for each sample. Samples with >/=1 crystal/hpf were classified as positive for either CaOxDi or struvite. Manual results were a consensus between the microscopists. Automated crystal classification starts with an unclassified (CRY) category followed by specific classification if satisfactory morphologic features are present.

Results: There were 48/300 and 41/300 positive samples for CaOxDi and struvite, respectively. There were 9 false-negative and 3 false-positive CaOxDi results and 3 false-negative and 11 false-positive struvite results. Positive/negative agreement between manual and automated evaluation was 81%/99% for CaOxDi and 93%/96% for struvite. There were 2/48 CaOxDi and 22/41 struvite positive cases without automated specific classification that were reported as CRY. Provided images allows rapid and easy specific recognition.

Conclusions: The SDx provides clinically valid CaOxDi and struvite characterization of dog and cat urine.

C-34: QUALITATIVE EVALUATION OF SQUAMOUS AND NONSQUAMOUS EPITHELIAL CELL NUMBERS WITH AN IN-CLINIC AUTOMATED URINE SEDIMENT MICROSCOPY SYSTEM

Arthur R Alleman, Graham Bilbrough, Jeremy Hammond, Donald J McCrann, Christine Myrick, Dennis B DeNicola

Background: Manual microscopic characterization or urinary epithelial cells is difficult in many practice situations. The recently introduced IDEXX SediVue DxTM (SDx) Urine Sediment Analyzer performs automated gentle centrifugation and microscopic analysis of urine. Seventy high-resolution microscopic images are processed with veterinary-specific convolution neural network image evaluation software.

Objective: To compare SDx results to manual review of SDx images for the presence or absence of significant squamous (EPI) and nonsquamous (NEPI) epithelial cell numbers.

Methods: A mixture of 300 canine and feline urine samples from a database of 1056 manually cataloged urine samples previously processed by the SDx were selected to assure adequate representation of common urine formed elements. Two experienced microscopists, blinded to image selection and SDx performance (SW1.0.0), characterized all 70 images for each sample. Samples with ≥1 cell/hpf were classified as positive for either EPI or NEPI. Manual results were a consensus between the microscopists.

Results: There were 85/300 and 41/300 positive samples for EPI and NEPI, respectively. Positive/negative agreement between manual and automated evaluation was 68%/98% for EPI and 76%/76% for NEPI. There were 27 false-negative and 4 false-positive EPI results and 10 false-negative and 63 false-positive NEPI results. Of these, 26/31(84%) EPI and 37/73(49%) NEPI manual and SDx sets of results had differences of less than 1 cell/hpf at the cutoff threshold between positive/negative buckets. Considering the imprecision of these analyses, clinical interpretation was considered similar in these cases.

Conclusions: The SDx provides clinically valid automated EPI and NEPI characterization of dog and cat urine.

C-35: QUALITATIVE EVALUATION OF ERYTHROCYTE AND LEUKOCYTE NUMBERS WITH AN IN-CLINIC AUTOMATED URINE SEDIMENT MICROSCOPY SYSTEM

Arthur R Alleman, Graham Bilbrough, Jeremy Hammond, Donald J McCrann, Christine Myrick, Dennis B DeNicola

Background: Blood cells are some of the most common formed elements in urine; manual microscopic identification in many practices is difficult. The recently introduced IDEXX SediVue DxTM (SDx) Urine Sediment Analyzer performs automated gentle centrifugation and microscopic analysis of urine. Seventy high-resolution microscopic images are processed with veterinary-specific convolution neural network image evaluation software.

Objective: To compare SDx results to manual review of SDx images for the presence or absence of significant RBC/WBC numbers.

Methods: A mixture of 300 canine and feline urine samples from a database of 1056 manually cataloged urine samples previously processed by the SDx were selected to assure adequate representation of common urine formed elements. Two experienced microscopists, blinded to image selection and SDx performance (SW1.0.0), characterized all 70 images for each sample. Samples with >/= 6 cells/hpf were classified as positive for either RBC or WBC. Manual results were a consensus between the microscopists.

Results: There were 104/300 and 71/300 positive samples for RBC and WBC, respectively. Positive/negative agreement between methods was 72%/95% for RBC and 82%/91% for WBC. There were 29 false-negative and 9 false-positive RBC results and 13 false-negative and 20 false-positive WBC results. Of these, 30/38(79%) RBC and 20/33(61%) WBC manual and SDx sets of results had differences of only few cells on either side of the cutoff threshold between positive/negative buckets. Considering the imprecision of these analyses, clinical interpretation was considered similar in these cases.

Conclusions: The SDx provides clinically valid automated RBC and WBC characterization of dog and cat urine.

C-36: INFLUENCE OF DIFFERENT COLLECTION TECHNIQUES ON URINALYSIS IN DOGS

Cathy Trumel, Laurene Plante, Romain Huve, Florence Palanche, Margot Diemer, Anne Geffre, Rachel Lavoue

Background: Cystocentesis is recommended for urine sampling because it decreases contamination of the genitourinary tract and facilitates interpretation. This prospective study aimed to investigate the influence of different urine collection techniques on physical and chemical urinalysis endpoints and on sediment examination.

Methods: Three urine specimens were collected from each dog by free-catch (FC), free-catch after genital cleaning (FCC) and cystocentesis (CYS), in that order. Influence of the collection method on urine specific gravity by refractometry, chemistry by urinary strips and sediment examination were evaluated by ANOVA and binary logistic regression. A p-value

Results: 62 dogs were included. Median delay between FC and CYS was 30 minutes. Specific gravity, glucose, bilirubin, protein and nitrite reactive pads, and mean count of casts were not influenced by the collection technique. Peroxidase activity, pH and mean count of RBC were higher by CYS. Mean counts of leukocytes and epithelial cells were higher by FC.Peroxidase activity, RBC, and WBC were lower with FCC than with CYS probably because of the absence of iatrogenic blood contamination. WBC and epithelial cells were lower with FCC than with FC likely secondary to less genitourinary tract contamination.

Conclusion: Although predicted, these results are useful for clinicians who need to repeat urinalysis in the follow-up of their patients or in case of pain, hemostatic disorders and more generally for ethical purposes. FCC could be recommended for routine canine urinalysis as it is not invasive and could give less pre-analytical alterations.

C-37: PLATELETS DOWNREGULATE THE EXPRESSION OF EMT-TRANSCRIPTION FACTORS IN CANINE OSTEOSARCOMA AND MAMMARY GLAND TUMOR CELL LINES

R.C. Silva, S.C. Bulla, P. Badial, M.M. Monobe, L.A. Anai, E.M. Terra, K. Lunsford, and C. Bulla

Diagnostic Pathology Focused Scientific Session I

December 4, 2016
1:40 PM – 1:50 PM

ULTRASTRUCTURE OF AIR-CONDUCTING MUCOSA OF DOGS WITH CHRONIC RESPIRATORY DISEASE SUSPECTED OF PRIMARY CILIARY DYSKINESIA

Ileana Miranda, Anibal Armien

Objective: This study aims to evaluate the air-conducting mucosa of dogs with chronic respiratory disease suspected of PCD that were submitted for ciliary ultrastructural evaluation.

Methods: Samples of respiratory mucosa from 15 dogs submitted to the Ultrastructural Pathology Unit at the Minnesota Veterinary Diagnostic Laboratory were evaluated by light and transmission electron microscopy (TEM). Fourteen biopsies and necropsy in one case were performed.

Results: PCD was only confirmed in the necropsied dog, which presented with chronic otitis, bronchopneumonia, hydrocephalus and ultrastructural abnormalities in 84% of the assessed cilia, including absence of dynein arms and microtubular changes. All other 14 cases showed only non-specific alterations in the minority of the evaluated cilia and were classified as SCD.

Conclusion: Ciliary ultrastructural analysis can confirm a PCD diagnosis if specific abnormalities exist. TEM remains an important investigation in Veterinary Medicine, as no other specific test for PCD in dogs has been standardized yet.

December 4, 2016
1:50 PM – 2:00 PM

DIAGNOSIS OF COLLAGEN TYPE III GLOMERULOPATHY USING PASH/MASSON'S TRICHROME COMBINATION (PASH/MT) AND PICOSIRIUS RED (PSR) STAINS

Anne Burnum, Brittany McHale, Uriel Blas, Cathy Brown

Collagen type III glomerulopathy (collagenofibrotic glomerulopathy, CG), is a rare familial disease of young dogs that causes proteinuria and progressive loss of renal function. Definitive diagnosis requires transmission electron microscopy (TEM) to confirm expansion of the mesangial matrix and subendothelial space by abnormal type III collagen. However, a presumptive diagnosis can be achieved with light microscopy alone through the use of histochemical stains. CG was diagnosed in a 12 week old male Pug that was one of 3 littermates with acute renal failure. At necropsy, there was generalized subcutaneous edema. Renal sections were evaluated with a panel of stains used for renal biopsies (H&E, periodic acid Schiff-hematoxylin [PASH], Masson's trichrome [MT], Jones methenamine silver, and Congo red [CR]). Glomeruli were globally expanded by deposits of eosinophilic material in the mesangium and capillary wall. The deposits were blue with MT and stained with Jones silver, consistent with collagen. Based on these results, differentials included CG and glomerulosclerosis (GS). Picrosirius red (PSR) and PASH/MT combination stains were used to differentiate type IV collagen from type III. The deposits were blue with PASH/MT and orange-red with green birefringence using PSR, ruling out GS and supporting a diagnosis of CG. For comparison, these stains were repeated on renal tissues from an age-matched control dog and on two dogs with TEM-verified diagnoses of CG or focal segmental GS. Final confirmation of CG was obtained by TEM. This diagnostically challenging case demonstrates the utility of histochemical stains for the diagnosis of glomerular disease.

December 4, 2016
2:00 PM – 2:10 PM

PULMONARY ALVEOLAR CAPILLARY DYSPLASIA IN PUPPIES

Katie J Barnes, Csaba Galambos, Kurt Williams

Respiratory distress is a significant cause of morbidity and mortality in neonatal puppies. In spite of the frequency of respiratory disease in young puppies there is little published literature describing non-infectious neonatal pulmonary pathology in veterinary medicine. In contrast, non-infectious interstitial lung disease in human neonates is well documented, with more than 13 entities described. Alveolar capillary dysplasia (ACD) with misalignment of pulmonary veins (MPV) is a rare developmental lung disorder in human infants resulting in severe pulmonary hypertension, hypoxia, and death shortly after birth. Herein we describe the pathology of an interstitial lung disease in five puppies (3 female, 2 male; median age of 4 days) with features similar to ACD/MPV. Lungs from affected puppies were evaluated histologically and using immunohistochemistry for a-smooth muscle actin (SMA), pancytokeratin, and CD31. The lungs had diffusely markedly thickened alveolar septa, medial hypertrophy of pulmonary arteries, and centrally located, dilated thin-walled veins similar to MPV in humans. Pancytokeratin labeling of alveolar epithelial cells highlighted the underdeveloped pulmonary acini. Numerous spindle-shaped SMA-expressing cells were present around small vessels and capillaries, and were often detected beneath the alveolar lining epithelium throughout the thickened interstitium. CD31 labeling identified capillary profiles within the alveolar interstitium; many of the capillaries were dilated and dysplastic, with double capillary layers frequently noted. Based on the current study, we conclude that pulmonary developmental abnormalities with similarities to human pediatric lung diseases are present in the domestic dog, and may be the cause of early life death in puppies.

December 4, 2016
2:10 PM – 2:20 PM

FATAL RANAVIRUS INFECTION IN A GROUP OF CAPTIVE MELLER'S CHAMELEONS (TRIOCEROS MELLERI)

Lauren B Peiffer, Nathan Pate, Kathleen Gabrielson, Allan P Pessier, Ellen Bronson, Lisa Mangus

Ranaviruses are well-documented, world-wide pathogens of fish and amphibians that are increasingly recognized as causes of morbidity and mortality in reptiles. This case series documents the first known occurrence of ranavirus-associated disease in chameleons. In November 2015, a group of five Meller's Chameleons from the Maryland Zoo in Baltimore died and were submitted to Johns Hopkins University veterinary pathology service for postmortem examination. All animals presented in thin body condition. The first four animals did not have any other significant gross findings, while the fifth animal exhibited mild transudative coelomic effusion and petechial hemorrhages affecting the tongue and kidneys. Microscopically, there was multifocal necrosis most notably affecting the spleen, liver, kidney, adrenal tissues, and nasal cavity. Moderate to abundant numbers of basophilic intracytoplasmic viral inclusions were present in the liver and nasal mucosa. All animals exhibited varying degrees of necrotizing rhinitis with secondary bacterial and, in one animal, fungal infection. Intravascular bacterial colonies were also observed in the liver of one chameleon, suggestive of intercurrent bacteremia. Samples of liver were sent to the San Diego Zoo Institute for Conservation Research, where samples were positive on Taqman qPCR for a portion of the ranavirus major capsid protein (MCP) gene. Sequencing of portions of the neurofilament-like and MCP genes further identified the virus as a member of the *Frog Virus 3* (FV3) group. This series demonstrates that *Ranavirus* should be considered a differential in lizards that present with sudden death, rhinitis, skin lesions, and splenic / hepatic necrosis.

December 4, 2016
2:20 PM – 2:30 PM

DEVELOPMENT OF QUANTITATIVE REAL-TIME PCR ASSAYS TO DETECT MYCOBACTERIUM SPP. IN ZEBRAFISH (DANIO RERIO)

Danielle M Meritet, Donna M Mulrooney, Micheal L Kent, Christiane V. Löhr

Background: Infections with *Mycobacterium* spp. are common in zebrafish research colonies. As co-morbidities can influence biological responses to stimuli including experimental treatment and conditions, detection of *Mycobacterium* ssp. infections in zebrafish colonies prove essential.

Methods: We developed quantitative real-time PCR assays to detect the three *Mycobacterium* species most commonly identified in laboratory zebrafish by targeting sequences of the mycobacterial heat shock protein 65 gene. Zebrafish were experimentally infected with *M. marinum*, *M. chelonae* and *M. haemophilum.* The experiment was terminated at 12 weeks post infection.

Results: Simplex PCR assays are both highly specific and sensitive for fresh frozen samples, and highly specific and moderately sensitive for formalin-fixed paraffin-embedded (FFPE) samples. We evaluated two different sampling techniques for FFPE samples for whole, sagittal sectioned zebrafish. DNA quantity and DNA purity was equal for both paraffin cores targeting granulomas containing bacteria and scrolls from the entire fish. The diagnostic sensitivity of cores was superior to scrolls for *M. chelonae* and *M. haemophilum*, but not *M. marinum* likely due to widespread granuloma formation with *M. marinum* versus more localized infections with *M. chelonae* and *M. haemophilum*.

Conclusions: The assays are specific, sensitive, cost effective, and rapid and ideally suited to diagnose common *Mycobacterium* spp. infections in laboratory zebrafish.

December 4, 2016
2:30 PM – 2:40 PM

SUBCUTANEOUS PANNICULITIS-LIKE T-CELL LYMPHOMA IN DOGS

Erica Noland, Stefan Keller, Matti Kiupel

Canine nonepitheliotropic cutaneous T-cell lymphomas (NETCL) are poorly characterized and may be confused with other conditions. In humans, a number of distinct subtypes of NETCL have been recognized, including indolent subcutaneous panniculitis-like T-cell lymphoma (SPLTL). We describe 5 dogs with subcutaneous T-cell lymphomas similar to SPLTL. All cases were characterized by proliferations of small to intermediate and sometimes large sized, CD3 positive T-cells that had infiltrated the subcutis in a lace-like pattern and frequently rimmed adipocytes. No epitheliotropism was observed and neoplastic cells were often karyorrhectic and there were regions of extensive necrosis. Heavy infiltrates of histiocytic cells with prominent phagocytosis masked the lymphoid neoplastic cell population in some sections. A clonal T-cell receptor gamma gene rearrangement was determined by PCR in 4 of the 5 cases. The mean age was 8.5 years (5.5 to12 years). No breed or sex predilection was observed.

Two dogs presented with an acute onset of multiple skin masses, 2 dogs had solitary masses with subsequent development of multiple smaller masses within 17 to 60 days of diagnosis, and 1 dog had a solitary mass. In at least two dogs, the masses were erythematous and/or ulcerated. Locations included shoulder, neck, and ventral abdomen. While two dogs were euthanized following diagnosis, one dog treated with chemotherapy (CCNU) survived 7 months post diagnosis. While SPLTLs are less aggressive lymphomas in humans, their biological behavior in dogs remains uncertain. SPLTL is a distinct entity in dogs and needs to be accurately diagnosed to better determine clinical behavior.

December 4, 2016
2:40 PM – 2:50 PM

COMPARISON OF HISTOLOGIC MARGIN STATUS IN LOW-GRADE CANINE MAST CELL TUMORS EXAMINED BY RADIAL AND TANGENTIAL SECTIONS

Milan Milovancev, Duncan Russell, Camila B Dores

Background: Histologic margin status may predict recurrence in cutaneous malignancies. However, there is a poor understanding of how sectioning technique influences margin outcome.

Objective: The aim of this study was to compare histologic margin status in canine mast cell tumors sectioned by both radial and tangential techniques. The hypotheses were that tangentially sectioned neoplasms are more frequently positive (ie. neoplastic cells at the inked surgical margin), and that complete tangential margins are associated with longer histologic tumor-free margins (HTFMs).

Methods: Eighteen inked surgical margins from canine cutaneous mast cell tumors (Kiupel low grade, Patnaik Grade II) were prospectively collected. All masses were excised with curative-intent, wide surgical excision. Following a complete radial section, tangential sections of the adjacent inked surgical margin were embedded and sectioned for routine histopathology. Tangential margins were graded as negative or positive (clusters with more than 10 mast cells in a 40μm radius).

Results: Tangential sections detected significantly more positive surgical margins than radial sections (8/18 (44%) vs. 2/18 (11%), $p = 0.0256$; Chi-square test). HTFM length was greater in the negative tangential margins (12.6 ± 6.0mm; range 0 – 19mm) compared to the positive tangential margins (5.5 ± 5.2mm; range 0 – 13mm, $p = 0.0176$; unpaired t-test). Tangential section margin categorization was weakly positively correlated to HTFM length ($r2 = 0.304$; $p = 0.0176$; Pearson correlation coefficient).

Conclusion: These data indicate that sectioning technique may influence histologic margin status. Tangential sections could be a more sensitive indicator of incomplete histologic margins.

December 4, 2016
2:50 PM – 3:00 PM

ZEBRA ALERT! EASTERN EQUINE ENCEPHALITIS IN PUPPIES IN THE MIDWEST AND NORTHEAST

Caroline Andrews, Jodie Gerdin, Jon Patterson, Scott D. Fitzgerald

Eastern equine encephalitis virus (EEEV) is an alphavirus within the family Togaviridae that is classified as a select agent and is capable of causing mortality in humans and a number of veterinary species. The virus is spread by mosquitoes and has recently been postulated to overwinter in snakes. Herein we describe three cases of EEEV in puppies in Michigan and New York. Two puppies were euthanized following an acute history of seizures and obtundation. A littermate of one of these puppies died two weeks earlier following a history of anorexia and fever. All three puppies lacked significant gross lesions at autopsy and tested negative for rabies virus. In all three puppies, histologic examination revealed a necrotizing, neutrophilic and lymphoplasmacytic meningoencephalitis with strong positive immunohistochemical labeling for EEEV antigen in neurons and glial cells. The diagnosis of EEEV was confirmed by PCR in one puppy and cell culture in the other two dogs. EEEV is rare in dogs and pathologic descriptions have only been reported from puppies in the Southeast. The clinical and pathologic features of the cases described here are similar to those previously reported. The initial clinical signs of EEEV in puppies are typically nonspecific, including anorexia, fever, and diarrhea, but rapidly progress to severe neurologic disease characterized by seizures and recumbency. Although rare, EEEV should be considered as a differential diagnosis for neurologic disease in puppies, especially after more common etiologies, such as canine distemper, rabies, and toxoplasmosis, have been ruled out.

December 4, 2016
3:30 PM – 3:40 PM

CLINICOPATHOLOGIC FEATURES OF CANINE LINGUAL T-ZONE LYMPHOMA

Lauren Harris, Kelly L Hughes, Emily D. Rout, Janna A. Yoshimoto, Claire M Cannon, Paul R. Avery, E.J. Ehrhart, Anne C. Avery

Background: Canine T-zone lymphoma (TZL) is a subtype of canine T-cell lymphoma characterized by unique cytomorphologic features and histologic pattern, immunophenotypic loss of CD45 expression, and an indolent clinical course. Dogs with TZL typically present with single to multiple lymphadenopathy and/or lymphocytosis. We report an unusual presentation of this disease involving the tongue.

Objectives: To characterize the clinicopathologic features of a novel extranodal presentation of TZL recently identified in the tongue of dogs.

Results: Twelve dogs presented to the Colorado State University Clinical Immunology Laboratory and Diagnostic Medical Center between 2006 and 2016 for evaluation of multifocal raised erythematous lingual masses. Seven dogs presented with concurrent peripheral lymphadenopathy or lymphocytosis, four presented with both peripheral lymphadenopathy and lymphocytosis, and one presented with neither peripheral lymphadenopathy nor lymphocytosis. All dogs were diagnosed with lingual TZL based on variable combination of immunophenotyping via flow cytometry, cytology, histopathology, and immunohistochemistry. Interestingly, three cases were initially diagnosed as plasma cell tumors based on histopathology suggesting differentiation of these entities might be a diagnostic challenge. With variable treatment protocols six dogs exhibit complete resolution of the lesions.

Conclusions: This case series highlights a unique presentation of canine TZL and presents a new entity that should be considered as a differential for lingual neoplasia. The characteristic indolent behavior of TZL makes differentiation from other, more aggressive, round cell tumors especially important for accurate prognosis and treatment decisions. This report also demonstrates the utility of immunophenotyping via flow cytometry in the diagnosis of extranodal round cell neoplasia.

December 4, 2016
3:40 PM – 3:50 PM

SYSTEMIC MYCOBACTERIUM KANSASII INFECTION IN A DOMESTIC SHORTHAIR CAT

Su-Hyung Lee, Du-Min Go, Sang-Ho Woo, Jee-Yong Eun, Dae-Yong Kim

A 1-year-old, female, feline leukemia virus and feline immunodeficiency virus negative Domestic Shorthair cat presented to a private clinic with anorexia, depression , weight loss, and multifocal facial, pinnal, and joint nodules. Medical treatment failed to improve the condition. Fine needle aspiration and biopsy of the pinna nodule revealed pyogranulomatous inflammation with a suspected infectious etiology. The patient's physical condition deteriorated and it died. Postmortem examination was performed and multifocal to coalescing firm nodules and ulcerations on the ear, periocular areas, nasal planum, oral cavity, laryngopharyngeal region, lung, tracheobronchial lymph nodes and mediastinal lymph node were observed. Impression smears of several nodules revealed myriad slender extracellular and intrahistiocytic bacterial rods. Histopathologic findings included nodular pyogranulomatous inflammation with or without necrosis. Large numbers of acid fast positive bacilli were observed. Multiplex PCR and DNA sequencing was conducted on fresh tissue and *Mycobacterium kansasii* was identified. Nocardia sp., *M. tuberculosis, M. avium*, *M. intracellulare* and *M. fortuitum* were nor detected, nor were protozoa with Giemsa and PAS stained sections. Systemic M. kansasii has been reported in cattle, infrequently in dogs, and a Chinese soft shell turtle, but has not been reported in cats.

December 4, 2016
3:50 PM – 4:00 PM

GENOMIC DETECTION OF CYTOLOGICALLY CONFIRMED MAST CELL TUMOR LYMPH NODE METASTASIS OTHERWISE NOT DETECTED BY CONVENTIONAL HISTOPATHOLOGY

Gregory A Krane, Barbara Davis, Sarah Beatty, Sarah Boston, Owen Skinner, Julia A Conway

An eight year-old, castrated Boxer presented with a recurrent interdigital forelimb mass 4.5 months after incomplete excision of a Patnaik grade II-Kiupel low grade mast cell tumor (mitotic index of zero). The cytologic diagnosis of the mass and corresponding prescapular lymph node was mast cell tumor with lymph node metastasis. Cytologically, the metastatic mast cells in the lymph node comprised 20-25% of the total cell population and exhibited moderate pleomorphism with increased multi-nucleation and mitoses. There was no detectable splenic, hepatic, or intra-thoracic involvement (abdominal ultrasonography, hepatic and splenic fine needle aspiration with cytology, three-view thoracic radiography). Partial foot amputation and lymphadenectomy followed. The histopathologic diagnosis of the mass was Patnaik grade II-Kiupel high grade mast cell tumor (seven mitoses in ten 400x fields). No metastatic mast cells were present in six histologic step-sections through the entire lymph node (H&E and Toluidine-blue stains). A quantitative nuclease-protection assay performed on mRNA extracted from formalin-fixed, paraffin-embedded sections of the mast cell tumor showed significant expression of Kit (15-fold increase over housekeeping genes), MAP2K1, MAP2K2, Vimentin, and CD11a. The patient's lymph node expressed low levels of Kit, which control lymph node did not express. Cytologic and genomic evaluations of the lymph node indicated metastatic spread of the mast cell tumor despite lack of histologic detection. This case highlights potential limitations of histopathology in evaluation of tumor metastasis, warranting further investigation into the capability of genomic analysis of tumors and lymph nodes to be a sensitive and useful adjunct diagnostic tool in such evaluations.

December 4, 2016
4:00 PM – 4:10 PM

CD31 IMMUNOREACTIVITY IN CANINE RENAL CELL CARCINOMAS

Tyler J Peat, Elijah Edmondson, Deidre M DuSold, Margaret A Miller, Jose A Ramos-Vara

CD31 (platelet endothelial cell adhesion molecule 1, PECAM-1) is considered a specific marker of endothelial neoplasms in both humans and dogs. However, CD31 immunoreactivity has been reported in some human non-endothelial neoplasms (carcinomas, sarcomas, mesotheliomas, and lymphomas). Immunohistochemical expression of CD31 in similar canine neoplasms has not been reported. Immunoreactivity for CD31 (mouse monoclonal antibody, clone M0823, Dako, Carpinteria, CA) was detected in 14 of 43 (33%) formalin-fixed, paraffin-embedded canine renal cell carcinomas (RCC): 5 of 16 (31%) papillary, 5 of 13 (38%) solid, and 4 of 14 (29%) tubular); and 2 of 2 (100%) metastatic RCC. The percent of neoplastic epithelial cells with membranous CD31 reactivity was scored as 0 (<1%), 1 (1-15%), 2 (16-30%), 3 (31-50%), or 4 (>50%). The intensity of immunoreactivity was scored as 1^+, 2^+, or 3^+. CD31 immunoreactivity was heterogeneous in the 14 positive primary tumors. The mean percent reactivity (MR) and mean intensity (MI) scores were 1.6 and 2.8, respectively, for papillary RCC; 2.6 and 2.6 for solid RCC; 2.2 and 3.0 for tubular RCC. The MR and MI were 1.0 and 2.0, respectively, in metastatic RCC. Patchy membranous and/or cytoplasmic CD31 expression was noted in cortical tubular epithelial cells in 25 (61%) of 41 samples with adjacent non-neoplastic parenchyma. In summary, CD31 reactivity in some canine RCCs expands the expression of this marker, once believed to be specific for endothelial cells.

December 4, 2016
4:10 PM – 4:20 PM

ZOLLINGER-ELLISON-LIKE SYNDROME IN A MEXICAN GRAY WOLF (CANIS LUPUS BAILEYI)

Miranda Frohlich, Nick Robl, Jason D. Struthers, Valerie M. Wong, Matti Kiupel

A 12-year-old, captive, male Mexican gray wolf (*Canis lupus baileyi)* presented for inappetence and weight loss. Abdominal ultrasonography revealed a thickened duodenum and peritoneal effusion. Two duodenal perforations were noted on exploratory celiotomy and repaired. Persisting clinical signs led to a second celiotomy, revealing a 4-cm mesenteric mass, which was diagnosed as a neuroendocrine tumor on histopathology. Over the next year, the wolf was maintained on famotidine (10 mg, PO, q24) and omeprazole (20 mg, PO, q24) with minimal clinical signs. Eighteen months after the initial presentation, deteriorating health prompted euthanasia. Necropsy revealed a firm, well-demarcated, 3.5 x 3 x 2 cm mottled tan to red mass expanding the right pancreatic limb; three similar masses within the mesentery adjacent to the pancreatic limbs; multifocal duodenal ulcerations; and a focal transmural duodenal perforation with septic fibrinosuppurative peritonitis. Other significant findings included chronic-active ulcerative esophagitis, generalized muscle atrophy, dehydration, and scant subcutaneous adipose stores. Histopathologically, the pancreatic mass was diagnosed as a neuroendocrine carcinoma. The parapancreatic masses were diagnosed as metastasis to lymph nodes. Immunohistochemistry of the pancreatic mass was strongly positive for gastrin and negative for glucagon, insulin, pancreatic polypeptide, serotonin, somatostatin, and vasoactive intestinal peptide. Antemortem serum gastrin level was 414 pg/mL (canine reference range: 20 - 104 pg/mL), corroborating hypergastrinemia. These findings support a diagnosis of a malignant pancreatic gastrinoma causing a Zollinger-Ellison-like Syndrome (ZES) and lymph node metastases. ZES has not been reported in wolves. Gastrinomas should be considered in wolves with duodenal ulceration or a pancreatic mass.

December 4, 2016
4:20 PM – 4:30 PM

CUTANEOUS AND VISCERAL HEMANGIOSARCOMA IN CATS: A RETROSPECTIVE STUDY OF 113 CASES

Yanet Velázquez-Jiménez, Daniel R Rissi

Background: Hemangiosarcoma (HSA) is a malignant endothelial neoplasm that occurs sporadically in cats.

Objective: We describe the pathological features of 113 cases of HSA in cats diagnosed at the DP and AVDL between 2008 and 2016.

Methods: Information was retrieved from necropsy and biopsy submissions.

Results: A total of 6,440 cases of neoplasms in cats were retrieved, with 113 (1.7%) diagnosed as HSA. The average age of affected cats was 10 years; domestic shorthaired cats were overrepresented (78% of cases) and no gender predilection was noted. Cutaneous HSA was diagnosed in 99 cases (87.6%) and visceral HSA was diagnosed in 14 cases (12.3%). Cutaneous tumors affected the trunk (29/99), head (31/99), limbs (21/99), mammary gland (3/99), and perianal region/tail (2/99). Tumor site was not informed in 13/99 cases. Visceral tumors affected the intestine (7/14), liver (6/14), mesentery (5/14), spleen (3/14), thorax (3/14), mesenteric lymph node (1/14), and brain (1/14). Tumors were multicentric in 8 of the 14 visceral tumors.

Conclusions: This study reveals that cutaneous HAS is more frequent than visceral HAS. Our results also suggest that the intestine is a predisposed site for the development of visceral HSA in cats.

December 4, 2016
4:30 PM – 4:40 PM

COEXPRESSION OF CD3 AND CD20 IN THREE CASES OF CANINE ENTEROPATHY-ASSOCIATED T-CELL LYMPHOMA, LARGE CELL TYPE (TYPE 1)

Erica Noland, Matti Kiupel

The majority of reported primary intestinal lymphomas in dogs are comprised of large T-cells that, according to the human World Health Organization criteria, are subclassified as enteropathy-associated T-cell lymphoma (EATL) type 1. Classically, T-cell lymphomas are thought to only express T-cell markers, i.e. CD3; however, there are growing numbers of reports of human peripheral T-cell lymphomas and cutaneous epitheliotropic T-cell lymphomas (CETCL) that coexpress CD20. Recently, coexpression of CD20 was reported in a CETCL in a dog. The clinical significance of these findings in either species has yet to be determined. We describe three cases of CD3+, CD20+ intestinal T-cell lymphoma in dogs. These lesions are characterized by proliferations of intermediate to large sized, CD3 and CD20 positive lymphocytes within the mucosal lamina propria and focally within the mucosal epithelium. These cells were Pax-5 negative similar to other reports of CD20+ T-cell lymphomas. PCR for rearrangement of the T-cell receptor gamma gene confirmed a monoclonal cell population in all 3 cases. Initial clinical signs included weight loss, inappetence, diarrhea, and/or vomiting. The mean age was 9 years (3 to 12 years). No breed predilection was observed. One dog was euthanized 11 days following diagnosis. One dog treated with CCNU survived 163 days. Interestingly, one untreated dog is alive after 597 days. Given the clinical implications, recognition of CD20+ T-cell lymphomas in order to limit misdiagnosis is important. While the clinical significance and prognostic value of expression of CD20 is unclear, it may allow for treatment with CD20 targeted therapy.

December 4, 2016
4:40 PM – 4:50 PM

DEVELOPMENT OF A NOVEL DIAGNOSTIC IMPLANT FOR DETECTION OF SPECIFIC IMMUNE RESPONSES IN MYCOBACTERIUM AVIUM PARATUBERCULOSIS (MAP) INFECTED CATTLE, MAP VACCINATED CATTLE, AND NAÏVE CATTLE

Tracy Lindquist, Saleh Albarrak, Shannon J. Hostetter, Doug E. Jones, Jesse M. Hostetter

Background: Current diagnostics for *Mycobacterium avium subspecies paratuberculosis* (MAP) cannot accurately identify cattle early in the course of infection. Development of a diagnostic assay detecting cell-mediated immune responses specific to the varying clinical stages of disease observed in MAP infection is needed.

Objective: The aim of this study was to develop a diagnostic tool that consistently identifies cell mediated immune responses specific to MAP infected cattle and MAP vaccinated cattle.

Methods: Using a permeable metallic implant containing MAP antigen emulsified in a collagenous scaffold, serologically positive, serologically negative, vaccinated, and naïve cattle were implanted subcutaneously. After 72-120 hours, the device was removed and the collagen center was collected. A portion of the collagen was formalin-fixed and the inflammatory cell infiltrate was observed histologically using a hematoxylin and eosin stain. The remaining collagen was harvested and cytokines were measured using a quantitative bead-based multi-analysis.

Results: Optimization of implant design as well as implantation technique, implant retrieval, and collagen collection was achieved. Cytokines (TNF-α, IFN-γ, and IL-10) were successfully measured from the collagen and the inflammatory infiltrate was observed within the collagenous scaffold via stained histological sections.

Conclusion: This new diagnostic tool can be used to successfully retrieve and measure specific cell-mediated immune responses occurring within an animal.

December 4, 2016
4:50 PM – 5:00 PM

ANGIOSTRONGYLUS CANTONENSIS INDUCED VERMINOUS MENINGOENCEPHALOMYELITIS IN A RED KANGAROO

Sonika Patial, Brooke A Delcambre, Peter DiGeronimo, Rudy Bauer

Angiostrongylus cantonensis is a zoonotic parasitic helminth that normally resides in the pulmonary arteries and the right ventricle of rats, the definitive host, in which it causes little disease. Humans, dogs, opossums, and various zoo animals are "accidental" hosts that acquire infection through the intermediate hosts (gastropods). However, no reports have described an infection in a Red Kangaroo. Here, we report verminous meningoencephalomyelitis caused by Angiostrongylus cantonensis in a 9-month-old male Red Kangaroo (Macropus rufus). This kangaroo first presented as lethargic, recumbent, and hypothermic with severe muscle wasting and diminished withdrawal reflexes in both hind limbs. Complete blood counts, chemistry panel, urine analysis, fecal exams, radiographs, and abdominal ultrasound were all unremarkable. Serology for toxoplasmosis and leptospirosis was negative. The kangaroo progressed to non-ambulatory paraparesis, declined mentally, and died within three weeks. Gross examination at necropsy showed mildly congested vessels and multifocal areas of dark brown discoloration, malacia and cavitation in the brain and the spinal cord. Histopathology revealed the presence of several transverse and longitudinal sections of nematode larvae in focal areas of cerebrum, cerebellum and the spinal cord, which were surrounded by extensive areas of rarefaction characterized by loss of gray or white matter, hemorrhage, spongiosis, neuronal necrosis, and gliosis. Interestingly, the eosinophilic response, commonly observed in human cases, was largely absent and the inflammatory response was minimal. Several nematodes were extracted from the fixed brain and spinal cord via dissection. The structural features and further analysis of these nematode larvae revealed at least one fifth-stage Angiostrongylus larva.

Diagnostic Pathology Focused Scientific Session II

December 6, 2016
1:30 PM – 1:40 PM

IMMUNOPHENOTYPIC CHARACTERIZATION OF CANINE SPLENIC FOLLICULAR DERIVED B-CELL LYMPHOMAS

Leah R Stein, Matti Kiupel, Cynthia Bacmeister

Background: Splenic follicular derived B-cell lymphomas belong to a subgroup of indolent lymphomas that arise on a background of lymphoid follicular hyperplasia. While both marginal zone lymphomas (MZLs) and mantle cell lymphomas (MCLs) have been reported in dogs, there is little information on their immunophenotypical characteristics.

Objective: The goal of this study was to characterize the immunophenotype of canine splenic MZLs and MCLs.

Methods: Thirteen MCLs and 35 MZLs were selected based on their morphologic features and tissue micro arrays were generated to evaluate expression of CD3, CD20, CD79a, Pax-5, Bcl-2, Bcl-6, CD10, MUM-1 and cyclinD1.

Results: MZLs were characterized by proliferating intermediate sized B-cells with abundant, lightly stained cytoplasm and mildly vesiculated nuclei with marginated chromatin and a large single, central nucleolus that formed homogenous cuffs around fading germinal centers. While MCLs also surrounded fading germinal centers, neoplastic B-cells were intermediate sized cells with scant cytoplasm and chromatin dense nuclei with variable degrees of indentation and angulation and inconspicuous nucleoli. Neoplastic cells in all cases were positive for CD45, CD20 and CD79a and negative for CD10, Bcl-6 and cyclin D1. All cases except one MCL and 3 MZLs were positive for Pax-5 and all except one MZL were positive for Bcl-2. Three MCLs and 27 MZLs were positive for MUM-1.

Conclusion: The observed immunophenotype for canine MZLs and MCLs is similar to their human counterparts with the exceptions that human MCLs overexpress cyclin D1 due to a translocation mutation. A similar mutation has not been reported in dogs.

December 6, 2016
1:40 PM – 1:50 PM

MALAKOPLAKIA MIMICKING NEOPLASIA IN THE URINARY TRACT OF AN ADULT CAT

Chiara Piccinelli, Marisa Ferreira, James E Hoare, Maurizio Longo, Claire S Taylor, Tiziana Liuti, Nicholas X Bommer, Paola Cazzini

An 8-year-old neutered female shorthaired cat presented with sudden worsening of a 3-week history of polyuria/polydipsia, dysuria, and lethargy. Abdominal ultrasound revealed a trigonal mass with extensive involvement of urethra and ureterovesical junctions resulting in bilateral hydronephrosis; the main differential was transitional cell carcinoma. A suction biopsy indicated marked septic inflammation, with no evidence of neoplasia. *Escherichia coli,* followed by a multi-resistant *Pseudomonas aeruginosa* were cultured from the urinary tract. Antibiotic treatment was initiated and bilateral subcutaneous ureteral bypasses were placed. Sequential traumatic catheterizations of the bladder demonstrated a transition from neutrophilic to granulomatous inflammation. Ultimately, fine needle aspirates of the mass revealed numerous, large, round cells with abundant, fine, purple cytoplasmic granules; occasional cells contained leukocytes, hemosiderin, and calcium-like crystals. Malakoplakia, or a granular cell tumor, were considered the most likely differential diagnoses. Histologically, the submucosa was expanded by numerous round cells with abundant, eosinophilic and PAS positive cytoplasmic granules (von Hansemann cells). Occasional intracellular and extracellular, 2 to 10 microns, basophilic structures were noted, which were positive for calcium (von Kossa and Alizarin Red stains), and frequently had a target-like appearance (Michaelis–Gutman bodies). Positive staining for CD18 confirmed histiocytic origin of the cells, supporting a final diagnosis of malakoplakia. Malakoplakia is a rare, granulomatous disease that can mimic neoplasia. Exact pathogenesis remains unclear, but impaired bactericidal function of macrophages is the proposed underlying mechanism. Only four cases have been reported in veterinary literature and, to our knowledge, this is the first cytological description of malakoplakia in domestic animals.

December 6, 2016
1:50 PM – 2:00 PM
CHANGES IN THE POSTMORTEM MICROBIOME OVER TIME
Kelsey E Lawrence, Khiem Lam, Renee Greer, Andrey Morgun, Natalia Shulzhenko, Christiane V. Löhr

Background: Postmortem microbiological analysis is important in diagnosing infectious morbidities and mortalities. To interpret findings from such analyses correctly, knowledge of the postmortem changes in the microbiome in tissues, over time is critical but is currently very limited. The goal of this study was to determine the effect of time and ambient temperature on the postmortem microbiome and bacterial translocation from the large intestine to the small intestine and extraintestinal sites.

Methods: The bodies of 20 healthy rabbits were kept either at 4°C or at ambient room temperature (20°C). Cecum, Ileum, kidney and lung were sampled at 0, 6 and 48 hours postmortem. Total DNA content was determined by qRT-PCR and microbiome diversity determined using 16S rRNA gene sequencing (Illumina MiSeq platform).

Results: Over time, total DNA, determined by qRT-PCR, decreased and bacterial DNA increased in tissue from bodies kept at 20°C. This trend was less pronounced and more variable at 4°C. Based on relative abundance of the microbiome, there was clear separation of the four tissues over all time and temperature points. Only cecal bacteria had sustained postmortem growth in all conditions. Interestingly, while cecal composition did not significantly change, there was an outgrowth of members of *Proteobacteria* in lungs and kidneys at specific times and temperatures. Bacterial translocation from intestine to extra-intestinal sites was not identified in the first 48h at 4C and was minimal at 20C.

Conclusions: Tissue-specific microbiomes change during the postmortem interval and this may affect outcomes from diagnostic testing.

December 6, 2016
2:00 PM – 2:10 PM
HEPATOSPLENIC T-CELL LYMPHOMA IN 16 HORSES
Adam Michel O Michel, William Vernau, Peter F Moore

Hepatosplenic T-cell lymphoma is an extra-nodal lymphoma mostly involving gamma-delta T-cells of splenic red pulp origin. It has only been described in a single horse. We assessed the clinicopathologic findings of 16 cases of equine hepatosplenic lymphoma that presented to the UC Davis Veterinary Medical Teaching Hospital from 1986 to 2016. The median age on presentation was 26-years and males (13/16) were more frequently affected than females. All horses had a rapid clinical course. Eleven horses had complete blood counts and 9 were anemic (median hematocrit 24%) with 5/11 having an increased MCV. Ten of 11 horses were thrombocytopenic (median platelet count 20,900/ul). Ten horses had serum biochemistry done with increased SDH present in 5/10, increased bilirubin in 9/10 and increased GGT in 10/10. Neoplastic lymphocytes were medium to large and variably erythrophagocytic in all cases. Horses had defining involvement of the splenic red pulp and hepatic sinusoids, along with intravascular involvement of lungs (15/16), bone marrow (8/16) and kidneys (8/16) in the absence of overt leukemia. Neoplastic lymphocytes expressed CD3 in all cases and when done (6/16), all had cytoplasmic Granzyme B immunoreactivity. Our findings suggest that equine hepatosplenic T-cell lymphoma is a rare, high-grade lymphoma of older horses, involving cytotoxic, erythrophagocytic T-cells. The clinicopathologic characteristics of this disease in horses are similar to those described in dogs and humans. Horses present with anemia, severe thrombocytopenia and cholestasis. Besides splenic and hepatic involvement, horses frequently have intravascular involvement of multiple other organs in the absence of overt leukemia.

December 6, 2016
2:10 PM – 2:20 PM

CARDIAC PATHOLOGY OF EXERCISE-ASSOCIATED SUDDEN DEATH IN RACEHORSES

Santiago Diab, Federico Giannitti, Jorge García, Francisco Uzal

Racehorses with exercise-associated sudden death often show no significant gross lesions to explain the cause of death. Death is then typically assumed to be related to a cardiac abnormality and, hence, the term "exercise-associated sudden cardiac death" (EASCD) is used. The hearts of 27 EASCD racehorses (group 1) and 27 racehorses euthanized after catastrophic musculoskeletal injuries (group 2) were thoroughly examined histologically following the same protocol. Lesions were categorized as follows: 1- Non-inflammatory, including myofibrillar degeneration, contraction band necrosis and myofiber drop-out; 2- Inflammatory, including mural/valvular endocarditis, myocarditis and epicarditis; 3- Fibrosis; and 4- Miscellaneous. Lesions of any type were observed in 24 (89%) horses from group 1 and 26 (96%) from group 2. Non-inflammatory lesions were observed in 17 (63%) horses from group 1 and 9 (33%) from group 2. Inflammation was observed in 23 (85%) horses from group 1 and 25 (93%) from group 2. Fibrosis was present in 6 (22%) horses from group 1 and 5 (18.5%) from group 2. Miscellaneous lesions were recorded in 9 (33%) horses from group 1 and 10 (37%) from group 2. The main difference between groups was the presence of myofibrillar degeneration in the atrial myocardium of 10 (37%) horses from group 1 only. Although most cardiac lesions were similarly present in horses from both groups, suggesting that many of these may be of limited clinical significance, myofibrillar degeneration was consistently associated with EASCD. Further research to establish possible cause/s of EASCD in racehorses is warranted.

December 6, 2016
2:20 PM – 2:30 PM

ACTINOBACILLUS EQUULI CAUSING SEPTICEMIA IN GROW-FINISH SWINE

Rachel J Derscheid, Adam Krull

Background: *Actinobacillus equuli* subsp. *haemolyticus* is occasionally reported as a cause of septicemia in pigs, with very few reports from the United States. The rarity of clinical disease in swine due to *A. equuli* and its close relation to other members within the genus *Actinobacillus* make this a somewhat challenging diagnosis.

Methods: We isolated *A. equuli* from lesions in pigs from diagnostic cases at Iowa State Veterinary Diagnostic Laboratory and further typed them by 16S rRNA sequencing and toxin gene expression. We then compared these to isolates of *A. equuli* isolated from equine cases and *A. suis* from porcine cases.

Results: Results aligned with expected gene expression and confirmed the case isolate as *A. equuli* subsp. *haemolyticus.*

Conclusion: This case series includes the first report of septicemia due to *A. equuli* in grow-finish swine.

December 6, 2016
2:30 PM – 2:40 PM

AN OUTBREAK OF MONENSIN TOXICITY IN HORSES IN CALIFORNIA

Guillermo M Rimoldi, John M Adaska, Patricia C Blanchard, Robert B Moeller, Robert H Poppenga

At an equestrian facility hosting 42 horses, acute onset of clinical signs including weakness, seizures, inability to walk or stand and muscular fasciculation were detected in 5 animals, early on Saturday morning, following exposure to a new batch of pelleted feed the previous day. Two adult horses died by night and were submitted on Sunday morning to California Animal Health and Food Safety Laboratory System (CAHFS) in Tulare, together with 2 pelleted feed samples. No gross lesions were detected during necropsy. Histologically, mild to moderate, segmental skeletal muscle degeneration with scattered swollen sarcoplasmas with pale eosinophilia, loss of cross striation, occasional sarcoplasmic fragmentation and occasional hemorrhages were detected in multiple sections. Degenerative changes were highlighted with PTAH staining, revealing severe muscular damage, affecting large numbers of myofibers. Heart lesions were mild to moderate, were restricted to atria with no significant lesions detected in ventricular sections. Monensin exposure was confirmed through detection in stomach contents, livers and hearts. Feed analysis detected elevated monensin levels, 130 and 150 ppm, at approximately 4 times the maximum recommended dose for cattle. Horses are 20 times more sensitive than cattle (DL50s, mg/kg, cattle, 26; horses, 1.3). On the following days the remaining 3 affected horses died, no necropsy was performed. Over the following months, 2 more horses died and 6 others were euthanized with histories of behavioral changes, primarily aggression. No significant heart or skeletal muscle lesions were detected in most of the animals with exception of one, where chronic, atrium osseous metaplasia was detected.

December 6, 2016
2:40 PM – 2:50 PM

SEVERE NECROTIZING HEPATITIS AND PNEUMONIA DUE TO CANINE HERPESVIRUS INFECTION IN TWO UNRELATED 8- AND 10-WEEK-OLD PUPPIES

Drew Magstadt, Greg Stevenson, Eric Burrough, Jianqiang Zhang

Background: Two puppies, aged 8- and 10-weeks-old, were submitted one month apart to the ISU-VDL. The older puppy had a history of acute respiratory disease and spontaneous death. The younger puppy died suddenly with no reported clinical signs.

Methods: Necropsy, bacterial culture of the liver and lung, and viral PCR of the liver and lung.

Results: Findings were similar in both cases. At necropsy, lungs were moderately firm and non-collapsing. There was mild hepatomegaly. Microscopic findings included random areas of coagulative and lytic hepatocellular necrosis. Swollen hepatocytes at the periphery of necrotic foci often contained round to polygonal, eosinophilic intranuclear inclusion body. In the lung, there was random necrosis of alveolar septa with associated accumulation of fibrin and neutrophils. Occasional pulmonary epithelial cells contained intranuclear inclusions. No significant pathogens were recovered from the liver and lung. PCR for Canine Herpesvirus was positive; Canine Adenovirus and Canine Circovirus PCR were negative.

Conclusion: Canine herpesvirus 1 (CHV1) infection is most commonly observed in puppies less than 2 weeks of age and is associated with rapid deterioration, high mortality, and random foci of necrosis in multiple organs. Infection in older puppies and adult dogs is most often reported as a non-fatal upper respiratory disease. However, these cases suggest CHV1 should be considered a differential diagnosis in cases of acute fatal respiratory disease or sudden death in puppies up to 10 weeks of age.

December 6, 2016
2:50 PM – 3:00 PM

MOUSE HEPATITIS VIRUS-LIKE PRESENTATION OF PROLIFERATIVE TYPHLOCOLITIS IN IMMUNODEFICIENT SENTINEL MICE

Katherine Wasson, Denise M Imai, Kerriann M Casey, Stephen M Griffey

Granulomatous hepatitis, splenitis and serositis with necrosis and intralesional multinucleated cells (possible syncytia) were identified in 8 homozygous athymic nude sentinel mice submitted from 1 conventional and 3 high barrier vivaria over a 5 day period. These index cases also had proliferative typhlocolitis characterized by neutrophilic, plasmacytic and granulomatous inflammation with mucosal hyperplasia and lamina proprial multinucleated cells. Because syncytia are considered a diagnostic feature of mouse hepatitis virus (MHV) infection, extensive testing was performed on the index cases, the cohort heterozygous nude sentinels and additional sentinels from corresponding vivaria. All serologic and qRT-PCR results were negative for MHV. Mice were also negative for other routinely surveyed pathogens, including *Helicobacter* spp.. Special stains (Gram, acid-fast, Steiner's silver) and ultrastructural evaluation of lesions and affected tissues failed to identify any potential pathogens. Routine rodent health screening of the production colony of these athymic nude sentinel mice remains negative for all potential pathogens tested. Based on these findings, the granulomatous lesions in the index cases were considered regional extensions of proliferative typhlocolitis and an infrequent manifestation of the enteric disease. This also suggests that the formation of multinucleated cells in homozygous athymic nude mice may represent a generalized response to chronic immune stimulation and should not be confused with pathognomonic lesions of MHV.

December 6, 2016
3:30 PM – 3:40 PM

RENAL HEMANGIOSARCOMA WITH PULMONARY METASTASIS IN A MARE

Katherine Hughes, Vikki HL Scott

Background: Hemangiosarcoma is an uncommon tumor in horses, although previously documented cases include neoplasms arising at cutaneous, ocular, vulvar, vertebral body and multicentric sites, some with metastasis. A thirteen year-old female Shire horse had presented a year previously with a history of intermittent mild serosanguinous nasal discharge and poor body condition. Evaluation at that time had not revealed the cause of the clinical signs, which subsequently improved. A year later the mare re-presented with a short history of bilateral mild, persistent, epistaxis and intermittent dullness. The mare deteriorated acutely, exhibiting abnormal mentation and worsening epistaxis, and died prior to further evaluation.

Methods: Gross and histopathological post mortem investigations were undertaken.

Results: The left kidney was expanded by a very large, poorly demarcated, and unencapsulated mass which was friable, dark red and cavitated. The lung lobes were expanded by myriad multifocal red, moderately firm, well-demarcated, unencapsulated foci, ranging from 3 mm to 10 mm diameter. Histopathology of renal and pulmonary sections revealed an unencapsulated neoplastic mass composed of spindloid cells arranged around blood filled spaces. The cells frequently exhibited oval nuclei with two prominent basophilic nucleoli. The neoplastic cells exhibited numerous mitotic figures and a moderate degree of anisocytosis and anisokaryosis. The cells exhibited positive punctate immunohistochemical cytoplasmic staining for factor VIII–related antigen and CD31. The final diagnosis was renal hemangiosarcoma with pulmonary metastasis.

Conclusions: Although uncommon, metastatic hemangiosarcoma should be considered as a possible differential diagnosis for horses presenting with a chronic history of intermittent epistaxis and weight loss.

December 6, 2016
3:40 PM – 3:50 PM

RENAL HISTOPATHOLOGY FOR NECROPSIED DOGS WITH FANCONI SYNDROME AFTER EXPOSURE TO JERKY PET TREATS

Jennifer Jones, Rachel Cianciolo, Cathy Brown, David Rotstein, Olgica Ceric, Lee Anne Palmer, Lauren Carey, Mary Nabity, Renate Reimschuessel

FDA has received over 5200 reports of canine illness after jerky pet treat (JPT) consumption, with over 1100 reported canine deaths. Approximately 7% of the reports to FDA indicate Fanconi-like syndrome (FS). The death rate for FS is low, and most dogs with FS reported to FDA have recovered with supportive care and JPT withdrawal. Vet-LIRN collaborated with noted renal pathologists to evaluate clinical data, feeding histories, and renal tissue from 61 necropsied dogs reporting various pre-mortem illnesses and JPT consumption. Only 11 of 61 dogs fit our case definition of clinical FS having pre-mortem normoglycemic glucosuria and/or a positive urine FS screening test. Two pathologists (RC, CB) from the International Veterinary Renal Pathology Service examined renal tissue and assigned a primary renal lesion to 10 of the 11 dogs. One dog was excluded due to tissue autolysis. Findings included: chronic kidney disease (CKD) (n=4), acute tubular necrosis (ATN) (n=3), immune complex glomerulonephritis (n=1), focal segmental glomerulosclerosis (n=1), and renal atherosclerosis (n=1). CKD and ATN were the most common lesions. Although this study initially evaluated approximately 5% (n=61) of the reported canine deaths after JPT consumption, the dogs with pre-mortem FS and kidney lesions (n=10) only represent approximately 2.5% of FS cases reported to FDA and approximately 1% of all reported deaths. It is not clear whether the described primary renal lesions represent those that may exist in dogs with clinical FS after JPT consumption. Vet-LIRN continues gathering data to help detect a root cause for the reported illnesses.

December 6, 2016
3:50 PM – 4:00 PM

BILATERAL PHEOCHROMOCYTOMA IN THE ADRENAL GLANDS OF AN 11 YEAR OLD BOER GOAT

Maureen T. O'Brien, Rodolfo Madrigal, George A. Stoica

An 11-year-old Boer goat wether presented to the Texas A&M Large Animal Emergency Services for a one-day history of recumbency due to hindlimb weakness. The wether had been hospitalized two weeks prior for lethargy, weight loss, and recumbency. While hospitalized, treatment included florfenicol for intermittent pyrexia, iron supplementation for anemia, and moxidectin for possible parasitism. Clinical evaluation upon return to the emergency service revealed decreased to absent reflexes bilaterally in the hindlimbs with lack of nociception and decreased reflexes in the forelimbs. Prognosis was poor due to clinical signs and euthanasia was elected. Relevant necropsy findings included bilateral adrenal medullary masses as well as bilateral stifle osteoarthritis, mucous membrane pallor, and poor body condition. Histology revealed bilateral adrenal medullary pheochromocytomas, with local cortical invasion and intravascular metastases. Additionally, one of the adrenal glands had central adrenal vein thrombosis associated with a locally extensive area of pleomorphic neoplastic cells, suggesting focal intra-tumoral heterogeneity as a result of microenvironment alteration (hypoxia) subsequent to the thrombus. Pheochromocytomas arise from chromaffin cells within the adrenal medulla and are uncommonly reported in goats. This case highlights a tumor with interesting histologic features that may have contributed to the intermittent pyrexia and weight loss reported clinically.

December 6, 2016
4:00 PM – 4:10 PM

TUMOR MARGIN ASSESSMENT USING A MODIFIED MOHS PROCEDURE IN AN ATLANTIC BOTTLENOSE DOLPHIN (TURSIOPS TRUNCATUS) WITH ORAL SQUAMOUS CELL CARCINOMA

David Rotstein, Lydia Staggs, Jeremy Sunseri

A 36 YO, F Atlantic bottlenose dolphin (*Tursiops truncatus*) had a single papillary mass on the right side of its mouth. The mass was observed in 2013 and initial biopsies indicated epithelial hyperplasia. The growth steadily increased in sized. Several other papillary masses grew and were biopsied. Some were still hyperplastic, but one mass was diagnosed as carcinoma-in-situ in 2015. Serial biopsies were taken from the original mass during this time, and all of these samples failed to show any histological evidence of neoplasia. Then, in the winter of 2016, a biopsy showed carcinoma-in-situ which was subsequently treated with CO_2 laser. Neoplastic cells were not observed at the borders of two additional excisions, but regrowth occurred and a mixture of carcinoma-in-situ and invasive squamous cell carcinoma was observed. Neoplastic cells were present at the margin. To attempt complete excision, a modification of Mohs Micrographic Surgery was performed in an effort to obtain complete margin control given the narrow surgical margins. The excised tissue, in a "double pacman" formation was mapped, inked, and processed routinely. Surgical margins were incomplete and additional surgery was performed. Mohs surgery could be useful in excisional biopsies for species which are not amenable to multiple capture, sample, and release.

December 6, 2016
4:10 PM – 4:20 PM

H1N1 INFLUENZA IN MINK KITS

Christine Watson, Kathy Toohey-Kurth, Hugh Hildebrandt, John Easley, Christopher Booth, Mary Lea Killian, Jan Shivers, Phillip Bochsler, Kathleen Deering, Doug Lyman, Amaranthan Muthuswamy, Melissa Behr

In spring 2016, an outbreak of respiratory disease in Wisconsin mink kits less than 1 month of age was observed in 4 mink operations in Northern Wisconsin. High mortality (20%) of young kits resulted in several submissions to WVDL. Six kits, 1-3 days of age were submitted on 05/10/2016, and divided into two groups. Interstitial pneumonia was noted histologically in both groups with fluid, fibrin and neutrophils in alveoli. Tracheal swabs from both groups were positive by polymerase chain reaction (PCR) for Influenza A; liver and lung samples from both groups were culture-positive for *Salmonella* Dublin. A second submission of four 4-week-old mink was received at WVDL on 05/24/2016.

They had died acutely or after a brief illness with tachypnea, and despite treatment with fluids, meloxicam and Baytril®. Grossly, lungs were diffusely dark red or red-brown, or had multifocal to coalescing, opaque, irregular, blotchy dark red foci interspersed with salmon pink pulmonary parenchyma. Histologically, widespread interstitial pneumonia was noted, characterized by fibrin, fluid, necrotic debris and PMNs in alveoli. Also, 25-50% of epithelia lining nasal passages were necrotic or ulcerated, with edema or lymphocytic-plasmacytic inflammation of the submucosa. Mink were negative for Aleutian Disease by PCR and positive for influenza A by PCR. Immunohistochemistry for Influenza A demonstrated nuclear staining in bronchial epithelium as well as in bronchial submucosal glands. Next generation sequence analysis of several isolates was done at the National Veterinary Services Laboratories (NVSL), and showed that they were all pandemic H1N1.

December 6, 2016
4:20 PM – 4:30 PM

PATHOLOGY AND IMMUNOHISTOCHEMICAL FINDINGS OF DISSEMINATED SALMONELLA TYPHIMURIUM INFECTION IN MODENA PIGEONS

Monique Franca, Joanna Echenique, Stabler Stabler, Hulimangala L Shivaprasad

Background: *Salmonella* Typhimurium infection in pigeons is typically associated with hepatitis, splenitis and encephalitis. We report the pathology and immunohistochemical findings of disseminated *Salmonella* Typhimurium infection in Modena Pigeons.

Methods: Three deceased juvenile Modena pigeons were submitted to the California Animal Health and Food Safety Laboratory System (CAHFS), Tulare Branch for necropsy. These pigeons presented neurological signs prior to euthanasia. Necropsy was performed and samples were collected for histopathology, bacteriology and immunohistochemistry.

Results: Gross lesions observed included yellowish exudate in the brain effacing the lateral cerebral and optic regions, mildly enlarged and friable spleen and pale nodule in the right caudal lung lobe. Microscopically, the pigeons had disseminated lesions of bacterial infection and septicemia including fibrinoheterophilic sinusitis, pneumonia, hepatitis, myocarditis, splenitis, nephritis, iridocyclitis, meningoencephalitis, meningomyelitis, osteomyelitis of the long bones and cranium, and otitis interna. Rod-shaped bacteria were observed within the lesions in many organs. *Salmonella* Typhimurium var Copenhagen was isolated in pure culture from brain, liver, lung and intestine samples. Immunohistochemistry using a *Salmonella* group B polyclonal antibody revealed a few to numerous *Salmonella* in all tissues with lesions.

Conclusion: In conclusion, we report disseminated *S.* Typhimurium infection with unusual lesions in Modena pigeons. Otitis interna is rarely reported in avian species and it was previously reported in turkey poults infected with *Salmonella*enterica*arizonae. S.* Typhimurium most likely spread from the brain to the inner ear through the vestibulocochlear nerve.

December 6, 2016
4:30 PM – 4:40 PM

PHAEOHYPHOMYCOSIS IN ZEBRAFISH

Laura Baseler, Christiane V. Löhr, Jan Spitsbergen

In studies of spontaneous and induced neoplasia in over 20,000 zebrafish at Oregon State University, systemic infections caused by pigmented fungi were observed on histologic examination of 5 fish. The zebrafish ranged from three to nineteen months old and represented multiple genetic strains. Gross lesions were variable and included poor body condition and cutaneous ulcers. Histologically, granulomatous inflammation or discrete granulomas surrounding variable numbers of fungal hyphae were present in one or more of the following organs: swim bladder lumen and wall, kidney, spinal cord, liver, intestine and gonads. Fungal hyphae were most numerous in the swim bladder. Hyphae were pigmented brown, frequently septate, rarely branched and had nonparallel walls approximately 3 µm in width, consistent with phaeohyphomycosis. Phaeohyphomycosis typically presents as an opportunistic infection caused by one of multiple genera of pigmented fungi that are often ubiquitous in nature. It is suspected that immunosuppression, induced by administered treatments, stress or genetic modification, predisposed these zebrafish to a water-borne fungal infection. Fungi were not detected in the cutaneous ulcers and vascular invasion was not observed in any of the affected organs. Therefore, the digestive tract and pneumatic duct were the presumed port of entry, while the swim bladder served as the initial target organ. This case demonstrates that zebrafish are susceptible to phaeohyphomycosis and that infection of the swim bladder likely results in secondary invasion of adjacent tissues through direct local extension, rather than vascular dissemination.

December 6, 2016
4:40 PM – 4:50 PM

CANINE OSTEOSARCOMA CELLS CAN LACK ALKALINE PHOSPHATASE ACTIVITY

Amy L. MacNeill, Alex Byas, Lauren Harris, Allison Vilander, Deanna Dailey, Kristen Weishaar

The goal of this study was to determine if vitamin D deficiency was present in a dog with osteosarcoma that lacked alkaline phosphatase (ALP) activity. A fine needle aspirate of a bony lesion was diagnosed as a malignant neoplasm. Differential diagnoses included osteosarcoma, hemangiosarcoma, other sarcomas, and metastatic carcinoma. Unstained cells incubated in BCIP/NBT were negative for ALP, making osteosarcoma less likely based on this parameter. .The limb was submitted for histology and diagnosed as osteosarcoma. Neoplastic cells were positive for vimentin and negative for cytokeratin, von Willebrand Factor, and CD18, supporting the diagnosis of osteosarcoma. ALP is decreased in some human osteosarcoma explants from poorly differentiated tumors. In culture, human osteoblasts can be induced to produce ALP after incubation in calcitriol. To determine if the same effect is observed in canine osteoblasts, cells were incubated in calcitriol for 48 hours, collected, and incubated in BCIP/NBT. In two isolates of cultured canine osteosarcoma cells, the number of cells that were positive for ALP increased following incubation with calcitriol in a dose-dependent manner. Low serum [calcitriol] is associated with an increased risk of cancer in dogs. This patient's serum 25-hydroxy-vitamin D concentration was 39.1 ng/mL (reference interval; 100-150 ng/mL). Although the low [vitamin D] may be a coincidental finding, it is intriguing to consider that it contributed to the lack of ALP activity in this patient's cells. These data suggest it would be prudent to determine serum [vitamin D] in osteosarcoma patients with unstained neoplastic cells that are negative for ALP.

December 6, 2016
4:50 PM – 5:00 PM

CANINE DISTEMPER: DETECTION OF A NOVEL GENOTYPE CAUSING DISEASE IN WILDLIFE

Jenny P Pope, Debra L Miller, Matthew C Riley, Eman P Anis, Rebecca P Wilkes

Canine distemper virus (CDV) is a common cause of a multi-systemic disease in both domestic dogs and wildlife species, including raccoons and foxes. Outbreaks of CDV in domestic dogs in eastern Tennessee have occurred since 2012, and it was determined that these outbreaks resulted from a novel genotype of CDV. We hypothesized that this virus is also infecting area wildlife and may be a source of the virus for these outbreaks in dogs. From 2013–2014, necropsies were performed and tissues collected from raccoons (*Procyon lotor*; *n* = 50) and gray foxes (*Urocyon cinereoargenteus; n* = 8) for CDV testing. A real-time, reverse-transcription polymerase chain reaction was used to document presence of CDV in tissue samples and a portion of the virus was subsequently sequenced for phylogenetic analysis. A high percentage of wildlife, both with (86%) and without clinical signs (55%), tested positive for CDV, with the majority (77%) testing positive for the novel genotype. Microscopic findings, including syncytia in the lungs and viral inclusions in urothelium, astrocytes, neurons, and bronchiolar epithelium, were also consistent with canine distemper. Minimal inflammation in the central nervous system of affected animals is indicative of the acute neurologic form of the disease. Pneumonia and parasitism were also commonly found in CDV-infected animals. Based on these results, CDV appears to be prevalent in East Tennessee wildlife. Subclinical or clinically-recovered shedders are a potential source of this novel genotype for domestic dogs, and this genotype is genetically distinct from vaccine strains.

Diagnostic Pathology Posters

D-01: MULTICENTRIC GLIOMA IN THE SPINAL CORD AND FOREBRAIN OF A BOSTON TERRIER

Whitney M Zoll, Ember D Epperson, Ji-Hey Lim, Sheila Carrera-Justiz

An 11-year-old neutered male Boston terrier presented to the University of Florida's College of Veterinary Medicine Neurology service with a one-week history of progressive proprioceptive ataxia, muscle tremors and cervical pain. An MRI showed an intramedullary spinal cord lesion at the level of the fourth cervical vertebra and multifocal lesions within the right aspect of the thalamus and midbrain. The patient died spontaneously at home and was submitted for necropsy. Grossly, the brain was unremarkable. Within the left dorsal aspect of the C4-C5 spinal cord, there was a 1.5 x 0.4 x 0.4 cm soft red mass. Upon sectioning, the mass was dark red, glistening, well demarcated, and it markedly displaced the ventral median fissure and midline to the right. Microscopically, the spinal cord mass was expansive, well demarcated, densely cellular, and composed of stellate to spindle cells forming sheets supported by a fine fibrovascular stroma. There were areas of vascular proliferation along the periphery of the mass and glial cells psuedopalisading around areas of necrosis and hemorrhage. Within the right hippocampus and thalamus were poorly demarcated foci of increased cellularity composed of fibrillary astrocytes with rare mitoses. The mass within the spinal cord was diagnosed as a high-grade astrocytoma and the masses within the hippocampus and thalamus were diagnosed as medium-grade astrocytomas. This is a rare phenomenon that to the authors' knowledge has only been reported in the human literature.

D-02: BACTERIAL VENTRICULITIS, MENINGOENCEPHALITIS, AND BRONCHOINTERSTITIAL PNEUMONIA IN A FERRET (MUSTELA PUTORIUS FURO)

In Joong Kim, Mellisa Nau, James W. Carpenter, Brian Lubbers, Ada G. Cino-Ozuna

A 2-month-old, male neutered ferret with history of intermittent generalized seizure and mucoid diarrhea was necropsied. On gross examination, the animal had a thin body condition (1/5) with pale mucous membranes and severe flea infestation. The brain had an area of edema and purulent exudate in the meninges at the left occipital lobe. The lungs were not deflated and were diffusely edematous, firm, and congested. Histologically, the lateral, third, and fourth ventricles of the brain contained numerous degenerate and non-degenerate neutrophils, few macrophages, and fewer lymphocytes with cellular debris and multiple colonies of gram-positive cocci that extended into the subventricular cortex. The meninges were diffusely expanded by edema and multifocal moderate lymphocytes and plasma cells, and lesser neutrophils with cellular debris. In the lungs, alveolar septa were diffusely congested and expanded by few to moderate neutrophils, edema, and fibrin. Approximately 40% of alveoli and some airways contained moderate alveolar macrophages and few neutrophils with edema, fibrin, and hemorrhage. Bacterial culture revealed abundant Streptococcus castoreus, low numbers of Stenotrophomonas maltophilia, and Pseudomonas sp. from the brain; and low numbers of Streptococcus zooepidemicus and unidentified gram-negative bacteria from the lung. Streptococcus castoreus was recently reported from a bitten skin wound in a European beaver (Castor fiber) submitted from a wildlife park. Streptococcus zooepidemicus is a mucosal commensal and opportunistic pathogen of cats, dogs, pigs, horses and human. To the authors' knowledge, this is the first report of Streptococcus-induced ventriculitis and encephalitis in a ferret, associated with bronchointerstitial pneumonia.

D-03: COPPER-ASSOCIATED HEPATOPATHY IN A FRIESAN HORSE
Linda Huang, Kurt J Williams

Copper hepatopathies are well described in sheep and some breeds of dogs; copper-associated hepatopathies are very rare in horses. A 20-year-old Friesian gelding had a 4 month history of decreased appetite and persistently elevated liver enzymes. GGT levels over the 4 month period ranged from 62-98 U/L (ref range: 5-24 U/L) while AST ranged from 568-717 U/L (ref range: 175-340U/L). The horse was housed in a facility with 35 other horses and was the only animal with elevated hepatic enzyme activity. The liver appeared normal during ultrasound-guided Tru-cut biopsy. Histologically the liver had moderate periportal bridging fibrosis, biliary hyperplasia with periportal lymphohistiocytic hepatitis, 'pigment granulomas', individual hepatocellular necrosis, and diffuse prominent hepatocellular cytoplasmic copper accumulation. Hepatic copper concentration was determined to be 1796.94µg/g Dry Matter (ref int: 12.00-175.00µg/g) using inductively coupled plasma-mass spectroscopy confirming the diagnosis of copper-associated hepatopathy. In sheep, copper hepatopathy usually develops secondary to excess dietary copper intake. Primary copper-associated hepatopathies in dogs have been reported in mostly in Bedlington terriers and Labrador Retrievers. Secondary copper hepatopathy in the current case is unlikely, as horses are resistant to dietary copper toxicity and this was the only horse on the farm to have biochemical evidence of underlying liver disease. A single case of copper-associated hepatic cirrhosis has been previously reported in a Friesian horse in the Netherlands. A mechanism for excessive copper accumulation in the Friesian horse liver is not known. A potential genetic basis for hepatic copper accumulation in Friesian horses remains to be evaluated.

D-04: A CASE REPORT: LARGE B-CELL LYMPHOMA WITH PLASMA AND MOTT CELL DIFFERENTIATION IN A YOUNG DOG
Alexandra M Dieterly, Grant Rezabek, April White, Laura Nafe, Melanie Breshears, Theresa Rizzi

Background: A 2.5 year-old, German Shepherd dog presented for one week of severe weight loss, vomiting, and diarrhea. Cytopathology, histopathology, PCR for antigen receptor rearrangement, and immunohistochemistry culminated in a diagnosis of splenic and hepatic large B-cell lymphoma with plasmacytic and Mott cell differentiation.

Objective: We describe clinical findings, cytology, histopathology, PCR testing, and immunohistochemical staining of this unusual neoplasm.

Methods: Diagnosis was achieved through cytopathological examination of fine needle aspirates and histopathology of spleen and liver, PCR testing, and immunohistochemistry for lymphocyte surface markers.

Results: Cytology revealed a round cell population with intracytoplasmic inclusions, consistent with Russell bodies.Histopathology of spleen and liver revealed neoplastic lymphocytes with scant amounts of globular, deeply eosinophilic cytoplasm, often with an intracytoplasmic perinuclear clear zone and single or multiple nuclei with marginated chromatin. The majority of neoplastic cells were positive for CD45, CD20, MUM-1, and IgM. In contrast to similar reported neoplasms, neoplastic cells were diffusely negative for CD79a and Pax-5. PCR for antigen receptor rearrangement revealed a clonal IGH3 population.

Conclusion: Cytology, histopathology, and clonality results culminate in a diagnosis of large cell, B-cell lymphoma with plasmacytoid and Mott cell differentiation, which has been reported only a few times in dogs. This poster presents the salient histologic, cytologic, and immunohistochemical staining features. Additionally, the neoplasm was diffusely negative for CD79a and Pax-5, which is unusual and may represent receptor loss as reported in people with plasmablastic lymphomas and Classical Hodgkin's lymphoma, respectively.

D-05: OSTEOSARCOMA IN A MATSCHIE'S TREE KANGAROO

Whitney M Zoll, Majorie Bercier, Jim Wellehan, William F Craft

A 16-year-old intact female Matschie's tree kangaroo presented to the University of Florida for lethargy, ataxia, and abnormal mentation. A head tilt was observed during neurologic examination. A CSF tap was blood contaminated and negative by PCR for *Angiostrongylus cantonensus*. A CT scan showed a mass in the right cervical region at the level of C1 extending into the foramen magnum. The tree kangaroo was treated with antiparasitics and antibacterials. The following day, marked dysphagia was noted. The patient was euthanized and submitted for necropsy. Gross examination revealed a 4 x 2.5 x 1 cm firm, tan mass that infiltrated and replaced the skeletal muscle of the right lateral neck at the level of the C1. The mass extended rostrally, infiltrating into the right ventrolateral basioccipital bone and extended into the cranial vault forming a 1.5 x 1 x 0.8 cm light tan, firm, and nodular mass. The mass replaced portions of cranial nerve XII and was associated with areas of hemorrhage along the ventral surface of the medulla oblongata. Histologic examination revealed pleomorphic neoplastic mesenchymal cells that formed interlacing streams and sheets of cells that surrounded osteoid multifocally. The mitotic index was variable ranging from seven (extracranial mass) to 50 (intracranial mass) mitoses per ten 400x fields. Multifocal vascular metastasis was observed. Within the neuropil of the medulla oblongata there were multifocal areas of necrosis and axonal degeneration with numerous spheroids. This is the first report of an osteosarcoma in a Matschie's tree kangaroo.

D-06: HISTOCHEMICAL ANALYSIS OF EOSINOPHILIC CRYSTALS IN A LARYNGEAL MYXOSARCOMA

Eunju Choi, Makoto Asakawa, Marina McConkey, Andrew D Miller, Jeanine Peters-Kennedy

A 2-year-old, female spayed, Labrador Retriever dog presented to the Cornell University Hospital for Animals for laryngeal mass resection. Histopathology revealed loose, haphazardly organized neoplastic spindle cells separated by myxomatous matrix. Embedded within the neoplasm were occasional radiating eosinophilic crystalline spicules. A preliminary diagnosis of laryngeal myxosarcoma with Charcot-Leyden crystals (CLCs) was made.

Histochemical staining including Ziehl-Neelson (ZN) for CLCs, Masson's trichrome, periodic acid Schiff (PAS), Alcian blue, and Okajima for hemoglobin were applied. The crystals stained red with trichrome, precluding collagen. PAS and Alcian blue were negative for the crystals, but highlighted the matrix supporting the diagnosis of myxosarcoma. ZN was negative ruling out CLCs. Okajima was positive for the crystals along with the erythrocytes. These results confirmed that these were hemoglobin crystals.

In vivo hemoglobin crystal formation is rare, other than in the rat, possibly representing an intermediate state of hemoglobin breakdown. Hemoglobin from all other species crystallizes *in vivo* only if structurally abnormal. In this case, no hematologic abnormality was observed. CLCs are common in human diagnostic cases often associated with an eosinophilic infiltrate such as asthma or parasitic infection. The crystal morphology in this dog was similar to CLCs described in the human; however, histochemical stains were not supportive. Comparison of the protein sequence of CLCs, also known as galectin-10, to protein sequences of common veterinary species revealed no protein sequence similar to that of human CLCs. This and the lack of reports of CLCs in the veterinary literature suggest that CLCs may be human specific.

D-07: TISSUE AUTOLYSIS USING HISTOLOGICAL CRITERIA IN HORSE TISSUES

Nanny Wenzlow, Dan Neal, Maureen T Long

Background: The goal of this study was to evaluate various histological criteria in brain, liver and muscle tissue to aid in the estimation of the post-mortem interval (PMI) in horses. Currently, only a combination of very few methods are used to estimate the PMI in animal or humans with a lack of accuracy. The ability to estimate the PMI with sufficient accuracy in field cases would be a significant contribution to investigations when foul play is suspected.

Material and Methods: Brain, liver, and skeletal muscle from 9 freshly euthanized horses, were held at 22°C and 8°C for 72h. Tissues were sampled at T0h, T1h, T2h, T4h, T6h, T12h, T24h, T36h, T48h, T60h, and T72h. Histologically, criteria for autolysis were evaluated for each tissue, at each temperature and at each time point.

Results: At 22°C, in liver tissue, hepatocyte individualization and the separation of bile duct epithelial cells from the basement membrane were the most field represented criteria. This was followed by disruption of myofiber continuity, increased eosinophilia and loss of striation in muscle tissue. Milder changes were observed in tissues held at 8 °C. In brain tissue, neuronal nuclear swelling decreases initially for 36h before increasing again untill 72h at 22 °C. The most striking observation in the brain at 8°C, was the progressive shrinkage of neuronal cytoplasm.

Conclusions: Liver and skeletal muscle tissue showed the most predictable criteria of autolysis over 72h after death and these results could aid to estimate the PMI in horses in field situations.

D-08: IMMUNOHISTOCHEMICAL CHARACTERIZATION OF B-CELL AND T-CELL LINEAGE OF BOVINE LYMPHOMA CASES

ANAMIKA GUPTA

Background: Lymph node pathology is useful in diagnosing inflammatory and neoplastic diseases in animals. However, by lymph node aspirate cytology and histopathology alone, it is difficult to predict the prognosis and design treatment regimens.

Objective: The present study was done to sub-classify bovine lymphomas by immunohistochemistry.

Methods: The study was conducted on 32 cases of bovine lymphadenopathies. Impression smears were prepared from lymph nodes and other tissues and stained with Wright's technique for cytology. Tissue samples were processed for routine histopathology. Immunohistochemistry was performed by using bovine-specific primary antibodies against B-cell (CD20, CD79, p-27) and T-cell (CD3, CD4, CD8) lymphocytes.

Results: Two cases of bovine lymphoma were suspected on the basis of peripheral blood smear examination and gross lesions in peripheral and visceral lymph nodes. Cytology revealed presence of pleomorphic lymphocytes with increase nuclear to cytoplasmic ratio. Histopathology of the first case showed pleomorphic small neoplastic lymphocytes in several lymph nodes. In second case, lymph nodes showed presence of monomorphic lymphoid cell infiltration with metastases in lungs. First case showed positive cytoplasmic reactivity of CD3 and CD8 with mild reactivity of CD4, whereas, some areas revealed positive reactivity of p-27 and the case was diagnosed as mixed T-cell and B-cell lymphoma. Second case showed positive reactivity of CD3 with mild reactivity of CD4 and the case was diagnosed as T-cell lymphoma. However, no reactivity of CD20 and CD79 was observed in either of the cases.

Conclusion: Immunohistochemical characterization of lymphomas is clinically importantas some may reveal mixed lineage.

D-09: LEUKOENCEPHALOMYELITIS AND POLYNEURITIS IN A MALTESE: ASSOCIATED WITH AUTOIMMUNE RESPONSE TO THE MYELIN?

Wen-Ta Li, Hui-Wen Chang, Chian-Ren Jeng, Fun-In Wang, Victor Fei F Pang, Zhi-Jing Wu, Chen-Hsuan Liu

Background: Acute disseminated encephalomyelitis (ADEM) is an inflammatory demyelinating disorder affecting the central and peripheral nervous systems in the human, and can be associated with a T cell-mediated autoimmune response to myelin basic protein, triggered by previous infection or vaccination (post-infectious and post-vaccinial forms). To our knowledge, no animal case has been reported.

Case presentation: A one year-old male Maltese showed mild ataxia and then became paraparesis. The patient died suddenly, and necropsy was performed. Samples for histopathology were preserved and processed routinely for slide preparation. DNA and RNA were extracted from the formalin-fixed, paraffin-embedded tissue blocks.

Results: Microscopically, leukoencephalomyelitis characterized by perivascular cuffing, axonal degeneration, demyelination, and astrogliosis were noted. The nerve roots of spinal cord, sciatic nerve, and nerve fascicles around larynx and adrenal glands were multifocally infiltrated by mononuclear inflammatory cells. Immunohistochemical stainings for canine distemper virus (CDV), pseudorabies virus (PRV), and rabies virus were performed, and polymerase chain reaction (PCR) using primer sets targeting CDV, PRV, *Toxoplasma gondii*, *Neospora caninum*, *Leishmania* spp., and *Encephalitozoon cuniculi* were also performed. However, all results were negative. The possible infectious etiologies were ruled out by negative results of PCR and immunohistochemical stainings. The possibility of non-infectious diseases, including granulomatous meningoencephalitis, necrotizing meningoencephalitis, necrotizing encephalitis, and acute polyneuritis, is less likely because of the dissimilar lesion distribution.

Conclusion: The leukoencephalomyelitis and polyneuritis are highly suspected as a result of autoimmune reactivity against the myelin, similar to the ADEM in humans.

D-10: MUCOPOLYSACCHARIDOSIS TYPE I IN THE DOBERMAN PINSCHER

Allison C Vilander, Crystal Cooley, Sushan Han, Rachel Han, Urs Giger, A R Moore

A 3 year-old intact male Doberman Pinscher presented with a poor body condition (3/9), ataxia, head tremor, decreased anal tone, positional nystagmus, and decreased withdrawal reflexes. Complete blood count, blood chemistry panel, and coagulation panel showed non-regenerative anemia, increased liver enzymes, and elevated D-dimers. Abdominal ultrasound and CT scan revealed a hyperechoic liver, multiple portosystemic shunts with thrombosis, abdominal effusion, and compression of the spinal cord at C4-C7 with narrowing of the spinal canal consistent with cervical vertebral myelopathy. CSF analysis revealed intracytoplasmic basophilic inclusions within mononuclear cells; these inclusions were PAS positive suggesting a lysosomal storage disease. Due to the severity of the hepatopathy, the patient was euthanized and submitted for necropsy. On gross exam, there were numerous abdominal vascular shunts, a thrombus partially occluding the portal vein, and a nodular liver. Histopathology revealed large foamy macrophages that contained PAS-positive vacuoles in the liver, lung, heart, kidneys, skeletal muscle, brain, and cervical spinal cord. A urinary metabolic screen showed a strongly positive mucopolysaccharidosis (MPS) spot test. Additional studies indicated high concentrations of urine dermatan sulphate and a complete lack of serum alpha-L-iduronidase activity with normal to increased beta-glucuronidase and arylsulfatase B activities, indicative of MPS I. MPS I is a lysosomal storage disease that leads to accumulation of the glycosaminoglycans due to an absence of the enzyme alpha-L-iduronidase. It has been reported in Plott hounds, Rottweilers, and Boston Terriers. This is the first report of MPS I in a Doberman Pinscher.

D-11: CAUSE AND MANNER OF DEATH OF VETERINARY FORENSIC CASES FROM METROPOLITAN AREAS IN TAIWAN

Chen-Hsuan Liu, Wei-Hsiang Huang

In recent years, veterinary forensic pathology has gained more and more attention in Taiwan. During 2011-2015, 72 cases which were reported as suspected animal abuse, were submitted to the School of Veterinary Medicine, National Taiwan University, as a forensic necropsy. Specific subsequent laboratory examinations were performed based on the gross necropsy findings. Cause of death and manner of death were designated after a comprehensive evaluation of submission information, pathology findings, and results of laboratory examinations. All of the 72 cases were from the Taipei and New Taipei metropolitan areas, and included 35 cats and 37 dogs. The cause of death was determined in 52 cases, while the manner of death was determined in 45 cases. For manner of death, 40% of cases were categorised into non-accidental injury, followed by undetermined (38%), natural causes (15%), and accidental causes (7%). The types of non-accidental injury included blunt force trauma, poisoning, sharp force trauma, asphyxia, thermal injury, neglect, and entrapment. Undetermined cause of death and manner of death were attributed to insufficient information from the crime scene and the state of decomposition of the corpse.

D-12: ENTEROCOLITIS CAUSED BY YERSINIA PSEUDOTUBERCULOSIS INFECTION IN A BOER GOAT

PANKAJ KUMAR

A 6 year old, female Boer goat was presented for necropsy to the Kansas State Veterinary Diagnostic Laboratory with a clinical history of diarrhea followed by recumbency, vocalization and rapid death. Grossly, the mucus membranes of the oral cavity were diffusely white. The abomasum was filled with liquid, reddish brown ingesta that contained numerous adult Haemonchus worms. There were multifocal petechiae within the mucosa of cecum and spiral colon and the lumen contained numerous whipworms. Histologically, the intestinal lesions were most prominent in the distal jejunum and ileum. The mucosa was multifocally eroded and contained numerous microabscesses consisting of large colonies of gram-negative coccobacilli surrounded by large numbers of neutrophils. These erosions and microabscesses were either superficial or extended deep into the mucosa but did not extend into the submucosa. In sections of the large intestine, there were small scattered foci of neutrophilic inflammation in the mucosa but the large colonies of bacteria present in the small intestine were not evident. The prominent colonies of gram-negative bacteria associated with microabscessation in the intestinal mucosa were consistent with yersiniosis due to *Y. pseudotuberculosis*. The diagnosis was confirmed by bacterial isolation from the small intestine and mesenteric lymph nodes. *Y. pseudotuberculosis* is a well-recognized zoonotic food-borne pathogen and infected animals are a potential public health risk. In goats, it has been associated with enterocolitis, mesenteric lymphadenitis, septicemia, placentitis/abortion, mastitis and ocular disease. In a recent study from California, goats were the most frequently reported species with clinical yersinosis.

D-13: A PLAMACYTOID UROTHELIAL CARCINOMA OF URINARY BLADDER WITH SYSTEMIC METASTASIS AND DISSEMINATED INTRAVASCULAR COAGULATION IN A BEAGLE DOG

Susanne Je-Han J.-H. Lin, Fun-In Wang, Jih-Jong Lee, Lei-Ya Wang, Yi-Jiun Chen, Hui-Wen Chang, Chian-Ren Jeng, Chen-Hsuan Liu, Victor Fei F. Pang

Background: Plasmacytoid urothelial carcinoma (PUC) is a rare neoplasm that has been described in human literature. To our knowledge, no cases have been reported in animals. The neoplastic cells have a unique discohesive plasmacytoid appearance and the tumor usually warrants a poor outcome. We report a canine case of PUC of urinary bladder with systemic metastasis and disseminated intravascular coagulation.

Case Description: An 11-year-old, male intact beagle showed hematuria and urinary incontinence for a year and was euthanized. At necropsy, extensive purpura hemorrhagica was observed in the skin of the lower abdomen. The urinary bladder exhibited generalized thickening with multiple small, mucosal hemorrhagic nodules.

Results: Microscopically, the bladder was diffusely invaded by discohesive neoplastic cells extending from the mucosa to the muscularis. Cells had a plasmacytoid appearance with a distinct cell border, eccentric nuclei, and abundant cytoplasm. Cells were positive for pan-CK and weakly positive for E-cadherin, but negative for vimentin, mum-1, MHC II, CD79a, CD3, and S-100. Neoplastic cells and tumor emboli were found in multiple organs. In the skin, lung, liver, and spleen, there were hematomas and intravascular fibrin thrombi.

Conclusion: This may be the first report of PUC in animals which based on the morphology, immunohistochemical staining pattern (reduced E-cadherin), and aggressive behavior bore similarity to human cases which show a unique discohesive morphology with reduced expression of E-cadherin and aggressive behavior similar to those reported in humans. Diagnosis may be problematic because of the plasmacytoid appearance which highlights the importance of confirmatory immunohistochemistry.

D-14: PULMONARY/ AIR SAC CARCINOMA IN A WHITE-BELLIED SEA EAGLE

Susanne Je-Han J.-H. Lin, Bao-Rong Wang, Hui-Wen Chang, Chen-Hsuan Liu, Fun-In Wang, Victor Fei F. Pang, Chian-Ren Jeng

Background: Primary respiratory neoplasms are uncommon and infrequently documented in pet or captive birds. Herein we report a carcinoma originating from the air sac and lung in a white-bellied sea eagle (*Haliaeetus leucogaster*).

Case Presentation: A 34-year-old white-bellied sea eagle had decreased appetite and activity for about 3 weeks without improvement after treatment with antibiotics and fluid therapy. The animal exhibited excessive panting and was unable to walk before its death. At necropsy, a large, infiltrative, soft mass was present in the right dorsal coelomic cavity extending to the right caudal lung lobe, right posterior thoracic air sac, vertebral column, right kidney, and testis.

Result: On microscopic examination, the mass was poorly demarcated and infiltrated the lung, air sac, vertebral column, kidney, and testis. It consisted of compact sheets of cuboidal to polygonal neoplastic epithelial cells that occasionally formed a tubulopapillary pattern. A high degree of anisocytosis and anisokaryosis with frequent karyomegaly was present. Some neoplastic cells had cilia on cell surface. Immunohistochemistry for cytokeratin and TTF-1 was positive.

Conclusion: In this case, the neoplasm's origin as air sac or lung could not be determined based on the degree of infiltration of the respiratory system, but histopathological findings and immunohistochemical stains did narrow the source. Respiratory neoplasia has been reported rarely in psittacines, but the occurrence in other aviary species is not known. To our knowledge, there is no published report of respiratory neoplasia in this species.

D-15: CYSTIC THYMOMA IN A RED PANDA
Susanne Je-Han J.-H. Lin, Yan-Syue Lai, Hui-Wen Chang, Chen-Hsuan Liu, Fun-In Wang, Victor Fei F. Pang, Chian-Ren Jeng

Background: Thymoma is the neoplasm arising from or displaying differentiation to thymic epithelial cells, and exhibits some features such as perivascular spaces, and the presence of benign lymphocytic infiltration. Cystic degeneration of the perivascular space in thymoma is reported and should be differentiated from thymic cyst. Herein we report a red panda (*Ailurus fulgens*) with thymoma that shows multiple cystic structures.

Case Presentation: A 19-year-old, female, captive red panda had a clinical history of respiratory distress and pleural effusion. Necropsy showed a well-circumscribed, multi-lobulated mediastinal mass found within the cranial thorax and ventral to the lung, measuring approximately 4.5 x 2 x 3 cm. Cutting sections of the mass exhibited pale parenchyma with a mesh-like appearance.

Result: Histologically, the well-circumscribed mass comprised sheets of mature small lymphocytes admixed with anastomosing ribbons and trabeculae of epitheiloid or spindloid epithelial cells. They show positivity of CD3 and pan-cytokeratin, respectively. Numerous, diffusely scattered cystic spaces with variable shapes and sizes were also noted among the neoplasm. Most of the cystic structures were cystic degeneration and coalescense of perivascular spaces that were devoid of epithelial lining, while the others were some dilated, large blood/lymphatic vessels lined by endothelium and showed positivity of factor VIII.

Conclusion: On the basis of the mass location, the histological findings, and the immunohistochemistry, a thymoma with diffuse cystic degeneration and multifocal dilated blood and lymphatic vessels was diagnosed. The lack of epithelial lining of the cystic structure rule out the multilocular thymic cyst.

D-16: PRIMARY MENINGEAL RHABDOMYOSARCOMA OF THE SPINAL CORD OF A YOUNG DOG WITH NEUROMELANOCYTOSIS AND MULTIPLE CUTANEOUS NEUROFIBROMAS: A CASE REPORT.

Laura L Hoon-Hanks, Chad B Frank, Elijah F Edmondson

A 7-week-old, male, black Labrador retriever dog presented for postmortem examination following progressive hindlimb paralysis and abnormal skin development. An intradural mass was found within the T9-T11 vertebral canal that was highly compressive and infiltrative into the underlying spinal cord. Microscopic analysis revealed features compatible with rhabdomyosarcoma (RMS; embryonal spindle cell type) which was confirmed with positive immunohistochemical staining of desmin and negative smooth muscle actin within the tumor sarcoplasm. The adjacent spinal cord also had numerous melanin-containing cells arranged in small nodules predominantly within the grey matter (micronodular neuromelanocytosis). The left lateral thorax had multifocal dermal masses, histologically consistent with diffuse neurofibromas. Primary meningeal RMS has not been previously described in dogs but is well documented in the human literature and is believed to arise from primitive mesenchymal cells. In human patients, RMS is the most common soft tissue sarcoma of childhood and adolescence and has been associated with concurrent congenital/genetic diseases. Similarly, in dogs a majority of RMS occur before 2 years of age, but a relationship between juvenile RMS and other congenital abnormalities has not been recognized. In this case, the constellation of (1) intradural spinal RMS, (2) multiple cutaneous neurofibromas, and (3) micronodular neuromelanocytosis represents a unique case presentation with unclear etiology. Overall, primary canine meningeal RMS of the spinal cord has not been previously reported and represents a novel differential for spinal tumors of young dogs. Moreover, such cases should be assessed for the presence of additional congenital abnormalities.

D-17: PRIMARY CEREBELLAR LYMPHOMA WITH HODGKIN'S LYMPHOMA-LIKE MORPHOLOGY IN A CAT

James K Chambers, Yuka Yoshino, Taichi Nakamori, Hiroyuki Nakayama, Kazuyuki Uchida

Feline intracranial lymphomas mostly affect the brainstem, cerebrum or meninges, however primary cerebellar lymphoma has not been reported. A 4-year-old cat exhibited neurological symptoms such as wobbling, right head tilt, and intention tremor, and magnetic resonance imaging revealed a mass lesion in the cerebellum. The cat was seropositive for feline leukemia virus (FeLV) antigen and feline immunodeficiency virus antibody. The cat died 5 months after initial presentation, and no neoplastic lesions, besides the cerebellar mass, were observed at necropsy. Histologically, the structure of the cerebellar medulla was obscured. In this area, large atypical round cells with abundant amphophilic cytoplasm were observed and they infiltrated into the cortex, meninges, and choroid plexus. These neoplastic cells showed severe anisokaryosis and anisocytosis. Neoplastic cells with single large inclusion-like nucleoli resembling

Hodgkin cells, and those with symmetrically arranged nuclei resembling Reed-Sternberg cells were often found. Lymphocytes, macrophages, and reactive astrocytes were admixed with the neoplastic cells. Neoplastic cells were positive for FeLV gp85/gp70, CD20, BLA36, vimentin, p16, p53, and Pax5, and negative for CD3, CD79a, and Iba1 by immunohistochemistry. Multiplex PCR for detecting immunoglobulin heavy chain gene rearrangement revealed monoclonal proliferation of B-lymphocytes. In the present feline case, the results indicated that the neoplasm was of B-cell origin, and that the neoplastic cells were comparable to those of Hodgkin's lymphoma in humans. This report describes feline primary cerebellar B-cell lymphoma that consisted Hodgkin's lymphoma-like neoplastic cells with FeLV protein expression.

D-18: PROFOUND BASOPHILIA IN A CAT WITH FELINE GASTROINTESTINAL SCLEROSING FIBROPLASIA

Harold W Tvedten

Feline gastrointestinal eosinophilic sclerosing fibroplasia was diagnosed by histopathological examination of lymph node, small intestine and spleen harvested at necropsy from a cat which had profound basophilia of 60 % (13.4 x 10^9/L). Both feline eosinophils and basophils have reddish granules on routine H+E histological sections. Feline basophils seen in H+E stained histological sections were reported by Fairley 2013 to have a clear cytoplasm while eosinophils have eosinophilic cytoplasm. Both have red colored granules though basophils had more distinct granules and eosinophils have more indistinct granules. Eosinophils tended to have bi-lobed nuclei while basophils had more polymorphonuclear shaped nuclei. Different histological stains were used to try to differentiate eosinophils versus basophils. The Luna stain appeared to illustrate well which granulocytes were eosinophils versus basophils. Eosinophils had red granules with Luna stain, while granulocytes with large, unstained and refractile granules were interpreted to be basophils. With this stain there were a moderate number of basophils seen in the intestinal sections, though granulocytes were mainly eosinophils. There were mainly eosinophils in the lymph node section and only occasional basophils. While in the spleen, half or more of the granulocytes were basophils, reflecting the blood basophilia. Luna stain is considered specific for eosinophils and appeared to be the most useful stain.

D-19: POSSIBLE VIRAL-INDUCED T-CELL LYMPHOMAS/LEUKEMIAS IN CAPTIVE NON-HUMAN PRIMATES

Shanny Hsuan H. Kuo, Wen-Ta Li, Chia-Da Hsu, Pao-Jung Wang, Jun-Cheng Guo, Victor Fei F. Pang, Hui-wen Chang, Chen-Hsuan Liu, Chian-Ren Jeng

Background: Lymphomas in primates can be either spontaneous, viral-induced, or associated with carcinogens. Although spontaneous cases are uncommon, there are several well-known viruses capable of inducing lymphoproliferative diseases. Simian T-cell leukemia/lymphoma virus, for instance, presents high seroprevalence among primates and is associated with aberrant T-cell proliferations. Transmission may occur via membrane exposure, vertical infection, and sexual intercourse with possible interspecies transmission.

Case presentation: During 2012-2015, five non-human primates of two distinct species in Taipei Zoo died sequentially. Affected species included four Cheirogaleus medius (3 females and 1 male), and a male Nycticebus pygmaeus. Three of the animals presented severe alopecia. The remaining animals were clinically normal. Age ranged from 10-17 years. Necropsy was performed and tissues were routinely processed for histopathology.

Results: Histopathologically, neoplasm cells were present in all primates and affected hematopoietic tissues, liver, kidney, gastrointestinal tract, and skin. Dense sheets of monotypic small to medium-sized lymphoid cells were seen causing variable structural effacement of the affected organs. The neoplastic cells were intensively labeled for CD3 and compatible with a T-cell origin.

Conclusion: All of the affected animals were housed adjacently and all diagnosed with T-cell lymphomas/leukemias. It is speculated that such serial neoplastic development between two species was caused by a specific viral infection, with evidences pointing to interspecies transmission.

D-20: SYNCHRONOUS MEDULLARY THYROID CARCINOMA AND PARATHYROID CARCINOMA IN A BEAGLE DOG

Shanny Hsuan H. Kuo, Chian-Ren Jeng, Victor Fei Pang, Hui-wen Chang

Background: Thyroid gland tumors constitute 1.2%-4.0% of all canine tumors with a breed predilection for Beagles, Boxers, and Golden Retrievers. In dogs, medullary thyroid carcinomas and parathyroid carcinomas are relatively rare. Here we describe synchronous medullary thyroid gland carcinoma and parathyroid gland carcinomas in an 11-year-old, female spayed, Beagle.

Materials and Methods: Routine CT scanning for metastasis due to a previously diagnosed mammary gland tumor (MGT) revealed abnormal masses in both thyroid glands with concurrent retropharyngeal lymphadenopathy, and pulmonary metastasis. The animal also presented with hypercalcemia. Thyroidectomy and parathyroidectomy were performed and samples evaluated histologically.

Results: Histopathologically, there were two distinct neoplastic cell populations. The first one (pattern A) featured pale eosinophilic, polygonal cells arranged in nests subdivided by a delicate fibrovascular stroma with capsular and vascular invasion. Multifocally blended with the afore-mentioned cells was a second neoplasm (pattern B) comprised of basophilic polygonal cells forming acini and glandular structures. Pattern A cells were positive for TTF-1, Chromogranin-a, and NSE, whereas pattern B cells were positive for PTH, Chromogranin-a and NSE, and negative for TTF-1.

Conclusion: Canine parathyroid gland carcinomas with concurrent thyroid gland tumors are rare. The thyroid gland and parathyroid gland tumors and the previously diagnosed MGT are thought to have occurred independently. Our report describes the histopathologic and immunohistochemical findings of a confirmed case of synchronous thyroid C-cell carcinoma and parathyroid carcinoma in a beagle dog.

D-21: PROLIFERATIVE LESIONS ASSOCIATED WITH POXVIRUS IN WILD RODENTS IN TEXAS

Carolyn L Hodo, Joanne L Mansell, Jessica E Light, Sarah A Hamer

Background: *Poxviridae* is a diverse family of viruses, many of which are pathogenic to animals and humans. Many poxviruses of the Old World have rodent reservoirs, but the only North American poxvirus known to be associated with wild rodents is *Volepox* and lesions in wild rodents are poorly described. Ectromelia, or mousepox, is a well-known disease of laboratory mice but has not been described in wild populations.

Objective: We report a poxviral infection characterized by proliferative epidermal lesions without systemic involvement in a wild northern pygmy mouse (*Baiomys taylori*).

Methods: In central Texas field research, we observed several rodents with proliferative lesions on the tail or feet and collected one for pathologic examination. Following necropsy, tissues were processed for histopathology and electron microscopy. DNA was extracted and subjected to PCR and DNA sequencing of a metalloproteinase gene, and the sequence was compared with poxviruses in GenBank.

Results: At necropsy, four pedunculated epidermal masses ranged from 0.4 to 0.8 cm in diameter on the tail and both hind feet. Microscopically there was a proliferative and necrotizing dermatitis, with the masses comprised of markedly hyperplastic epidermis with hyperkeratosis, ballooning degeneration, and keratinocytes containing eosinophilic intracytoplasmic inclusions. Electron microscopy revealed pox virions within the inclusions. The sequence was distinct but most similar to a clade containing *Raccoonpox* and *Skunkpox* viruses.

Conclusion: The proliferative nature of the lesions is unlike any previously described manifestation of poxvirus in North American rodents. Ongoing molecular work aims to sequence additional genes to more fully characterize this virus.

D-22: MANUAL STRANGULATION ON A STRAY CAT—THE LINK BETWEEN THE SUSPECT'S BEHAVIOR AND THE VICTIM

Wei-Hsiang Huang, Chen-Hsuan Liu

In late December 2015, a missing stray cat received public attention when a suspect allegedly strangulated the cat according to a witness. Surveillance video footage of the incident surfaced, showing the cat was strangulated by both hands of the suspect without struggling signs as well as kicking the left abdomen post mortem. The body was placed in a travel bag and then transferred and stored in a motorcycle trunk. The suspect confessed that he had abandoned the travel bag for several hours and then transferred it to the trunk. The body was submitted to our institute for forensic necropsy. Radiographs revealed suspected luxation of C7. On necropsy, the body was at an early bloat stage, and the fur was soaked and attached by plant fibers. A bruise at the middle area of the neck and a small abrasion at left submadibular skin was revealed after shaving the hair. Soft tissue hemorrhage was noted around the left submadibular area, especially surrounding the left jugular vein. Suspect hemorrhage was noted in the left abdominal wall and at subdural space of C7;petechiae were noted on the scalp. Histopathologically, left jugular vein wall tears were significant, along with parenchymal crush and mininal hemorrhage of left submandibular lymph node. Spinal nerve around C7 and adjacent soft tissues demonstrated significant hemorrhage. Subpleural pulmonary emphysema and extreme dilation of bronchioles were noted. Necropsy findings were compatible with the suspect's confession and the video footage, elucidating the mystery why the cat didn't struggle.

D-23: ZYGOMATIC ARCH PAROSTEAL OSTEOSARCOMA IN THREE DOGS
Randi M. Gold, Fabiano Oliveira, Roy R. Pool

Parosteal osteosarcoma is a relatively rare bone tumor in animals characterized by slow growing and locally invasive masses having predilection sites that include the surface of long bones. Tissue or histological sections from three dogs with zygomatic arch masses were presented to the biopsy service of Texas A&M University for evaluation. Case 1 presented for chronic sneezing. CT of the skull revealed a small lesion in the frontal lobe of the cerebrum and a bony proliferation on the axial border of the left zygomatic arch. Case 2 was noted to have a non-painful swelling of the left zygomatic arch one month prior to presentation. CT showed minimal invasion of the zygomatic arch with rostral osseous proliferation along with several other bony proliferations on the right and left frontal bones and caudal to the globe of the right eye. Case 3 had increased bone density centered on the left zygomatic arch noted on CT. Minimal additional case history was available. As is the case in humans, these tumors were histologically characterized by well differentiated fibro-osseous and chondroid components that radiate outward from an undisturbed cortex. Cellular atypia and mitotic figures were uncommon. Parosteal osteosarcomas have previously been reported to occur in the skulls of dogs and cats with to our knowledge only one being reported to originate from the zygomatic arch. Surgical resection is recommended as while these tumors are initially histologically benign, they can invade the underlying bone structure, and undergo progressive malignant transformation into osteosarcoma.

D-24: ANATOMOPATHOLOGICAL FINDINGS OF RESPIRATORY SYSTEM IN MARSH DEER (BLASTOCERUS DICHOTOMUS).
Pedro Enrique E. Navas-Suárez, Jose Mauricio M. Barbanti Duarte, Eliana Reiko Matushima, José Luiz L Catão Dias

Background: The marsh deer (*Blastocerus dichotomus*) is the largest South American deer. It is known that excessive hunting, conversion of wetlands for agriculture and dams are the most important causes of their population size declining.

Methods: This study describes the gross and histological lesions of the respiratory tract in a population of marsh deer and its associated risk factors. Over a 20-year period

(1995-2015), 68 free ranging and individuals of an ex situ conservation project marsh deer from southcentral Brazil, were autopsied.

Results: Frequent gross lesions included pulmonary edema (60.3%) and congestion (38.0%). The most common histological lesions were pulmonary congestion (92.6%), edema (72.1%) and pneumonia (60.3%). The major patterns of pneumonia were interstitial (73.2%) and bronchopneumonia (41.5%); in twelve cases these two patterns were observed concurrently. In twenty cases (48.8%) of pneumonia there was vascular and perivascular infiltration of lymphocytes, plasmatic cells, and few histiocytes and neutrophils. In thirteen cases (31.7%) of pneumonia was observed multiple intra alveolar larvae of metazoan parasites. In seven cases (41.2%) of bronchopneumonia was observed intralesional cocci and coccobacilli colonies.

Conclusions: Circulatory disturbances were highlighted as the most common anatomopathological finding in marsh deer. A vascular and perivascular lymphoplasmacytic infiltration is a common finding observed in malignant catarrhal fever and could be the cause, however, it was not possible to identify the etiologic agent . To our knowledge, this is the first comprehensive assessment of the pathology of the respiratory system in marsh deer.

D-25: ANATOMOPATHOLOGICAL FINDINGS OF RESPIRATORY SYSTEM IN BROWN-BROCKET DEER (MAZAMA GOUAZOUBIRA).

Pedro Enrique E Navas-Suárez, José Luiz L Catão Dias, Adriana Marques Joppert da Silva, Eliana Reiko Matushima

Background: The brown brocket deer (*Mazama gouazoubira*) is classified among the Small Solitary Forest Ruminants (SSFR) highlighting the difficulty of studying these animals due to their secretive nature and the closed habitat. Further research for species conservation including the evaluation of the sanitary aspects such as, infectious and noninfectious diseases and their relationship with wildlife-livestock interface.

Methods: This study describes the gross and histological lesions of the respiratory tract in a population of brown brocket deer and its associated risk factors. Over a 20-year period (1995-2015), 120 deer from São Paulo state, Brazil, were autopsied.

Results: Frequent gross lesions included pulmonary edema (57.5% n=50/87) and congestion (51.7% n=45/87). The most common histological lesions were pulmonary congestion (75.6% n=68/90), edema (63.3% n=57/90) and pneumonia (53.3% n=48/90). The major patterns of pneumonia were interstitial (70.8% n=34/48) and bronchopneumonia (18.8% n=9/48). Sixty-one animals (50.8% n=61/120) presented trauma, the most common causes were: vehicle-collision (24.6% n=15/61) and canine attack (23.0% n=14/61). It is remarkable that the majority of the traumatized deer were adults (59.0% n=36/61) in good body condition (73.8% n=45/61), which could be interpreted as a loss of reproductively viable and apparently healthy individuals resulting in a great impact in terms of natality.

Conclusions: Pneumonia is one of the main anatomopathological finding in pathology of ruminants, and is considered as a major morbidity and mortality process both in captive and free-ranging deer. To our knowledge, this is the first comprehensive assessment of the pathology of the respiratory system in brown brocket deer.

D-26: HEPATOBLASTOMA AND MIXED ENDOCRINE PANCREATIC CARCINOMA IN A CYNOMOLGUS MACAQUE FETUS

Chee Bing Ong

Background: Hepatoblastoma, the most common pediatric liver malignancy, is an uncommon tumor that arises from embryonic and fetal hepatocytes. Hepatoblastoma is a rare tumor in domestic animals and has been reported in the dog, alpaca, horse, mouse, cat and sheep.

Methods: The current report describes the gross, histopathological and immunohistochemical characterization of a hepatoblastoma and mixed endocrine pancreatic carcinoma in a Cynomolgus macaque fetus.

Results: Neoplastic cells of hepatoblastoma are arranged in primitive acinar formation often forming pseudorosettes, separated by extramedullary hematopoietic cells. The neoplastic cells are strongly positive for pancytokeratin, cytokeratin 18 and alpha-fetoprotein, and moderately positive for beta-catenin. The mixed endocrine pancreatic carcinoma was in vicinity of the pancreas and duodenum, and is composed of neoplastic epithelial cells arranged in cords, tubules and papillary structures supported by fibrocollagenous stroma. These neoplastic cells are positive for pancytokeratin, beta-catenin, cytokeratin 18 and scattered small clusters of cells are positive for chromogranin A and neuron specific enolase.

Conclusion: Histopathological and immunohistochemical characteristics of the neoplasms are consistent with a hepatoblastoma and mixed endocrine pancreatic carcinoma. In human and mouse, hepatoblastomas have high prevalence of deletion mutation in β-catenin gene. β-catenin gene mutation was not detected in this animal. Natural occurring hepatoblastoma has not been reported in Cynomolgus macaque based on the authors' knowledge, and this is the first reported case of hepatoblastoma in this species. Antenatal investigation via non-invasive modalities may be useful for future surveillance of fetal tumors in macaques and domestic animal species.

D-27: METASTATIC GRANULOSA CELL TUMOR AND SECONDARY SEPTICEMIA IN A FLORIDA MANATEE (TRICHECHUS MANATUS LATIROSTRIS)

David S. Rotstein, Martine de Wit, Ray Ball

An approximately 50 YO, 368 cm and 1190 kg female Florida manatee was found dead in a holding pool. She had a decline in body condition, lethargy,and abnormal ECG approximately 5 months prior to death. Treatment was initiated with enalapril and after initial improvement, her condition worsened. Gross necropsy findings included bilateral ovarian masses with adhesions to the body wall, peri-uterine masses, multiorgan congestion, thickened ventricular muscle, and ascites with fibrin. Significant histopathologic findings were observed in the urogenital, cardiovascular, hepatobiliary, digestive, musculoskeletal, and hematolymphatic system. In the left ovary, there was a luteoma, and in the right ovary, an invasive granulosa cells tumor was observed with mesenteric and serosal implantation as well as metastasis to the axillary lymph node. Cystic endometrial hyperplasia of the uterine horns was present. Renal lesions included a membranous glomerulopathy, intravascular fibrin thrombi, tubular loss, ectasia, and proteinosis. Myocardial findings included hypertrophy and degeneration of the left ventricle and atrophy and fatty infiltration of the right ventricle and atrium. A thrombus with bacterial cocci was present in the left atrium and bacterial cocci were present within the capsule of the liver, gastric serosa, mandibular lymph node, and peritoneum. The manatee succumbed to septic peritonitis that either arose from the atrial thrombus or the necrotic regions within the granulosa cell tumor. Aged manatees provide an understanding of the life history and natural diseases for this species.

D-28: MODIFIED BIOPSY TECHNIQUE AND HISTOPATHOLOGICAL GRADES FOR CRESTY NECK IN SPANISH PUREBRED HORSE

Abelardo M Morales-Briceño, Alejandro E Escamilla-Sanchez, Aniceto M Mendez-Sanchez, Jose P Perez-Arevalo, Jose M Mendez-Angulo

Background: The study describes a modified biopsy technique and histologic grades for cresty neck in Spanish Purebred Horses as a diagnostic tool.

Methods: The study included 80 stallions and 40 mares from Andalusia and Extramadura, Spain that were normal or varying degrees of deformity. Six punch biopsy samples were obtained from the left and right dorsal neck region (cranial to caudal) with a depth of 2-3 cm including muscle. Histologic samples were assigned a grade based on the degree of lipomatosis and skeletal muscle involvement. The grade ranged from 0 to 5. Grade 0-normal, Grade 1-Small amounts of fat deposit with small vacuoles and unaffected skeletal muscle, Grade 2-coalescing fat vacuoles present in the intermyofibrillar space, Grade 3-moderate amounts of coalescing fat vacuoles and moderate intermyofibrillar fatty infiltration (moderate lipomatosis), Grade 4-Abundant coalescing fat vacuoles infiltrating the intermyofibrillar (marked lipomatosis), and Grade 5-fat vacuoles present and no muscle tissue present (severe lipomatosis).

Results: For the 120 horses, grades were as follows: Grade 0 (24%), Grade 1 (20%), Grade 2 (35%), Grade 3 (14%), Grade 4 (5%), and Grade 5 (2%).

Conclusion: The biopsy and grading methods provide a potential diagnostic tool for assessing cresty neck in Spanish Purebred horses

D-29: MUCOEPIDERMOID CARCINOMA IN AN ORANGE-WINGED AMAZON PARROT

Denise Lin, Sanjeev Narayanan, Melissa Nau, James W Carpenter, Mackenzie Hallman

A 33-year-old intact female orange-winged Amazon parrot (A*mazona amazonica*) presented for a growing mass over the right eye. Computed tomography scan revealed a heterogenous mixed soft tissue and mineral dense mass with lysis of the right frontal bone. Gross necropsy confirmed right frontal bone lysis leading into the infraorbital sinus with partial replacement by a soft tissue mass. Histologically, the tumor obliterated the cortical bone of the rostral skull and infiltrated the surrounding soft tissue, and was composed of islands and trabeculae of intermediate, epidermoid, and mucus-producing epithelial cells embedded in abundant loose fibrovascular stroma. The majority of cells were of intermediate type, characterized by polygonal shape, moderate eosinophilic lacy cytoplasm, and irregularly round nuclei with vesiculate chromatin and 1-2 prominent nucleoli. Rare epidermoid cells were present, characterized by lamellar cytoplasmic keratinization. Mucus-producing cells contained abundant clear cytoplasm and peripherally displaced nuclei (signet-ring cells), and were found singly or forming acinar-like structures. Toluidine blue and acid mucicarmine stains confirmed mucin production. Neoplastic cells stained positively for cytokeratin AE1/AE3 and negatively for vimentin on immunohistochemistry. Collectively, these findings were consistent with a diagnosis of mucoepidermoid carcinoma, with suspected origin from salivary or nasal mucous glands. Mucoepidermoid carcinomas are commonly found in the salivary gland in humans, and less commonly in other glands or the lungs.

D-30: MAMMARY CARCINOMA WITH METASTASIS TO MULTIPLE JOINTS IN A DOG

Maggie McCourt, Alix Dieterly, Paige Mackey, Shane Lyon, Theresa Rizzi, Jerry Ritchey

Background: In dogs, widespread metastasis of mammary carcinomas is common, however, metastasis to the synovium has not been previously reported. An 8-year-old, intact female, mixed breed dog presented for progressive lameness. Physical examination revealed joint effusion in the right elbow and hock joints bilaterally, left axillary and right popliteal lymphadenomegaly, and a subcutaneous mass within the mammary glands.

Objective: To characterize the cytologic, histopathologic, and immunohistochemical features of mammary carcinoma with synovial metastasis.

Methods: Cytology, histopathology, and immunohistochemical staining including pancytokeratin and CK 8/18 for epithelial cells and calponin for myoepithelial cells were performed.

Results: Cytologic examination of the mammary mass, enlarged lymph nodes, and affected joint fluid revealed a monomorphic population of loosely cohesive epithelial cells. Neoplastic cells exhibited high nuclear to cytoplasmic ratio, nuclear molding, and binucleation. Histopathologic examination was consistent with a mammary carcinoma forming intraductular acini and papillary projections surrounded by myoepithelial cells with invasive nests or individualized neoplastic epithelial cells. Similar neoplastic epithelial cells were present in lymphatics of affected lymph nodes and within the synovium of affected joints. Neoplastic luminal epithelial cells within the mammary carcinoma, synovium, and lymph nodes were positive for pancytokeratin and CK8/18.

The myoepithelial cells within the mammary carcinoma, affected lymph nodes, and synovium were positive for calponin.

Conclusion: The cytologic and histopathologic features along with immunohistochemical stains culminate in a diagnosis of mammary carcinoma with metastasis to the synovium and lymph nodes.

D-31: EQUINE OCULAR PATHOLOGY: A RETROSPECTIVE HISTOPATHOLOGICAL STUDY
Mariana M Flores, Fabio Del Piero, Perry L Habecker, Ingeborg M Langohr

Background: Ocular diseases are a common problem in equine medicine; however, few articles demonstrate the general prevalence of different ocular conditions affecting horses. Thus, the aim of this study was to investigate the occurrence of ophthalmic diseases in horses submitted to enucleation or necropsy.

Methods: The archives of the Pennsylvania Animal Diagnostic Laboratory System (PADLS - New Bolton Center) (1997-2011) and the Louisiana Animal Disease Diagnostic Laboratory (LADDL) (2003-2015) were searched for equine submissions with available ocular histology slides. All ocular slides were reviewed by two pathologists (M.M.F. and I.M.L.) and all case records were examined to determine the ophthalmic conditions and their accompanying clinical history.

Results: A total of 102 enucleations and 83 necropsies were included. The most common ophthalmic conditions were keratitis (36), traumatic injuries (32), equine recurrent uveitis (ERU) (29), ocular neoplasms (20), of which squamous cell carcinoma was the most common (11), and uveitis/endophthalmitis due to sepsis (19). Additional conditions included degenerative changes (9) (with predominance of retinal atrophy and detachment [5]) and congenital anomalies (4). Of the ERU cases, the majority was reportedly unilateral (17) and had another preceding or concomitant ocular condition (17), such as keratitis (10), neoplasia (2), and cataract (1).

Conclusion: ERU was the only pathologic condition that was occasionally associated with a second primary pathologic process. It is possible that ERU represents a non-specific albeit unique intraocular inflammatory response triggered by a variety of local inflammatory and neoplastic processes.

D-32: LISTERIOSIS IN TWO NEONATAL LLAMAS (LLAMA GLAMA)
Ian K. Hawkins, Marcia R.S. Ilha, Eman Anis, Rebecca P. Wilkes

Listeria monocytogenes is a common cause of disease among domestic ruminant species, however, among South American Camelids (llamas and alpacas) the prevalence is low, with only a few documented reports describing pathologic changes. Here we describe the clinical signs, gross findings, and histopathology in two neonatal llamas with listeriosis. Both crias were approximately fourteen days old and came from the same South Georgia herd of 50 animals. The animals presented with acute lethargy, fever, and rapidly succumbed to the infection within the same five day time frame. The dams did not exhibit clinical signs and remained healthy following the neonatal losses. At gross necropsy, the brains were congested with cloudy-beige meningeal surfaces, the umbilici were mildly swollen, and in one animal, a mild amount of fibrin was observed in the thoracic and abdominal cavities. Prominent histopathologic lesions included a suppurative meningoencephalitis, necrotizing hepatitis, nephritis, omphalophlebitis, and vasculitis. *L. monocytogenes* was confirmed in both animals with culture and real-time PCR (qPCR) on fresh and paraffin-embedded tissues. With the evidence of omphalophlebitis, the infection may have originated from the umbilicus, however, an in utero route or oral route of infection, remain possibilities as well. This presentation of multiple forms of listeriosis (septicemia and meningoencephalitis) in multiple animals from the same group is unusual, especially among South American Camelids.

D-33: GENERALIZED DEMODICOSIS WITH VISCERAL LYMPH NODE INVASION IN A DOG

Moges W. Woldemeskel, Ian K. Hawkins

Demodex mites are widespread arthropod ectoparasites among domestic mammals and demodicosis is one of the most common skin diseases of the dog. Two forms are represented in canines: localized demodicosis and the more severe, generalized demodicosis. However, systemic invasion of the infection is not known to occur. In this report, we documented the clinical, gross, and histopathologic findings in a dog with generalized demodicosis and subsequent spread of the infection to multiple internal lymph nodes. Grossly the animal was severely alopecic, emaciated, and several peripheral and internal lymph nodes were enlarged. Histopathology confirmed a severe dermatitis with abundant Demodex mites in the hair follicles. Multiple lymph nodes (peripheral and internal) were disrupted by a marked granulomatous lymphadenitis with numerous, viable, intralesional *Demodex* mites. In addition, viable *Demodex* mites were often observed in the subcapsular sinuses and peri-nodal, thrombosed lymphatic vessels, indicating lymphatic invasion; a focal degenerated arthropod parasite was also identified in the skeletal muscle. These histopathologic findings confirmed a disseminated visceral infection by *Demodex* primarily affecting the skin, peripheral lymph nodes, lymphatic vessels, and internal lymph nodes. To the authors' knowledge, this is the first report of canine generalized demodicosis with lymphatic invasion, thrombosis, and invasion of visceral lymph nodes documented in the literature.

D-34: IMMUNOHISTOCHEMICAL PROFILE OF 20 FELINE RENAL CELL CARCINOMAS

José A Ramos-Vara, Elijah F Edmonson, Tod Bass, Margaret A Miller, Dee M Dusold

Renal cell carcinoma (RCC) is an uncommon neoplasm in cats. We examined the immunohistochemical profile of 20 formalin-fixed, paraffin-embedded feline RCC: 11 chromophobe tubular carcinomas, 4 chromophobe tubulopapillary carcinomas, 1 chromophobe papillary carcinoma, and 4 anaplastic carcinomas. Immunohistochemical markers used were Pax8 (monoclonal BC12, Biocare, Concorde, CA), KIT (polyclonal, Dako, Carpenteria, CA), CD10 (monoclonal 56C6, Vector Labs, Burlingame, CA), cytokeratins (monoclonal AE1/AE3, Dako), and vimentin (monoclonal 3B4, Dako). A polymer-based immunoperoxidase procedure was used. Immunoreactivity was scored as percent positive cells (50%) and intensity of expression (weak, intermediate, or strong). Nineteen tumors (95%) expressed Pax8; 12 (60%), KIT; 15 (75%), CD10; 20 (100%), cytokeratins; 19 (95%), vimentin. Pax8 immunoreactivity (nuclear) was readily apparent, but varied in intensity within a section. KIT reactivity was diffuse cytoplasmic and fairly homogeneous within a section. CD10 immunoreactivity was predominantly in the apical border of tubular profiles and less commonly cytoplasmic. CD10 immunoreactivity was particularly strong in tubular lumens, less common/intense in areas with papillary differentiation, and absent in solid areas. Cytoplasmic cytokeratin expression was strong in 18 tumors, and weak in 2; the papillary portion of 1 tumor had distinct submembranous expression. Vimentin immunoreactivity, which ranged from diffuse to focal, was difficult to evaluate due to strong stromal immunoreactivity and its patchy expression in phenotypically similar neoplastic cells. As in canine RCC, Pax8 appears to be a very sensitive marker for feline RCC.

D-35: HISTOLOGIC LESIONS IN 207 CANINE PITUITARY GLANDS

Andrea L. Vanderpool, Margaret A. Miller, Joanna C. Scott-Moncrieff, David S. Bruyette

With increasing performance of transphenoidal hypophysectomy, pathology has become important in antemortem diagnosis of pituitary disease. Therefore, pituitary glands from 207 dogs (Indiana Animal Disease Diagnostic Laboratory archives, 2008-2015) were reviewed to determine the prevalence and nature of histologic lesions. Ten cases were hypophysectomy specimens; the remainder had been collected at autopsy. There were 115 female (81 spayed) and 92 male (59 castrated) dogs aged 1 day to 17 years (median, 9 years). Seventy-nine pituitary glands, including 36 that were considered unremarkable on initial postmortem examination, had adenohypophyseal proliferations of densely or sparsely granular (Periodic acid-Schiff) chromophobes. These were classified as hyperplasia (n=42) for lesions 5 mm diameter masses. The Gordon and Sweet technique was used to accentuate reticulin fibers for borderline (approximately 1-mm) lesions. The largest macroadenoma was 1.2 cm in maximal dimension. Dogs with proliferative lesions were older (2-17 years; median, 11 years) than dogs without proliferation (1 day to 15 years; median, 7 years). Breed or sex predilection for pituitary proliferation was not evident. Secondary neoplasms included 4 lymphomas, 3 metastatic carcinomas, 1 metastatic melanoma, and local extension of 2 ependymomas, 2 craniopharyngiomas, and 1 case each of gliomatosis, germ cell tumor, and meningioma. Hypophysitis was the diagnosis in only three dogs. Craniopharyngeal duct cysts (n=32) were common incidental lesions. In 87 dogs, the pituitary gland was within normal limits or had only microscopic craniopharyngeal duct cysts.

D-36: INTRA-ABDOMINAL LYMPHANGIOSARCOMA WITH PELVIC LIMB PARESIS IN A HORSE

Sarah E. Sykes, Susan J. Bender, Perry L. Habecker

Background: An 18-year-old Quarter horse gelding presented for evaluation of progressive hind limb lameness. Lameness examination revealed severe atrophy of the right quadriceps and a shortened cranial phase of the stride in that limb, consistent with paresis secondary to femoral nerve injury. Euthanasia and an autopsy were elected.

Results: Autopsy revealed marked atrophy of the right thigh muscles. Numerous light brown, thin-walled, cyst-like structures were embedded within the caudodorsal abdominal connective tissue. These structures measured up to 7 cm in diameter and contained clear fluid. The peri-neural connective tissue around the lumbar and sacral spinal nerve roots was edematous and yellow. The muscles surrounding the seventh lumbar vertebra were also yellow-brown and edematous.

Histopathology revealed a moderately cellular mesenchymal neoplasm that infiltrated the sublumbar and subsacral fibroadipose tissue and skeletal muscle, the dura mater of the lumbosacral spinal nerve roots, and the cortex of one kidney. Neoplastic cells lined bundles of collagen, forming disorganized and occasionally dilated vascular channels filled predominantly with small lymphocytes and free neoplastic cells. Less frequently, the neoplastic cells formed densely-packed, solid bundles. Neoplastic cells were fusiform, with abundant eosinophilic cytoplasm, and a paracentic nucleus. Pleomorphism was marked and the mitotic index was 27. The neoplastic cells exhibited mild to moderate Prox-1 nuclear labeling, and occasional LYVE-1 cytoplasmic labeling.

Conclusion: The gross, histologic and immunohistochemical findings were consistent with a lymphangiosarcoma. To our knowledge, this is the first reported case of an equine lymphangiosarcoma that surrounded the spinal nerve roots, resulting in pelvic limb paresis.

D-37: IDENTIFICATION OF MYXOZOA PARASITE IN THE CONNECTIVE TISSUE OF A CASCADURA FISH (HOPLOSTERNUM LITTORALE) IN TRINIDAD AND TOBAGO

Karelma Frontera-Acevedo, Carla Phillips

Background: A total of thirty cascadura (*Hoplosternum littorale*) fish collected from 5 different ponds were submitted alive to the Aquatic Animal Health Laboratory of the School of Veterinary Medicine, University of West Indies as part of monitoring and quarantine procedures of a commercial aquaculture farm. Cascadura fish are commercially important in Trinidad.

Methodology: Water from the five ponds was tested for quality. All fish were euthanized and necropsies were performed. Pooled samples from each pond were submitted for microbiology, parasitology, and histopathology.

Results: All water samples had low alkalinity, low hardness, and low pH. In general, fish from all ponds had evidence of low to moderate ecto- and endoparasitism. A section of connective tissue from the histopathologic samples of fish from the pond with the lowest pH contained an area of hemorrhage and numerous poorly staining oval to tear-drop shaped organisms. These organisms stained brightly acid fast, and some included two polar capsules and a polar filament. No other signs of infection were noted in these fish.

Conclusion: This is the first histopathologic description and report of myxozoan parasites in cascadura in Trinidad. Myxozoa can be pathogenic depending on species, target organs and especially in heavy infestations. They should be included in the differential for causes of morbidity and mortality in farmed cascadura, particularly when there is poor water quality which could negatively affect the immune system of the fish.

D-38: CHAETOMIACEAE FUNGI, NOVEL PATHOGENS OF EQUINE CEREBRAL PHAEOHYPHOMYCOSIS

Quinci Plumlee, Alix Dieterly, Courtney Meason-Smith, Aline Rodrigues-Hoffman, Brian Porter, Gabriel Gomez

Background: Mycotic encephalitis is rare in horses. Many previously unrecognized fungi, however, are emerging as potential pathogens in veterinary species and humans alike. One such group of rare fungi includes dematiaceous fungi of the Chaetomiaceae family (phylum Ascomycota, class Sordariomycetes). These fungi are rare causes of opportunistic neurotropic phaeohyphomycosis in humans but are not known to infect animals.

Objective: The aim of this study was to investigate equine hyphal mycotic encephalitis cases, characterize key histopathologic features, and classify causative organisms with molecular diagnostic techniques.

Methods: Seven cases were identified, reviewed, and histopathologically evaluated via H&E stain and special stains. Panfungal PCR targeting the fungal ITS2 region and DNA sequencing of the PCR products were performed on formalin fixed, paraffin embedded sections of affected brain. Results were BLASTed against published fungal genomes in the NCBI database.

Results: Four of the seven cases yielded good quality DNA sequences. Animals ranged from 8-22 years of age and presented with severe neurologic signs. Macroscopic lesions in the central nervous system were observed in two horses and included enlargement with red and yellow mottling of the midbrain and focal cerebral malacia. Major histologic findings were multifocal, discrete foci of necrosis, neutrophilic to granulomatous inflammation, marked vasculitis, and intralesional fungal hyphae. DNA sequences showed >98% homology with several species within the Chaetomiaceae family.

Conclusion: This case series is the first report of equine mycotic encephalitis caused by members of the Chaetomiaceae family, neurotropic dematiaceous fungi, previously only reported as rare emerging pathogens in immunocompromised people.

D-39: EXERCISE-ASSOCIATED SUDDEN CARDIAC DEATH IN RACEHORSES: CARDIAC HISTOLOGY PROTOCOL

Santiago Diab, Francisco Uzal

Racehorses that suffer exercise-associated sudden death (EASD) often have no gross lesions and, therefore, a thorough examination of the heart is warranted. A few studies have recorded the postmortem findings in cases of EASD in horses, but the cardiac examination and heart histology sampling vary greatly, and their results are difficult to compare and interpret. The goal of this cardiac histology protocol is to standardize the histological examination of the heart in cases of EASD to generate consistent data that can be better compared and interpreted over time and across institutions. Eleven sections of the heart are collected to include key areas of the conduction system as well as major components of the heart, including the pericardium, myocardium, mural and valvular endocardium, great vessels and coronary arteries. The following components are included in each section: Section 1, right ventricular free wall, right atrial wall, right coronary artery, tricuspid valve. Section 2, pulmonary artery valve, right ventricular outflow tract, pulmonary artery. Section 3, right atrial appendage. Section 4, sinoatrial node region. Section 5, left atrial appendage. Section 6, left ventricular free wall, left atrial wall, left coronary artery, left atrioventricular valve. Sections 7 and 8, left ventricular papillary muscles. Section 9, atrioventricular node region. Section 10, interventricular septum. Section 11, aortic semilunar valve, aorta, left ventricular outflow tract. These sections provide a wide representation of some of the most important structures of the heart.

D-40: INFLAMMATORY MAMMARY CARCINOMA IN A SHIH-TZU DOG

Soon Seek Yoon, Byeong Jae So, Ji-Youl Jung

Background: Inflammatory mammary carcinoma (IMC) is a locally advanced mammary cancer that is associated with aggressive behavior and a poor prognosis. Metastasis to the inguinal lymph node, muscle, lung, heart, uterus, and urinary bladder have been reported.

Objective: To describe a case of IMC with disseminated tumor embolism in a Shih-Tzu dog.

Methods: An 8-year-old intact female Shih-Tzu dog was referred to a regional animal hospital with 3-day history of anorexia, severe pain, and respiratory distress. A mastectomy had been performed 1 month earlier because of a mammary mass. Despite intensive care, the patient died the following day and was submitted to the Animal Disease Diagnosis Division in Animal and Plant Quarantine Agency for necropsy. Representative tissue samples were stained with hematoxylin and eosin and immunohistochemistry for microscopic examination.

Results: On gross examination, mammary nodules were not palpable and edema, erythema and ulceration of abdominal skin was observed. Other organs did not show evidence of metastases. Microscopically, severe diffuse inflammation, tumor cell proliferation and tumor emboli within dermal lymphatic vessels were present in the abdominal skin, and lymphovascular tumor emboli were disseminated to the kidney, lungs, and urinary bladder. The neoplastic cells were immunoreactive for cytokeratin 7, 8, and 19, but negative for vimentin.

Conclusion: Based on the histopathological features, this case was diagnosed as IMC with disseminated tumor embolism. There is limited data about the metastatic pattern of IMC. IMC is usually associated with a poor prognosis and is not responsive to chemotherapy.

D-41: EXPRESSION OF BETA-AMYLOID PRECURSOR PROTEIN IN POST-ANESTHETIC MYELOMALACIA IN A CLYDESDALE HORSE
Jessica S. Fortin, Joanne Kramer, Dae Y. Kim

A 2-year-old female Clydesdale horse was submitted for necropsy to the University of Missouri Veterinary Medical Diagnostic Laboratory. The horse had an elective surgical procedure for osteochondritis dissecans involving the intermediate ridge of the right tibia. Following surgery, the horse lost motor and sensory function in the hind limbs and was unable to stand. The status deteriorated and the horse was euthanized 20 h post-surgery. At necropsy, there were no significant observations. Histopathologic examination revealed mild to moderate acute myelomalacia of the spinal cord between T14 and S3. There were multifocal to coalescing areas of mild to moderate vacuolation, primarily in the dorso-lateral white matter, and mild multifocal hemorrhage in white and grey matter. A few neurons in the dorsal horns exhibited central chromatolysis and cytoplasmic eosinophilia. The lesions were bilateral, with one side being slightly more affected, and more severe in the distal lumbar and sacral spinal cord segments. Immunohistochemistry for beta-amyloid precursor protein (APP) was performed. APP has been used to demonstrate very early damage of traumatic brain injuries in human medicine and experimental studies using animal models. The APP immunohistochemistry revealed positively-labeled, multiple, segmentally-swollen axons (axonal bulbs) and a few neurons, and the positively-stained areas overlapped with the lesions seen on HE slides. APP immunohistochemistry may be a useful tool to study early traumatic and axonal injuries of central nervous system in veterinary medicine.

D-42: SUBLINGUAL PYTHIOSIS IN A CAT
Jessica S. Fortin, Michael Calcutt, Dae Y. Kim

Extracutaneous pythiosis is rare in cats. A biopsy specimen from a male, 2-year-old, domestic short-haired cat, was submitted for histopathologic evaluation to the University of Missouri Veterinary Medical Diagnostic Laboratory. The biopsy was from a multilobulated, sublingual mass measuring 2.5 x 2 x 1 cm, present for 3 months. No pharmacological treatment was attempted. Histopathological examination revealed severe multifocal to coalescing granulomatous inflammation, characterized by infiltration of large numbers of macrophages, epithelioid macrophages and occasional multinucleated cells surrounded by lymphocytes and plasma cells. Fungal-like hyphae and occasional degenerate eosinophils occurred centrally within the lesion. The hyphae were broad (3 to 7 mm-wide), non-parallel, occasionally septate and branching, staining faintly with a Gomori methenamine silver (GMS) stain. Based on the microscopic characteristics of the inflammation and intralesional hyphal structures, *Pythium insidiosum* infection was highly suspected and lagenidiosis was also considered as a differential diagnosis. The identification of the fungal-like organism as *Pythium insidiosum* was carried out by PCR and amplicon sequencing using DNA extracted from formalin-fixed paraffin-embedded tissue. The resulting sequence was >99% identical to multiple ITS1 sequences from *Pythium insidiosum* in GenBank. Only a few feline pythiosis cases have been reported and, when encountered, it caused diseases of the skin or gastrointestinal tract. This case presents an unusual manifestation of feline pythiosis, representing the first involving the oral cavity.

D-43: BROWN SKIN DISEASE IN PUERTO RICAN CRESTED TOADS
Denise Lin, Kelli Almes, Melissa Nau, David Eshar, James W Carpenter

Two endangered Puerto Rican crested toads (*Peltophryne lemur*) from the Sunset Zoo in Manhattan, KS, were submitted for necropsy at the Kansas State Veterinary Diagnostic Laboratory. On clinical examination, both were lethargic and had redness of the hind limbs and pelvic patch, dysecdysis, and thickened brown skin. Testing performed on conspecifics in the exhibit included skin scraping and chytrid PCR testing (with negative results) and skin cytology (which revealed a mixture of unidentified cocci bacteria and yeast). Supportive treatment was unsuccessful, and the toads were found dead within hours to weeks of initial presentation.

On gross necropsy, the toads were in good body condition with excess brown unshed skin on the ventrum and legs. Histologically, the most consistent findings were epidermal hyperplasia and hyperkeratosis with a thick, brown discolored layer of flattened keratinizing epithelial cells. Other skin findings included multifocal thickening of the dermal Eberth-Katschenko calcium layer and multifocal keratinocyte necrosis. Histochemical staining of the skin with Fontana Masson stain revealed a lack of melanin in the brown discolored epidermal layer, and Gram and GMS stains were negative for infectious organisms.

The lesions are consistent with the recently described idiopathic “Brown Skin Disease” of Puerto Rican crested toads. While the majority of cases have originated from the Toronto Zoo, other zoos have reported similar findings in their collections. This new emerging disease can potentially be a significant hindrance to the breeding and reintroduction of this endangered species.

D-44: PHARYNGEAL LIPOID PROTEINOSIS IN 2 DOGS
Wen-Ta Li, Bo Chen, Chian-Ren Jeng, Fun-In Wang, Victor Fei F Pang, Ching-Ho Wu, Hui-Wen Chang, Chen-Hsuan Liu

Background: Lipoid proteinosis (LP) is an autosomal recessive disorder associated with pathogenic mutations in extracellular matrix protein 1 (*ECM1*) gene, and characterized by diffuse deposition of glycoprotein with interspersed lipoid deposits in the skin, mucosa, and viscera. Skin and upper respiratory-alimentary tract (especially pharyngeal region) are the most common affected sites in the human. To our knowledge, no cases have been reported in animals. The current study describes the pharyngeal mass compatible with lipoid proteinosis in a Poodle and a Maltese.

Case presentation: Both dogs showed varying degree of acute respiratory distress episode (cough, dyspnea, and cyanosis), and a pharyngeal mass was noted and excised. The excised pharyngeal mass was preserved and processed routinely for histopathological examination. After surgery, the respiratory signs of both dogs were improved.

Results: Both pharyngeal masses showed similar pathological findings. There were varying amounts of an amorphous, acellular, eosinophilic substance with interspersed clear vacuoles surrounded by layers of fibrous connective tissue with small numbers of inflammatory cells. The eosinophilic substance was positive for Periodic acid–Schiff staining, but negative for Congo red and trichrome stainings.

Conclusion: The current study reported 2 canine cases of pharyngeal LP, and was aimed at promoting the *awareness of such disease in canine*. However, the pathogenesis of LP in canine is still poorly understood; hence further investigation to detect the mutations on *ECM1* gene in these diseased dogs is proposed.

D-45: CONGENITAL MALFORMATIONS SIMILAR TO PENTALOGY OF CANTRELL IN A NEONATAL DACHSHUND

Christina J Ramirez, Lisa G Shaffer

Background: One of seven neonatal dachsunds, a male, was born dead with severe congenital malformations. A similar congenital abnormality was anectdotally reported in the breeder's line. The stillborn puppy from the most recent litter was submitted for necropsy.

Objective: To describe the congenital malformations and relatedness to a human disorder.

Methods: Gross necropsy and histopathology of representative tissues were performed.

Results: Gross diagnoses included non-closure of the thoracic and abdominal wall with ectopic cordis and omphalocele/gastroschisis, caudal sternal cleft, imperforate anus, cleft palate, scoliosis, and syndactyly of the right 2nd, 3rd, 4th, and 5th digits of the forelimb. The heart appeared grossly and microscopically normal for a neonatal canine.

Conclusions: The puppy had severe malformations suggestive of a midline, developmental field complex similar to the human disorder Pentalogy of Cantrell. In humans Pentalogy of Cantrell may also include of cardiac abnormalities including transposition of the great vessels and patent ductus arteriosus. No cardiac abnormalities were present in the neonate; however, given the small size of the heart, subtle unidentified cardiac abnormalities may have been present. An inherited condition in this line of dachshunds cannot be excluded and continued follow-up may provide supportive evidence.

D-46: CYTOMEGALOVIRUS AS AN UNUSUAL CAUSE OF URINARY RETENTION AND ABDOMINAL MASS EFFECT IN AN SHIV-INFECTED RHESUS MACAQUE

M Kelly Keating, James Hayes, Tara Jones, David Garber

Background: Human immunodeficiency virus (HIV) continues to be a serious cause of world-wide morbidity and mortality. As HIV does not replicate in most non-human species, the lack of appropriate animal models is a major obstacle in the development of treatments and prevention measures. Asian macaques have become a widely-used model due to their permissiveness to infection leading to high viral loads, CD4+ T cell depletion and ability to contract opportunistic infections.

Objective: Describe an atypical case of a SHIV-infected rhesus macaque that presented for an abdominal mass effect secondary to a severely distended urinary bladder.

Methods: The rhesus macaque was infected with SHIVsf162P3 and exhibited high viremia over a course of 7 months. Tissues were collected at necropsy and routinely processed for histology and testing using special stains and immunohistochemistry.

Results: Gross examination of the animal revealed a distended urinary bladder with regional edema of pelvic inlet tissues and no evidence of urinary obstruction. Other findings included non-collapsing, tan lungs. Evaluation revealed widespread cytomegalovirus infection, involving the spinal cord and exiting nerve roots, pelvic nerves, urinary bladder, stomach and the lungs. SHIV-induced giant cell pneumonia and intraalveolar foamy material containing Pneumocystis species organisms were seen in the lungs. Other findings included myocardial fibrosis, gastroenterocolitis, and renal tubular degeneration.

Conclusion: Cytomegalovirus infection is one of the most common opportunistic infections seen in SHIV infected macaques, and can present with a wide range of symptoms. Atypical presentations of secondary infections, such as this, should be considered in this population of immunocompromised animals.

D-47: HOW STERILE ARE LESIONS OF GRANULOMATOUS AND PYOGRANULOMATOUS DERMATITIS AND PANNICULITIS IN DOGS? A NEXT GENERATION SEQUENCING APPROACH TO EVALUATE THE CANINE SKIN MICROBIOME.

FABIO B Rosa, Caitlin E Older, Courtney Meason-Smith, Jan S Suchodolski, Sonia Lingsweiler, Joanne Mansell, Aline R Hoffmann

Background: Dogs with sterile granulomatous and pyogranulomatous dermatitis (SGPD) are presented with multifocal skin nodules due to (pyo)granulomatous inflammation often involving the dermis and panniculus, with no obvious infectious organisms observed with routine or special stains. The objectives of this study were to investigate if bacterial or fungal infections could be the underlying cause of SGPD in dogs.

Methods: The study included 20 formalin fixed paraffin embedded (FFPE) skin samples and 12 fresh samples from SGPD dogs; and 10 FFPE and 10 fresh samples from healthy dogs. From these punch biopsies, the epidermis and follicles were separated from the dermis and panniculus. DNA extraction kits were used to isolate DNA from all samples. Microbial DNA was amplified using primers targeting the bacterial 16S rRNA V1-V3, and fungal ITS1-2 regions. The amplified DNA was utilized for next generation sequencing on an Illumina MiSeq instrument. The sequences were processed using QIIME.

Results: No differences in fungal or bacterial alpha diversity were observed between the SPGD and control dermis. Beta diversity analysis demonstrated differences between the bacterial communities in SGPD and control, but not fungal samples. The family Erysipelotrichaceae and the genera *Staphylococcus* and *Corynebacterium* were significantly more abundant in SGPD FFPE and fresh samples compared to controls, respectively.

Conclusion: The bacteria found to be more abundant in SGPD are common skin surface inhabitants, and likely secondary contaminants in SGPD cases due to disruption of the skin barrier and ulceration. This study gives additional evidence to support that SGPD lesions are likely sterile.

D-48: EXPRESSION OF CYTOKERATINS 7 AND 20 IN TWO FELINE UROTHELIAL CELL CARCINOMAS OF THE RENAL PELVIS

Fernanda Castillo-Alcala, Anna Kokosinska, Mark G Collett, Josepha DeLay

Urothelial carcinomas of the renal pelvis are rare tumours occasionally reported in domestic animal species. In cats, the most common site for primary urothelial neoplasia is the urinary bladder and 75% of urinary bladder urothelial carcinomas are positive for both CK7 and CK20; tumours originating from the renal pelvis are considered rare. This report investigated immunoreactivity for CK7 and CK20 in two aged cats diagnosed with unilateral urothelial cell carcinomas originating from the renal pelvis. Gross examination of the affected kidneys revealed pale, firm, infiltrative masses that extended from the renal pelvis into the adjacent parenchyma. In one of the cases, the tumour formed a discrete subcapsular nodule that slightly compressed the corresponding ureter. Affected kidneys had irregular indented cortices and showed mild to moderate hydronephrosis. Histologically, the tumours had an infiltrative growth pattern, with multiple nests and sheets of neoplastic urothelium extending into the renal medulla and cortex. In both cases there was an extensive desmoplastic reaction. CK7 immunoreactivity of neoplastic cells varied from sparse, faint to variably intense. Internal positive controls (urothelium and collecting ducts) demonstrated strong immunoreactivity for CK7. CK20 immunoreactivity of neoplastic cells in both tumours was negative, with variable reactivity among the epithelium lining residual medullary tubules. No gross or histological evidence of tumour metastases were present. Both cats had typical histological features of chronic renal disease. To the authors' knowledge this is the first study that attempts to characterize CK7 and CK20 immunoreactivity of urothelial carcinomas arising from the renal pelvis in feline patients.

D-49: NEUROPATHOLOGICAL CHANGES AND IMMUNOHISTOCHEMICAL FINDINGS IN 26 NATURAL CASES OF FELINE INFECTIOUS PERITONITIS IN CATS
Lorelei L Clarke, Daniel R Rissi

Background: Feline infectious peritonitis (FIP) is a common disease of cats affecting multiple organ systems with a wide variety of clinical signs. The brain is occasionally involved and associated neurologic signs have been described.

Objective: This retrospective case series systematically describes the gross, histological, and immunohistochemical findings in the brain of naturally occurring cases of FIP in domestic cats.

Methods: The University of Georgia Athens Veterinary Diagnostic Laboratory record system was searched for cases of FIP between 2005 and 2015. Submission forms and pathology reports were reviewed and archived histology slides were examined. Neuropathological changes were distributed into 3 broad categories: widespread superficial meningoencephalitis, widespread periventricular encephalitis, and rhombencephalitis. Cases that had not been tested or that had negative fluorescent antibody testing (FAT) results at the time of necropsy were submitted for immunohistochemistry (IHC) for FIP.

Results: Of the 26 cats that met the inclusion criteria, 85% presented with neurologic signs. Six cats (23%) had lesions only in the brain. Gross changes were reported in 58% of cases. The most common histologic pattern was widespread periventricular meningoencephalitis (46%), followed by rhombencephalitis (31%), and widespread superficial meningoencephalitis (23%). Overall, histological changes were similar regardless of lesion distribution. IHC confirmed the diagnosis of FIP in all 26 cases, even when FAT was negative.

Conclusions: This study describes multiple presentations and distributions of neuropathological changes associated with FIP in domestic cats. It also highlights the challenges of using FAT for diagnostic confirmation and suggests a need for IHC to confirm suspected cases.

D-50: EPIZOOTIC BOVINE ABORTION OUTBREAK IN EASTERN NEVADA CATTLE
Alisha T Massa, Kathleen A Potter, Dan Bradway

An approximately 180 days of gestation, black Angus fetus was submitted to the Washington Animal Disease Diagnostic Laboratory for abortion diagnosis. The herd from Eastern Nevada was experiencing a late term abortion storm. Previously, nine calves were submitted ranging from 230 days of gestation to full term. Stillborn and full term fetuses had no significant gross or histologic lesions. In the 180 day fetus, gross findings were severe diffuse splenomegaly and hepatomegaly with fibrinous peritonitis and effusion. Multifocal petechiae covered the thymus, esophagus, and trachea. On histopathological evaluation there was histiocytic, lymphoplasmacytic, and necrotizing inflammation in the spleen, liver, kidneys, heart, thymus, lungs, tongue, and lymph nodes. The thymus was hemorrhagic and depleted of lymphocytes. Fresh samples of spleen, liver, and kidney were tested for the novel deltaproteobacterium that causes epizootic bovine abortion via quantitative polymerase chain reaction. The cycle threshold value was low indicating a strong positive. Also called foothill abortion, naïve heifers and cows that graze on tick infected pastures prior to six months gestation are susceptible. Fetuses which become infected after six months gestation can mount successful immune responses and clear infection or be born weak with no apparent lesions. This case demonstrates that early third trimester fetuses are most diagnostic with characteristic gross and histologic lesions supportive of epizootic bovine abortion. While the tick vector is present in Eastern Nevada, this disease has not previously been reported in this area. The causative deltaproteobacterium may be spreading east within the pajahuello tick (*Ornithodoros coriaceus*).

D-51: LIGHT AND ELECTRON MICROSCOPY OF IDIOPATHIC PULMONARY FIBROSIS AND TYPE II PNEUMOCYTE CARCINOMA IN A WEST HIGHLAND TERRIER

Alexandra P.R. Armstrong, Ned Patterson, Frederic H. David, Daniel A. Feeney, Anibal G. Armien

A 17-year-old, neutered male, West Highland Terrier dog with a history of idiopathic pulmonary fibrosis was enrolled in the University of Minnesota Pulmonary Fibrosis study. The dog underwent a CT scan in November 2014 that showed moderate to marked pulmonary fibrosis. The dog was euthanized due to disease progression and presented for necropsy. Grossly, the lungs were mottled light pink and light brown, with reduced elasticity and multifocal, poorly demarcated firm areas. Additional findings included nodules within the medulla of both adrenal glands and severe mitral and tricuspid valvular degeneration. On histologic evaluation, there was variably mild to marked fibrosis throughout the lungs, affecting alveolar septal walls, peribronchial interstitium, bronchiolar interstitium, and perivascular stroma. There was scattered honeycombing of alveoli in more severely affected areas. There was a well-demarcated, unencapsulated nodule composed of a proliferation of presumed neoplastic type II pneumocytes, which also had a neuroendocrine pattern. On further evaluation, fibrotic areas were positive for collagen with Picrosirius red and negative for elastic fibers with Verhoeff-Van Gieson/Hart's method. With electron microscopy, the pulmonary interstitium was markedly expanded by collagen fibers. The nodules were composed of neoplastic type II pneumocytes. The final diagnosis was pulmonary fibrosis and type II pneumocyte carcinoma. Electron microscopy also allowed further classification of the pulmonary tumor and detailing of changes in the regions of fibrosis.

D-52: ABERRANT MIGRATION OF SETARIA DIGITATA TO THE SPINAL CORD OF A HORSE FROM SOUTH KOREA
Hyunkyoung Lee, Kyunghyun Lee, Kanghyun Baek, Choiej Choi, Hayoung Kim, ByungJae So

Background: Adult *Setaria digitata* are generally considered to be non-pathogenic in their natural hosts, but transmission of infective larvae through mosquito and other arthropod vectors to non-permissive hosts such as goats, sheep, cattle, horse and man, can result in serious and often fatal neuropathological disorders.

Objective: We describe a case of aberrant migration of *Setaria digitata* to the spinal cord of a horse from South Korea.

Methods: A 17-year-old male Thoroughbred horse suffering from hind limb ataxia for two months was presented to the Animal and Plant Quarantine Agency for necropsy. Sections were fixed in 10% neutral buffered formalin, embedded in paraffin, sectioned and stained with H&E. Genomic DNA isolated from formalin-fixed paraffin embedded tissue curls was used as a template for PCR targeting 12S rDNA conserved among nematodes. Amplified PCR products were sequenced to confirm the identity of the nematode.

Results: On gross examination, no significant lesions were observed. On histopathologic examination, infiltrates of eosinophils, lymphocytes, plasma cells, macrophages, and neutrophils were observed in the meninges of the brain and spinal cord. Cross-sections of nematodes, approximately 100-200 μm in diameter were observed in subarachnoid space of lumbar spinal cord. Nematodes were identified as *Setaria digitata* by sequencing.

Conclusion: Setaria digitata has been reported in cattle in South Korea. This is the first irst report of equine cerebrospinal nematodiasis with *Setaria digitata* in South Korea. Cerebrospinal nematodiasis should be a differential for horses with neurologic signs in endemic regions.

D-53: UNCLASSIFIED MALIGNANT ROUND CELL TUMOR IN A JUVENILE DOG: A CANINE COUNTERPART OF SMALL ROUND CELL TUMOR OF CHILDHOOD?
Ikki Mitsui, Masaki Michishita, Eri Fukazawa, Tetsuya Kobayashi

Malignant neoplasms rarely develop in juvenile dogs with exception of rhabdomyosarcoma, lymphoma, and nephroblastoma. Malignant neoplasms of children are more extensive and include leukemia, neuroblastoma, Ewing sarcoma/primitive neuroectodermal tumor, rhabdomyosarcoma, desmoplastic small round cell tumor, malignant rhabdoid tumor, Wilms tumor, and osteosarcoma. Cytomorphological similarities among these human tumors necessitate immunophenotyping by a panel of antibodies and genetic analysis on top of routine histopathology for definitive diagnosis. A 2-year-old intact male Border collie dog presented with right pelvic limb lameness. Computed tomography revealed a destructive mass in the right proximal tibia, enlargement of the right popliteal and multiple subiliac lymph nodes and multiple renal cysts. Despite amputation of the affected leg, the dog died 28 days after the initial presentation. Autopsy and histopathologic evaluation revealed sheets and nests of atypical round cell replacing retroperitoneal structures with lymphatic/vascular invasion and systemic metastasis. Suspected cause of death was acute renal failure due to bilateral ureteral compression/obstruction and diffuse alveolar damage. Reticulin silver stain, toluidine blue stain, and various immunohistochemical stains were performed to determine tumor cell origin. The tumor cells were positive for vimentin, NSE, nestin, inhibin alpha, c-KIT, desmin, and S100, while negative for cytokeratin AE1/AE3, EMA, E-cadherin, CD3, CD20, MUM1, Iba-1, CD163, CD99, Melan A, synaptophysin, chromogranin A, PGP9.5, insulin, glucagon, myoD1, muscle specific actin, WT-1, and GFAP. A cell of origin could not be determined, and there was no distinct correlation with known human round cell neoplasms. This case represents a rare event and supports implementing wider immunohistochemical screening.

D-54: DETECTION OF SKUNK ADENOVIRUS 1 (SKADV-1) IN AN AFRICAN PIGMY HEDGEHOG (ATELERIX ALBIVENTRIS)

Hiroo Madarame

A 34-month-old female pet African pigmy hedgehog (*Atelerix albiventris*) with ascites was necropsied. The main pathological lesion was cardiomyopathy and the cause of death was congestive heart failure. An incidental, segmental desquamating tracheitis with nuclear inclusion bodies was observed. Formalin-fixed paraffin embedded tissue sections of liver, spleen, kidneys, heart, lungs, trachea, ovaries, uterus, gastrointestinal tract, adrenal glands, pancreas, and spinal cord underwent PCR amplification and sequence analysis. The assessment of the DNA dependent DNA polymerase sequence of the extracted sample from the trachea found a 98% sequence alignment for skunk adenovirus 1 (SkAdV-1) (GenBank accession No. KP238322). Nested-PCR of a partial sequence of the adenoviral DNA polymerase gene was positive in the trachea and lungs. Electron microscopy demonstrated that nuclear inclusions of tracheal epithelial cells contained non-encapsulated, hexagonal, electron-dense particles, approximately 60 nm in diameter. SkAdV-1 is a novel adenovirus, isolated in Canada from a dead wild skunk suffering from acute hepatitis. Although SkAdV-1 was isolated from frozen liver tissue of the diseased skunk, Koch's postulates have not been fulfilled to link this virus to the histopathologic findings. It remains to be determined whether skunks act as reservoirs, amplifier hosts, or spillover hosts for SkAdV-1. In the case of the hedgehog, SkAdV-1 was suggested as a possible causative agent of respiratory tract infection. This is the first report of SkAdV-1 infection in animals other than skunks (*Mephitis mephitis*). SkAdV-1 should be monitored closely as a potential cause of global emerging and cross-species infection.

D-55: EXPRESSION ANALYSIS OF OBESITY RELATED FACTORS AND THEIR ASSOCIATION WITH AROMATASE EXPRESSION IN MALIGNANT CANINE MAMMARY TUMORS

Jong-il Shin, Ha-young Lim, Hyun-woo Kim, Byung-joon Seung, Jung-hyung Ju, Jung-hyang Sur

Background: For decades, the rate of overweight and obese people has risen in people and this condition has been associated with several diseases including cancer. Human researches have identified inflammatory mediators with roles in carcinogenesis, especially breast cancer. Obesity is increasingly problematic in dog populations, but relatively little research has been done concerning a relationship with cancer.

Objective: This study was designed to investigate the role of obesity in malignant canine mammary tumors (CMTs), by assessing the aromatase expression and the regulatory roles of immune-mediators such as cyclooxigenase-2 (COX2), prostaglandin E2 (PGE2), nuclear factor kappa beta (NF-kB), hypoxia inducible factor-1a (HIF-1a), and adipokines (leptin) in lean, optimal body weight, overweight, and obese animals.

Methods: Clinicopathological data, including the breed, body weight, body condition score, age, and spaying status were collected together with histopathological characteristics (histological types, grading, and lymphatic invasion). To determine the expression of aromatase, COX2, PGE2, NF-kB, HIF-1a, and leptin, immunohistochemistry was conducted in 60 samples of malignant CMTs. For statistical analysis, Statistical Package for Social Science software program, version 22.0 was used.

Results: CMTs from overweight and obese animals had significantly elevated levels of PGE2 ($p=0.022$), and aromatase expression significantly correlated with PGE2, NF-kB, and leptin($p=0.015$, $p=0.004$ and $p=0.001$, respectively). However, no significant difference was observed in terms of histopathological characteristics.

Conclusion: PGE2, a known obesity-related immune mediator, could be upregulated in malignant CMTs from overweight and obese animals. In addition, PGE2, NF-kB, and leptin influenced the expression of aromatase, as observed in women.

D-56: THE EXPRESSION OF THE M2 MACROPHAGE ANTIGEN CD163 AND MACROPHAGE RELATED-CYTOKINES IN CANINE MAMMARY TUMORS IS ASSOCIATED WITH OBESITY

Byung joon Seung, Ha young Lim, Jong il Shin, Hyun woo Kim, Jung hyung Ju, Jung hyang Sur

Background: Obesity is significant risk factor for human breast cancer and canine mammary tumor (CMT) development. Tumor- associated macrophages (TAMs) are an important component of leukocyte infiltration in tumors, and TAMs are increased in obese animals.

Objective: We analyzed the relationship between TAM polarization and obesity based on the body condition score (BCS) using select immunohistochemical stains.

Methods: The expression of CD163, which is M2 phenotype macrophage receptor, and the expression of cytokines secreted from TAMs such as interleukin (IL)-6, IL-10, and transforming growth factor-beta, were investigated by immunohistochemistry in the area surrounding TAMs in 60 cases of CMTs on the basis of the BCS. To evaluate the expression of CD163 and cytokines, computer-assisted image analysis software was used.

Results: CD163 was expressed in TAMs and cancer cells. On the basis of BCS, the expression of CD163 was significantly higher in overweight or obese dogs than lean or normal dogs (P =0.006). In contrast, the expression of IL-6 was significantly higher in lean or normal dogs than overweight or obese dogs (P =0.000).

Conclusions: CD163 expression is related to M2 phenotype macrophages, which promote tumor progression, while IL-6 expression around TAMs is related to M1 phenotype macrophages, which produce pro-inflammatory cytokines and help kill tumor cells. Therefore, obesity can affect tumor development by recruiting and polarizing macrophages, and overweight or obese dogs are expected to have a worse prognosis compared with that of lean or normal dogs.

D-57: A UNUSUAL CANCER IDENTIFIED IN AN ADULT MALE SEA LION (Zalophus californianus) STRANDED IN BRITISH COLUMBIA, CANADA

Stephen A Raverty, Martin Haulena, Karisa Tang, Abigail McClain

An adult male California sea lion live stranded in BC waters in poor body condition and was presented to the Vancouver Aquarium for assessment and rehabilitation. Along the right and to a much lesser extent left mandible, there was marked enlargement of the salivary glands with subcutaneous swelling throughout the fascial region. Cytology of lymph node aspirates disclosed carcinoma. Due to progressive deterioration in overall condition and poor response to supportive care, the animal was humanely euthanized. Necropsy confirmed generalized emaciation and revealed a mass within the right salivary gland which breached the capsule and invaded locally to the tonsils and regionally lymph nodes as well as abutted the mandibular body and ramus. Microscopically, the mass was an adenocarcinoma and locally invasive. Molecular studies to screen the tumor and other tissues proved positive for universal herpesvirus and gene sequencing identified Californian sea lions herpesvirus. Close evaluation of the reproductive and urinary tracts did not reveal any discernible tumor involvement. Despite the positive PCR result, lack of urogenital involvement and identifiable viral inclusions within examined tissues suggested viremia with spontaneous tumor development of the parotid gland. This is a rare tumor in stranded marine mammals and in this instance was likely spontaneous. The mass would have contributed significantly to impaired foraging and feeding, resulting in the decline in nutrition. To date, only sporadic cases of urogenital carcinoma have presented in the northeastern Pacific and this case may present a distinct and novel pathologic entity.

D-58: LAFORA'S DISEASE IN AN EPILEPTIC CHIHUAHUA WITH NECROTIZING CYSTITIS AND PULMONIC STENOSIS

Molly C. Friedemann, Maureen T. O'Brien, Melanie L. Puchot, Brian F. Porter, Carolyn L. Hodo

A seven-year-old, spayed female Chihuahua presented with a two-week history of hematuria and pollakiuria. The patient also had a history of chronic pulmonic stenosis and seizures controlled with phenobarbital. An abdominal ultrasound showed a severely thickened trigone, and the urinary bladder contained a large amount of mineralized material. Urine culture revealed >100,000 colony forming units/mL of *Corynebacterium urealyticum*. Despite hospitalization and treatment, the patient became anorexic and oliguric. A repeat ultrasound showed progressive hemorrhagic cystitis and bilateral renal pelvis dilation and partial urethral obstruction. Overnight the patient declined, went into cardiac arrest, and died. Relevant necropsy findings included severe necrohemorrhagic cystitis, hydronephrosis, hydroureters, and pulmonic stenosis. Histologic evaluation revealed severe fibrinonecrotizing cystitis with abundant bacterial colonies and widespread neuronal Lafora body-like inclusions throughout the cerebrum and cerebellum. Skeletal muscle also contained rare, subsarcolemmal aggregates of lightly basophilic material suggestive of Lafora disease. The Lafora bodies in the cerebellum and the subsarcolemmal material within skeletal muscle were PAS positive. Lafora's disease is a hereditary neurodegenerative disorder resulting from defects in enzymes responsible for glycogen metabolism. The condition is reported in humans and dogs and is a progressive disorder characterized by myoclonic epilepsy and eventual death. Lafora-like bodies are commonly seen in aged dogs; however, when abundant in a dog with a history of seizures, this finding is suggestive of Lafora's disease.

D-59: ADENOCARCINOMAS ARISING FROM APOCRINE GLANDS OF THE CHEEK AREA IN TWO BLACK-TAILED PRAIRIE DOGS (CYNOMYS LUDOVICIANUS)

Hirotaka Kondo, Haruka Sakashita, Hiroki Toge, Shinkichi Tsuruno

Some rodent species including prairie dogs exhibit olfactory communication, which is associated with characteristic scent-marking behavior using integumentary glands. Prairie dogs, presumably, have at least two glandular anatomic areas: the oral cheek and perianal. We describe two cases of adenocarcinoma of the cheek glands in domestic black-tailed prairie dogs (*Cynomys ludovicianus*). Case 1 was an 8-year-old, female that developed a solitary mass measuring 8mm in diameter on the cheek area. Case 2 was a 7.5-year-old, female that presented with a solitary mass measuring 9mm in diameter on the cheek area. Masses were surgically removed and were submitted to the Pathology Service at Synergy Animal General Hospital. Masses showed similar morphology and were composed of polygonal to columnar cells arranged in irregular acini and tubules with small amounts of intervening fibrous connective tissue. Neoplastic cells had distinct cell borders and moderate amounts of eosinophilic cytoplasm. Nuclei were ovoid to round with coarsely stippled chromatin and prominent nucleoli. Anysocytosis and anisokaryosis was marked.. There were 12 and 20 mitoses per 10 HPF in cases 1 and 2, respectively. Histologic findings were consistent with adenocarcinomas of apocrine gland origin. Recurrence of the mass was observed in

Case 2. Neoplasia is uncommon in this species, and to the best of our knowledge, this is the first report of adenocarcinomas arising from apocrine glands of the cheek area in prairie dogs.

D-60: AXIAL OSTEOSARCOMA IN FOUR AFRICAN HEDGEHOGS

Alonso Reyes-Matute, Adriana Méndez-Bernal

Neoplasms in African hedgehogs are a common occurrence and represent one of the main causes of disease. Osteosarcomas have been reported in skeletal and extraskeletal locations including mandible, ribs, and vertebra. Four hedgehogs with osteosarcoma submitted to the Pathology Department at Facultad de Medicina Veterinaria y Zootecnia, UNAM from 2014 to 2016 are reported. In two cases, the neoplasm arose from the skull (parietal bones) with extension along the skull; one case arose from the ribs with associated compression of the thoracic and abdominal cavity; the last case involved the vertebrae. Histological lesions were similar in all cases and consisted of well-demarcated nodules, rimmed by thin layers of fibrous tissue. Neoplastic cells were arranged in sheets of polyhedral to spindle-shaped cells with interspersed necrosis. Numerous trabeculae of osteoid were present throughout the tumors. No metastases were detected. The predominant histological pattern was osteoblastic, but a teleangiectatic-like pattern was observed in the vertebral osteosarcoma. Electron microscopy was performed in two cases, and malignant osteoblasts had features consistent with descriptions in other species including deposits of hydroxyapatite in osteoid. Axial osteosarcoma has been observed in this species in contrast to appendicular osteosarcoma in other species.

D-61: METASTATIC SUBCUTANEOUS PHEOCHROMOCYTOMA IN A CAT

Ashley Leisering, Brian Meyer, Mario Sola, William Wigle, Craig Thompson

A 14-year-old, spayed female Domestic Shorthair was presented with multiple, rapidly coalescing skin nodules, four months after a urinary bladder urothelial carcinoma was excised. On ultrasonographic examination, multiple nodules were detected in the abdomen. One of the skin nodules was aspirated and cytologically evaluated at the Clinical Pathology Laboratory of the Purdue University Veterinary Teaching Hospital. The smears consisted of cohesive clusters of epithelial cells with indistinct cell borders and abundant bare nuclei, most consistent with a neuroendocrine tumor. Due to the patient's deteriorating condition, humane euthanasia was elected. In addition to nodules in the subcutis, necropsy revealed multiple masses in the caudal abdomen, omentum, left kidney and adrenal gland, abdominal wall, mesentery, lungs, and brain. On histologic evaluation, the cells from all sites appeared similar with the neoplastic cells arranged in sheets and nests. The cells exhibited variably distinct cell borders and a faintly granular cytoplasm. The morphology, taken into consideration with the involvement of the adrenal medulla, raised the suspicion for a metastatic pheochromocytoma. Immunohistochemical evaluation revealed expression of PGP9.5 and endorphin by the neoplastic cells, whereas no reactivity was observed for met-enkephalin and uroplakins 2 and 3. These findings are most consistent with a metastatic pheochromocytoma involving the skin. Pheochromocytomas are rarely reported in cats and metastasis to the skin is uncommon in any species. To the authors' knowledge, this is the first report of a disseminated pheochromocytoma with subcutaneous metastasis in a cat.

D-62: SYNOVIAL HISTIOCYTIC SARCOMA IN AN EIGHT YEAR OLD CASTRATED MALE DOG

Jonathan M Bagwell, Heather R Herd, Brad L Njaa, Theresa E Rizzi

An eight year old, castrated male, mix-breed dog presented for intermittent lameness of the right hind limb with swelling and warmth over the metatarsus. Radiographs showed soft tissue swelling. Initial response to NSAIDs and antibiotics was short-lived, followed by continued swelling, lameness, and right popliteal lymph node enlargement. A two centimeter mass was palpated proximal to the metatarsal pad on the plantar surface. Repeat radiographs showed mild lytic and proliferative changes in the diaphysis of the third metatarsal. Aspirates of the swelling contained a population of pleomorphic round cells with round to oval to irregularly shaped nuclei that have coarse chromatin and large, prominent nucleoli. Binucleate and multinucleate cells were frequently observed. Cytoplasm was moderately abundant, basophilic, and occasionally contained variable numbers of small eosinophilic granules. Lymph node aspirates contained an expanded population of intermediate lymphocytes. PCR for Antigen Receptor Rearrangement performed on the mass and the popliteal lymph node was negative. Histologically, the neoplasm was composed of pleomorphic spindle-shaped to discrete round cells with occasional myxoid differentiation, multiple areas of necrosis, and lymphocytic infiltrates consistent with a synovial cell sarcoma. Round cells were morphologically similar to the aspirates. Neoplastic cells had positive cytoplasmic immunoreactivity for CD18 (histiocytic origin), consistent with synovial histiocytic sarcoma. Popliteal lymph node impressions made after limb amputation contained neoplastic cells similar to the initial aspirate and a background of reactive lymphoid cells. The patient is clinically doing well two months after mid femoral amputation.

D-63: PULMONARY VENO-OCCLUSIVE DISEASE AND PULMONARY CAPILLARY HEMANGIOMATOSIS IN A YOUNG CAT

Tiffany L Jenkins

Pulmonary capillary hemangiomatosis and pulmonary veno-occlusive disease are rare causes of primary pulmonary hypertension in humans, and have only recently been reported in dogs. A 1-year-old male neutered Persian cat (*Felis catus*) presented for autopsy after sudden death following a grooming appointment. Physical examination and medical history prior to death were unremarkable. Grossly, the lungs were mottled red to pink and multifocally rubbery to firm. There was marked left-sided cardiomegaly, consistent with hypertrophic cardiomyopathy (HCM). Microscopically, there was multifocal to widespread expansion of pulmonary alveolar septa and perivascular and bronchiolar interstitium by numerous, densely arranged thin-walled congested capillaries. Multifocally, the proliferating capillaries formed large, expansive, nodular foci, corresponding to the grossly described red, firm foci. Proliferative capillaries multifocally infiltrated along smooth muscle bundles and extended into the walls of respiratory bronchioles. Additionally, rare veins, identified using Verhoeff van Gieson (VVG) elastin stain, were occluded by intraluminal fibrous stroma and endothelial lined vascular spaces. The gross and microscopic changes were consistent with pulmonary capillary hemangiomatosis (PCH) with rare features of pulmonary veno-occlusive disease (PVOD), both rare primary pulmonary diseases previously undocumented in cats. The severe primary pulmonary disease was the presumed cause of death despite concurrent, and presumably unrelated, HCM. The incidence of PVOD and PCH in the veterinary field is unknown at this time, likely due to these diseases being overlooked and/or misdiagnosed. Awareness of these pulmonary diseases within the veterinary community is imperative to better diagnose, ascertain pathogenesis, direct treatment, and investigate comparative models of PCH and PVOD.

D-64: KARTAGENER SYNDROME IN A WEST HIGHLAND WHITE TERRIER PUPPY
Charles Assenmacher

Background: An 18-week-old female intact West Highland white terrier puppy was presented for a recent history of coughing and increased respiratory effort with no response to treatment by the referring veterinarian. Initial assessment revealed a dull mentation, tachypnea, and a body temperature of 40.2°C. Crackles and wheezes were audible in all lung fields. Euthanasia and an autopsy were elected due to the poor response to initial treatments.

Results: The autopsy revealed a situs inversus totalis (thoracic and abdominal), as well as moderate bilateral lateral ventricle hydrocephalus, marked atrophy of the nasal turbinates and increased firmness of all lung lobes.

Histopathology revealed a moderate to severe multifocal acute to subacute fibrinosuppurative and necrotizing bronchopneumonia with myriad intralesional bacteria and a mild multifocal chronic-active rhinitis with bone loss and reactive new bone formation.

Conclusion: The presence of situs inversus along with the histologic findings of rhinitis, bronchopneumonia, and hydrocephalus are suggestive of a primary motile ciliary disorder termed primary ciliary dyskinesia (PCD). PCD is typically an autosomal recessive disorder that causes defects in the action of cilia lining the upper and lower respiratory tract (including the middle ear), the fallopian tubes in females, flagella of spermatids in males, and the ependymal cells of the brain and thus predisposes to chronic rhinitis/sinusitis, bronchitis and bronchiectasis, hydrocephalus, and infertility. This condition is also known as “Immotile cilia syndrome” or Kartagener’s syndrome and has been described in many species, including humans, rodents, pigs, cats, a horse and numerous breeds of dogs.

D-65: CLOISONNE KIDNEY IN A RABBIT

Krystal J Vail, Michelle Whitehead, Sharman Hoppes, Raquel R Rech

An adult, male, wild Eastern Cottontail rabbit (*Sylvilagus floridanus*) presented to Texas A&M University Veterinary Medical Teaching Hospital for trauma of unknown origin. On gross examination, the renal cortices were bilaterally and diffusely dark brown, stopping abruptly at the corticomedullary junction. The medulla was diffusely slightly red. Histologically, approximately 60% of the proximal tubular basement membranes had segmental thickening by dark brown pigmentation on both H&E and unstained sections. Thickened basement membranes showed moderate staining for Perl's iron stain. Cloisonne kidney is a pigmentary condition most commonly described in goats and rarely reported in sheep, horses and a lion. The condition is not associated with clinical disease or impaired renal function. It is thought that repeated episodes of intravascular hemolysis result in deposition of ferritin and hemosiderin on proximal tubular basement membrane.

D-66: DERMATITIS WITH ASSOCIATED SARCOPTES SCABEI, CORYNEBACTERIUM SP., AND STAPHYLOCOCCUS HYICUS INFECTION IN A MINIATURE POT-BELLIED PIG (SUS SCROFA DOMESTICUS)

Elizabeth A Kieran, Michael J Dark, Heather S Walden, Elizabeth A Nelson

A 4 month old, male miniature pot-bellied pig (*Sus scrofa domesticus*) presented to the University of Florida Veterinary Medical Center for sudden onset dyspnea and unresponsiveness. Initial exam findings included obtunded mentation, cyanosis, bradycardia, bradypnea, hypothermia, hypoglycemia, poor body condition, and severe exudative dermatitis with lichenification; despite supportive therapy spontaneous death occurred 36 hours after hospital admission. Post-mortem gross evaluation of the skin identified widespread, marked to severe lichenification, multifocal to coalescing erosions, numerous tan crusts forming broad superficial mats, serous to greasy exudate oozing from all skin surfaces, and patchy reddening of the ventral body wall. Histologic evaluation of the skin identified marked to severe, chronic dermatitis with marked orthokeratotic and parakeratotic hyperkeratosis with multifocal intracorneal pustules, numerous 300 x 400 micrometer intracorneal mites (morphology consistent with *Sarcoptes scabiei*), and numerous intracorneal coccobacilli. Aerobic culture of the skin yielded heavy growth of predominately *Corynebacterium* sp.; *Staphylococcus hyicus* was also cultured. A skin sample was also submitted for parasite identification, confirming *Sarcoptes scabiei* infestation. *S. scabiei* infestation is a highly contagious disease of economic importance in production swine operations. The mite causes mechanical injury to the skin in addition to irritant effects induced by secretions and excreta and hypersensitivity reactions to mite antigens. A recent study exploring scabies infection in a porcine model has demonstrated scabies impacts host skin microbiota, with a shift from normal to pathogenic bacterial populations and notably identified *Corynebacterium* sp. infection associated with crusted skin lesions and associated mite microbiota.

D-67: A DIVERGENT MORPHOLOGIC FEATURE OF CUTANEOUS PLASMA CELL TUMORS IN DOGS

Brittany McHale, Uriel Blas-Machado, Fabiano Oliveira, Daniel Rissi

Background: Cutaneous plasma cell tumor (CPCT) is a common neoplasm of middle-aged to older dogs. Tumors can be single or multiple and occur predominantly on the pinna, digits, lip, and oral cavity, although other sites can also be affected.

Objective: CPCT typically offers no diagnostic challenge. However, a subset of tumors with pseudofollicular arrangement of neoplastic cells may make it difficult to differentiate from epithelial neoplasia. Here we describe 5 cases of canine CPCT with pseudofollicular arrangement.

Methods: Cases originated from the diagnostic service at the AVDL and AD. Pathology reports were reviewed and histology slides were examined. Replicate sections were submitted to immunohistochemistry for multiple myeloma oncogene 1 (MUM-1) and pancytokeratin AE1/AE3 (PCK).

Results: Mean age of affected dogs was 11.4 years and multiple breeds and sites were affected. Histologically, neoplastic cells were arranged in sheets, packets, and scattered pseudofollicular structures having central accumulations of blood or eosinophilic material admixed with neoplastic cells and hemosiderin-laden macrophages. Because pseudofollicular structures resembled neoplastic acini, epithelial neoplasia was cogitated as a possible differential diagnosis in some of these cases. Tumors were immunopositive for MUM-1 and immunonegative for PCK.

Conclusion: CPCT with pseudofollicular morphology may resemble epithelial neoplasia with acinar or follicular differentiation and raise questions about tumor histogenesis.

D-68: PULMONARY ARTERITIS ASSOCIATED WITH HYDROPHILIC POLYMER EMBOLI FOLLOWING INTRAVENOUS CATHETERIZATION IN A PET RABBIT

Laura L Bassel, Amanda Mansz, Jeff L Caswell

A 9-year old lop-eared pet rabbit died unexpectedly 2 days after bilateral medial canthoplasty surgery. No complications were reported during the surgery but the rabbit had a decreased appetite following discharge. Significant findings on gross postmortem examination included a moderately dilated right atrium. Histological examination of lung demonstrated extensive acute neutrophilic vasculitis affecting the small to medium-sized pulmonary arteries. The lumens of the affected arteries were filled with ribbon-like, pale pink material that was birefringent with polarized light, positive with periodic acid–Schiff reaction, and unstained using Masson's trichrome. Our analysis of the material was consistent with hydrophilic polymer frequently used to coat intravascular devices including catheters and stents. It was concluded that the marked vasculitis was induced by the intraluminal foreign material and caused pulmonary hypertension leading to dilation of the right atrium and was the likely cause of death. Hydrophilic polymer embolism is a recognized complication of interventional endovascular procedures in humans. Hydrophilic polymer may dissociate from the catheter and embolize to distant sites where it can be associated with significant iatrogenic injury including death. To the authors' knowledge this has not been reported in animals, but veterinary pathologists should be aware of the highly characteristic morphology.

D-69: NASAL PYTHIOSIS IN A SHEEP FROM TEXAS

Paula R Giaretta, Roy Pool, Glaucia D Kommers, Sara Lawhon, Caitlin E Older, Philippa Gibbons, Rodolfo Madrigal, Raquel R Rech

An 18-month-old, Dorper ewe presented to the Texas A&M University Veterinary Medical Teaching Hospital for a two-week history of dyspnea. On physical examination, bilateral facial swelling, mucopurulent nasal discharge, and a 4x2 cm ulcerated and necrotic area of the rostral dental pad were noted. A CT showed an expansile, heterogeneous, soft tissue mass within the rostral nasal cavity, causing leftward deviation of the nasal septum, thickened nasal mucosa, and areas of osteolysis in the rostral vomer and incisive bones. A nasal septal biopsy consisted of multifocal eosinophilic granulomas in the submucosa, with abundant 3-5 µm in diameter, nonparallel walled, irregular branching, and GMS positive hyphae, surrounded by granular and eosinophilic Splendore–Hoeppli material. Areas of caseous necrosis, vascular invasion, and thrombosis were observed. Due to poor prognosis, euthanasia was elected without a necropsy. PCR of DNA extracted from formalin-fixed paraffin-embedded samples using ITS3 and ITS4 primers targeting the Internal Transcribed Spacer (ITS) 2 region resulted in a single positive band, sequenced and classified as *Pythium insidiosum* with more than 98% identity match for ITS3 and ITS4. *Rhizopus* sp. was cultured from the nasal mucosa collected 12 hours post mortem but was considered a contaminant in this case. Immunohistochemistry was positive for a rabbit polyclonal anti-*P. insidiosum* antibody and negative for a mouse monoclonal anti-*Rhizopus* antibody . The diagnosis of pythiosis was based on the morphology, PCR results, and immunohistochemical identification and to the author's knowledge, is the first report of the disease in a sheep in the United States.

D-70: IMMUNOREACTIVITY OF CANINE LIPOSARCOMA TO MUSCLE AND BROWN ADIPOSE MARKERS

Sarah E Stevens, Elise EB LaDouceur, Jason Wood, Christopher M Reilly

Rhabdomyosarcoma, liposarcoma, and hibernoma (a benign neoplasm of brown fat) share some overlapping histologic and immunohistochemical features. Specifically, human liposarcomas and canine hibernomas have both been reported to express some muscle markers via immunohistochemistry (IHC), which is commonly used to diagnose rhabdomyosarcoma. The overlapping histologic and immunohistochemical features of liposarcoma, rhabdomyosarcoma, and hibernoma in dogs may present a diagnostic challenge, with important differences in prognosis and treatment. By definition, rhabdomyosarcomas and liposarcomas are malignant, with rhabdomyosarcoma having much higher potential for metastasis. In contrast, hibernomas are benign and pose low risk of recurrence. Recently, it has been reported that the expression of muscle markers by hibernomas can support this diagnosis over other tumors of fat. This study investigated the expression of muscle markers and uncoupling protein 1 (UCP1) by canine liposarcoma in 25 cases using IHC for muscle antigens desmin, myogenin, and α-smooth muscle actin (α-SMA), and the brown fat marker UCP1. Oil red O histochemistry was performed to confirm the presence of lipid and the diagnosis of liposarcoma. Histologic subtyping identified 15/25 well-differentiated, 5/25 pleomorphic, 3/25 myxoid, and 2/25 dedifferentiated subtypes. 23/25 expressed UCP1, 7/25 expressed α-SMA, 7/25 expressed desmin, and 3/25 expressed myogenin. Data from this study helps clarify the immunohistochemical profile of liposarcoma and suggests that there is considerable overlap in the expression of several muscle antigens and UCP1 between neoplasms of brown fat, white fat, and skeletal muscle origin. This overlap may limit their utility in the differentiation of these groups of neoplasms.

D-71: ACUTE MYELOID LEUKEMIA IN A TOGGENBURG GOAT

Jolie A Demchur, Susan J Bender, Koranda A Walsh

Acute Myeloid Leukemia (AML) with granulocytic differentiation and extramedullary dissemination in a 4-year-old Toggenburg wether is described. The goat exhibited marked weight loss over three months with a profound non-regenerative anemia, elevated white blood cell count, and generalized peripheral lymphadenopathy. The goat was euthanized with a presumptive diagnosis of myeloid leukemia, although acute lymphoproliferative disease could not be ruled out. The CBC demonstrated 38% unclassified blast cells, 8% mildly toxic neutrophils, 2% metamyelocytes, 2% progranulocytes, and 10% small lymphocytes. On postmortem examination the peripheral, thoracic and abdominal lymph nodes were enlarged and partially to completely effaced by soft, bulging, pale green tissue interspersed with small hemorrhagic foci. Nodules of similar pale green tissue were throughout the pancreas. Cytology of the lymph nodes revealed abundant neoplastic round cells in varying stages of myeloid maturation (myeloblasts, promyelocytes, myelocytes, bands, and few mature segmented neutrophils). Histologically, sheets of similar neoplastic cells effaced the nodal architecture, often exhibiting blast morphology. More mature cells had band-shaped nuclei and cytoplasmic granules. The neoplastic population was identified within the bone marrow, pancreas, spleen, adrenal gland, liver, kidney, and cecum. A diagnosis of AML with neutrophilic differentiation is assigned based on cell morphology, numbers of blasts in circulating blood, and positive immunohistochemical staining for neutrophil elastase. AML occurs sporadically in most domestic species; however, there is only one other report of caprine AML. Dissemination of neoplastic cells to form solid extramedullary tumors is rare in humans, but is most often diagnosed in patients with myeloproliferative neoplasms.

D-72: BRAIN LESIONS IN A CALF WITH HISTORY OF PERINATAL ASPHYXIA AND APPLICATION OF FLUORO-JADE C

Kenji Koyama, Natsuko Fukumoto, Ken-ichi Watanabe, Noriyuki Horiuchi, Tomomi Ozawa, Yoshiyasu Kobayashi

Background: Perinatal asphyxia and subsequent brain injury remain as important problems in the cattle industry. However, no detailed studies describing these conditions have been conducted to date.

Methods: A 3-day-old, Japanese black, female calf showed respiratory distress immediately after delivery with human assistance. The calf was treated with artificial respiration and an injection of respiratory stimulant, and started to breath on her own after 10 minutes of birth. The calf showed astasia and consciousness disturbance, and died at postnatal day 3. Routine histopathological examination was carried out. In addition, a neuropathological examination using Fluoro-Jade C (FJC), a neuronal degeneration marker, was performed.

Results: Histologically, there was bilateral laminar cortical necrosis in the deep layer of the cerebrum. Neuronal necrosis was also observed diffusely in the corpus striatum and partially in the hippocampus and Purkinje cells. Giemsa-stained sections revealed loss of Nissl bodies in affected cells. In FJC sections, positive reactions were observed in necrotic cells. In FJC sections, positive reactions were observed in necrotic cells. Histological diagnosis was polioencephalomalacia.

Conclusions: The distributional pattern of the lesions was consistent with that of global ischemia and interpreted as representing secondary changes to a hypoxic/ischemic status during perinatal asphyxia. FJC seemed to be applicable for a neuropathological examination on calves with history of perinatal asphyxia, and was useful to detect the distributional pattern of the lesions.

D-73: MULTIPLE TUMORS IN A GUINEA PIG (CAVIA PORCELLUS)
Celic Berenice B Montoya Ménez, Adriana Méndez Bernal, Angélica Rodríguez

A 6-year-old Guinea pig (Cavia porcellus), was brought for postmortem evaluation at the Department of Pathology of the College of Veterinary Medicine of UNAM from the Veterinary Specialty Hospital of Wildlife and Clinical Ethology of the same institution. The animal was humanely euthanized due to severe wasting and a non-resectable mass in the right prescapular region. Histologically, the prescapular mass was diagnosed as a poorly-differentiated, high-grade subcutaneous sarcoma. The mass was evaluated by transmission electron microscopy (TEM) and diagnosed as fibrosarcoma. In addition to this tumor, other neoplasms of different cell origins were present: a pancreatic neuroendocrine tumor, a thyroid adenoma, and a neuroendocrine tumor in a mesenteric lymph node. Other lesions seen included an endometrial polyp and ovarian cysts. The aim of this case is to describe multiple neoplasms, most of endocrine or neuroendocrine origin, in a single Guinea pig, and their possible relationships. It also highlights the importance of using TEM as a diagnostic tool in poorly differentiated neoplasms.

D-74: CUTANEOUS SPINDLE CELL SARCOMAS IN A KOI CARP (CYPRINUS CARPIO)
Laura E Rice, Josue Delgado, Raquel R Rech, Michelle Whitehead, Sharman M Hoppes, Brian F Porter

An adult female Koi carp (Cyprinus carpio) presented to the Texas A&M University Zoological Medicine service for multiple raised, ulcerated, erythematous cutaneous masses along the dorsum of two months duration. These lesions were biopsied, and histologic examination revealed an unencapsulated, mildly infiltrative, poorly demarcated, and densely cellular neoplasm that elevated the epidermis and expanded the dermis. Neoplastic cells were spindled with indistinct cell borders, arranged in bundles, steams, and interlacing fascicles on a fine fibrovascular stroma. Neoplastic cells had scant eosinophilic cytoplasm and a single oval to cigar-shaped nucleus with finely stippled chromatin and one to two prominent nucleoli. Anisocytosis and anisokaryosis were moderate, and two mitotic figures were seen in ten 40x fields. PCR testing for Koi (cyprinid) herpesvirus and Koi poxvirus was negative. Undifferentiated cutaneous spindle cell sarcoma was diagnosed, and humane euthanasia was elected due to the poor prognosis. On gross postmortem examination, four 3 x 1.5 to 5 x 3 cm, dark red, raised, firm, cutaneous masses were along the dorsal midline. Histology of the four masses confirmed a diagnosis of cutaneous spindle cell sarcoma. Cutaneous spindle cell sarcomas are rarely reported in koi, and the gross and histologic appearance of the dermal masses resembles that of Hikui disease, a disfiguring skin disease that has been recently described in this species.

D-75: A REVIEW OF THE CAUSE AND MANNER OF DEATH OF SUSPECTED ANIMAL ABUSE CASES IN TRINIDAD AND TOBAGO, 2008-2015

Karelma Frontera-Acevedo, Lana A Gyan, Rod B Suepaul

Background: Forensic veterinary pathology is becoming increasingly important in investigation situations of suspected animal abuse. It is thought that the submission of suspected cases of animal abuse has increased over the past seven years in Trinidad and Tobago.

Methodology: A search for owner or law-enforcement suspected animal abuse necropsy cases submitted at either the University of West Indies Veterinary Hospital or the Ministry of Agriculture, Land and Fisheries, Veterinary Diagnostic Laboratory for the period of 2008-2015 was performed. Based on pathology records and submission information, cases were categorized by signalment and cause of death (COD) or manner of injury (MOI) if the animal was euthanized.

Results: A total of 84 cases of suspected animal abuse were submitted in the years 2008-2015. Although the submission of these cases decreased from 2008-2012, there has been an increase since then. The most commonly submitted animal species were dogs (74%) and sheep (7%). Of the 84 cases, only 22 (24%) had a confirmed COD/MOI. The most common confirmed COD/MOI was trauma (45%), followed by deliberate poisoning (27%). The majority of the unconfirmed COD/MOI involve suspected deliberate poisoning.

Conclusions: Owners are becoming increasingly aware of the importance of pathology to help diagnose suspected animal abuse cases, and are becoming more interested in trying to prosecute the suspects. Toxicological studies are currently limited, but they would be very useful in confirming cause of death for the majority of unresolved cases.

D-76: STERILE NEUTROPHILIC DERMATOSIS IN A DOG

Myeon-Sik Yang

Background: Canine sterile neutrophilic dermatosis is a rare skin disease, characterized by dermal edema with a moderate to severe perivascular or diffuse neutrophil infiltration, which is similar to Sweet's syndrome in humans.

Objective: The aim of this report is to present microscopic appearance of skin lesions diagnosed with sterile neutrophilic dermatosis.

Methods: The papule was dissected on occipital region of a 2-year-old Yorkshire terrier by routine surgical procedure. Imprint method for cytologic examination was used before removal. The dissected papule was fixed in 10% neutral buffered formalin. Fixed tissues were processed, embedded in paraffin, sectioned at 5-μm thickness, and stained with hematoxylin and eosin (H&E) for routine histopathology.

Results: There was no clinical sings and no abnormalities on chemistry and CBC examination. Imprint cytologic examination revealed a cellular sample that consisted of neutrophil, lymphocyte and mesenchymal cell. Histopathologic examination revealed that dermis contains a moderate neutrophilic infiltrate intermingled with lymphocytes, macrophages and eosinophils. Ulceration occurred in the center of the papule and neovessels containing neutrophils were observed in ulcerated lesion. Leukocytoclasia were not prominent and there was no vascular damage.

Conclusion: Based on histopathologic observation, the biopsied papule was diagnosed as sterile neutrophilic dermatosis

D-77: HEPATIC ENCEPHALOPATHY AS AN ATYPICAL PRESENTATION OF METASTATIC CARCINOMA IN A DONKEY

Amanda Anderson, MeeJa Sula

A 17-year-old, intact female jenny presented to the referring veterinarian with acute onset of neurologic abnormalities including aggression toward the owner. The jenny had a one-month history of colic and lethargy that was refractory to treatment with antibiotics and mineral oil. Serum biochemistry abnormalities were limited to elevated liver enzymes. Due to the presumptive diagnosis of hepatic encephalopathy the owners elected humane euthanasia and the body was submitted to the University of Tennessee College of Veterinary Medicine for postmortem evaluation. Gross necropsy examination revealed a markedly enlarged and friable liver with multifocal, irregular, firm, pale tan, umbilicated masses replacing nearly two-thirds of the hepatic parenchyma, and a severely distended stomach impacted with feed. Similar masses were present in the lung, multiple abdominal lymph nodes, pylorus of the stomach, and duodenum. Histologic examination revealed a densely cellular multilobulated neoplasm composed of islands and acini of polygonal cells that occasionally formed papillary structures surrounded by abundant desmoplasia. Throughout the cerebral gray and white matter were numerous pairs and small clusters of Alzheimer type II astrocytes. Microscopic diagnosis was carcinoma with regional and pulmonary metastasis, and hepatic encephalopathy. Primary differentials for the neoplasm included biliary, gastric, or neuroendocrine origin. The gastric impaction and chronic signs of colic were likely secondary to neoplastic invasion and obstruction of the pylorus. Neoplasms of non-lymphoid or melanocyte origin are rare in equids and although uncommon, hepatic encephalopathy may be an unexpected presenting complaint.

D-78: CUTANEOUS HIGH-GRADE MYXOFIBROSARCOMA WITH LOCOREGIONAL METASTASES IN A DOG

Valeria Cafe Marcal, Graham Stock, Melanie Dobromylskyj, Marina Scoffone, Jerome Benoit

Background: Myxofibrosarcoma is a malignant canine soft tissue tumor of fibroblast origin distinguished by the presence of an abundant myxoid matrix rich in mucopolysaccharides.

Objective: To describe the histological, immunohistochemical features and treatment protocols of a locally aggressive canine myxofibrosarcoma.

Methods: Formalin-fixed and paraffin-embedded biopsy samples were processed routinely and stained with H&E. Primary antibodies against vimentin (1:200), S-100 protein (1:800), α-SMA (1:400), desmin (1:100), CD18 (1:25), and CK (cytokeratin) AE1/AE3 (1:50) (Dako, Glostrup Denmark) using the biotin-free peroxidase method Envision Detection System (K5007, Dako, Glostrup Denmark) were applied. The 3,3'-diaminobenzidine tetrahydrochloride (DAB) served as chromogen. A 6 weeks metronomic chemotherapy course, consisting of daily administration of low-dose cyclophosphamide 10 mg/m2, PO, SID, (Specials Laboratory Ltd, UK) and meloxicam 0.1 mg/Kg, PO, SID (Metacam®, Boehringer Ingelheim, Germany) was administered. A palliative hypofractionated course of radiation therapy of four once-weekly 8Gy fractions prescribed to the planned target volume was given.

Results: The first course of chemotherapy was unsuccessful and the hypofractionated course of radiotherapy only provided some short-term tumor control with no remission. The tumors progressed and the dog was euthanized 12 months after initial clinical presentation. Neoplastic cells were positive for vimentin but negative for S-100, smooth muscle actin, desmin, CD18 and cytokeratin. Based on the myxoid and multinodular histological appearance, the presence of a curvilinear capillary pattern within tumor stroma and the immunohistochemical results, the masses were diagnosed as high-grade myxofibrosarcomas.

Conclusion: High-grade myxofibrosarcomas are clinically aggressive and difficult to diagnose and aggressive local treatment strategies are necessary.

D-79: HERPESVIRAL BLEPHARITIS ASSOCIATED WITH ORAL B CELL LYMPHOMA IN A CAT

Susanne Je-Han J.-H. Lin, Hui-Wen Chang, Chian-Ren Jeng, Fun-In Wang, Victor Fei F. Pang, Chen-Hsuan Liu

Background: Feline herpesvirus-1 (FHV-1) infection is commonly seen in domestic cats, which leads to feline viral rhinotracheitis and is also one of the major causes of feline ocular diseases manifested mainly by conjunctivitis and keratitis. Herein we report a cat with ocular FHV-1 infection associated with oral B cell lymphoma.

Case Presentation:A 13-year-old mixed cat showed left eyeball proptosis and third eyelid protrusion; an oral mass was also found during examination. Enucleation and oral mass biopsy were performed and specimens were submitted for pathological examination. DNA was extracted from paraffin embedded tissue and used for polymerase chain reaction (PCR).

Result: Microscopically, there was severe necropurulent and lymphoplasmacytic inflammation of the protruding nictitating membrane admixed with abundant intralesional hair fragments. Epithelial necrosis and intranuclear eosinophilic inclusion bodies were present, and the PCR result was positive for FHV-1 by using primers to amplify the FHV-1thymidine kinase gene. Keratoconjunctivitis, iritis, as well as hematoma in anterior chamber and corneal perforation were also observed. The biopsied tissue of oral mass consisted of sheets of large B cells confirmed by anti-CD79a immunohistochemistry.

Conclusion: Herein we report a cat with oral B cell lymphoma and FHV-1infection leading to third eyelid blepharitis, keratoconjunctivitis, hematoma in anterior chamber with corneal perforation as well. Typical intranuclear inclusion body was observed and the PCR result showed positivity of FHV-1. A multifactorial etiology combining neoplasm, viral, secondary bacterial infection is considered.

D-80: DUODENAL DUPLICATION CYST IN A BOVID

Ian T. Sprandel, Adam W. Stern

Background: Cystic duplications of the gastrointestinal tract are uncommon congenital abnormalities that occur anywhere in the alimentary tract from the tongue to the anus. Inclusion criteria consists of a location adjacent to the intestinal lumen, a wall composed of smooth muscle, and a lining of intestinal-like mucosa. While most commonly described in humans, gastrointestinal duplication cysts have also been described in foals, goats, dogs and a cat.

Objective: This case study describes the gross and histological findings of the first reported incidence of a gastrointestinal duplication cyst in a bovid.

Methods: A postmortem examination with routine histopathology was performed on a 290 kg castrated male bovid that was submitted to the Veterinary Diagnostic Laboratory at the University of Illinois at Urbana-Champaign after a 48-hour period of deteriorating health. The animal bloated immediately prior to death.

Results: Gross examination revealed a 20 cm aspect of the duodenum that was markedly dilated (15 cm in diameter) and fluctuant on palpation. On cut surface, the duodenal wall contained a cystic cavity that was entirely filled with a crystal-clear, viscous fluid. The cavity did not communicate with the adjacent duodenal lumen. Histologically, the wall of the cyst is composed of smooth muscle with a thin, underlying lamina propria and lined with a flat to minimally folded (no villous structures) layer of intestinal epithelium consisting of a monolayer of columnar epithelial cells with numerous interspersed goblet cells.

Conclusion: This is the first reported case of a gastrointestinal duplication cyst in this species.

D-81: SATELLITE TAG DEPLOYMENT, DETACHMENT AND LOSS OF A SOUTHERN RESIDENT KILLER WHALE (L95) IN BRITISH COLUMBIA, CANADA

Stephen A Raverty, Brad Hanson, Paul Cottrell, David S Rotstein, Sophie Dennison, Tracey Goldstein, Tori McKlveen, Lisa Spavin, Joseph Gaydos

To better define critical habitat and determine areas transited by the endangered southern resident killer whales, a satellite tagging program has been initiated. An adult male L95 was tagged February 23, 2016 with detachment within 5 days and the animal was recovered dead, March 30, 2016. A comprehensive necropsy was undertaken to determine a cause of death, assess the animal for possible evidence of human interaction and screen for recognized pathogens of concern. The carcass was recovered floating off the west coast of Vancouver Island and presented with advanced autolysis which hindered gross evaluation of multiple organs. Two puncture wounds at the base of the dorsal fin were consistent with prior satellite tag deployment and a tissue block was excised for imaging studies. Tissues were processed for histopathology and ancillary diagnostic studies. The animal was in fair to moderate body condition and there was diffuse fibrinous peritonitis with marked splenomegaly. Radiology and MRI studies confirmed retained tag petals and histopathology of the tag site wound and lungs disclosed mucormycosis with transmural vasculitis, tracking fasciitis and bronchopneumonia with florid intralesional fungal hyphae. There was heavy growth of Clostridium novyi and Cl septicum from the implant site with no aerobic isolates. PCR proved negative for Brucella spp and morbillivirus. Multiple killer whales have been previously tagged with little untoward effect. Retained barbs have been observed post tag detachment in live and dead animals and this is believed to be the first wild whale to succumb with mucormycosis and retained tag petals.

D-82: RHABDOMYOLYSIS SECONDARY TO PRESUMPTIVE BLACK WIDOW SPIDER ENVENOMATION IN A CAT

Melanie A. Breshears, Rebecca S. Tims, Shane D. Lyon

A two-year-old domestic shorthair cat initially presented to the rDVM for mandibular subcutaneous edema accompanied by lethargy. The presumptive cause of localized edema was an arthropod bite or sting and edema resolved following treatment with injectable dexamethasone. Lethargy persisted, and a few hours later the cat vomited and ultimately collapsed. The cat was presented to OSU Veterinary Teaching Hospital for profound weakness that progressed to respiratory failure. Serum biochemical abnormalities indicated severe muscle injury accompanied by stress leukogram and stress hyperglycemia. Despite supportive care, including mechanical ventilation, the cat's condition deteriorated and euthanasia was followed by necropsy. Gross examination revealed lesions limited to ventilator-associated pulmonary changes and subtle pallor of some skeletal muscle groups. Histopathology of limb and epaxial muscles revealed acute, multifocal and monophasic myofiber degeneration and necrosis accompanied by minimal inflammation. Skin from the region of previous submandibular edema was characterized by a mild to moderate infiltrate of eosinophils and mast cells accompanied by regionally extensive necrosis of the adjacent superficial subcutaneous muscles. Based on the suspicion of an arthropod bite followed by muscle weakness and accompanying necrotizing myopathy, black widow spider envenomation was presumptively diagnosed. Black widow spider (*Latrodectus mactans*) venom contains a neurotoxin (alpha-latrotoxin) that binds nerve terminals and induces release of neurotransmitters, causing muscle spasms. Flaccid paralysis, proposed to be caused by toxin-induced motor end plate degeneration, follows muscle spasticity. The precise cause of acute myofiber injury is unclear, but is presumed to be secondary to prolonged myofiber contraction caused by toxin-induced neurotransmitter release.

D-83: PULMONARY MYCOSIS SECONDARY TO ENTERIC SALMONELLOSIS IN HORSES

Martha Hensel, Courtney Smith, Quinci Plumlee, Aline Rodrigues-Hoffmann, Raquel Rech

Background: Pulmonary mycosis secondary to enteric salmonellosis is often caused by *Aspergillus* sp. Diagnosis is based on histologic features with corresponding fungal culture, but the species is rarely identified.

Objective: The aim of this retrospective study was to use PCR and DNA sequencing to determine the etiologies associated with pulmonary mycosis in horses diagnosed with *Salmonella* sp. enteritis by culture or PCR.

Methodology: The Texas A&M College of Veterinary Medicine Teaching Hospital archived medical records were searched for cases of confirmed enteric salmonellosis with pulmonary mycosis. DNA was extracted from formalin-fixed, paraffin-embedded lung using the BiOstic FFPE Tissue DNA Isolation Kit (MO Bio, Carlsbad, CA). PCR amplification and DNA sequencing was performed for ITS2. Trimmed sequences were aligned to NCBI archived sequences using Basic Local Alignment Search Tool (BLAST).

Results: Four of the five horses had *Aspergillus* species including *fumigatus* and *flavus*. The fifth horse had *Curvularia spicifera*. A single horse had a dual infection of *Aspergillus fumigatus* and *Fusarium* sp.

Conclusion: PCR and DNA sequencing from FFPE for fungal agents provide a safe, quick and effective alternative to traditional culture methods. This study confirms *Aspergillus* species as the most common isolate and adds two additional agents including *Fusarium* sp. and *Curvularia spicifera*.

D-84: MALIGNANT MELANOCYTIC SCHAWNNOMA IN A CANINE

Ramon A. Wong-Saavedra, Ignacio C. Rangel-Rodriguez, Lucia A. Garcia-Camacho

A 12-year old male Golden retriever canine was presented to the Small Animal Teaching Hospital with a dark brown nodule on left pelvic member. At cytology evaluation, a mixed pleomorphic cell population is observed composed of epithelioid and spindle shape cells with abundant brown cytoplasmic granules. A diagnosis of mixed malignant melanoma was made. The tumor was surgically removed for histopathology evaluation. Microscopically, a partially well demarcated non-encapsulated tumor composed of melanin-laden pleomorphic epithelioid cells intermingled with pleomorphic fusiform small markedly elongated cells with scant to moderate eosinophilic cytoplasm displaying abundant fine melanin granules. The latter cells were predominantly arranged in whorls and palisades. Tissue bleaching was performed to fully evaluate the cell morphology, depicting predominantly spindle shaped pleomorphic cells with large ovoid nuclei portraying vesicular to fine chromatin and 2-4 round to angular nucleoli intermixed with fusiform cells with a thin cross-linked cytoplasm and wavy nuclei with fine chromatin and inconspicuous nucleoli. Both cell types are arranged in characteristic palisading and whirling with Verocay bodies which are more evident among cells with higher differentiation. In addition, blue-gray cytoplasm compatible with neuro-ectodermal origin is observed at trichrome stain. Taking together, a diagnostic of malignant melanocytic Schwannoma was made. Schawnnoma tumor cells are regarded to synthetized melanine since they are derived from neural crest. Electron microscopy is more reliable to rule out melanocytic Schwannoma from melanoma since immunohistochemistry markers are shared due to embryological origin. Currently, case reports of this variant of Schwannoma are scarce in veterinary.

D-85: MALIGNANT MESOTHELIOMA IN A CANINE

Ramon A. Wong-Saavedra, Ignacio C. Rangel-Rodriguez, Lucia A. Garcia-Camacho

An 8year old male, Afghan Hound dog was presented to the teaching hospital with abdominal distention. The clinical examination revealed multiple nodules along the cranial to middle aspect of the abdominal cavity. Due to worsening of clinical condition, euthanasia and necropsy were conducted at owner's request. Macroscopically, multiple dark green nodules were on the peritoneum and splenic and hepatic surfaces. Impression smears from the nodules and tissue samples from different organs were obtained to perform cytologic and histologic evaluation. Cytologically, groups of rounded cells displaying moderate anisocytosis, abundant cytoplasm with small projections along the membrane, and nuclei with fine to coarse granular chromatin and 23 nucleoli were observed. Findings were compatible with neoplastic mesothelial cells. Histologically, poorly demarcated, nonencapsulated tumors composed of pleomorphic epithelioid cells arranged as papillary projections on dense fibrovascular stroma were observed. The neoplastic cells were cuboidal to columnar with abundant eosinophilic cytoplasm showing scalloped border and round to oval nuclei with fine to coarse granular chromatin and round to angular multiple (24) nucleoli. The mitotic index was high. In addition, multiple foci of neoplastic cells with papillary projections were found throughout the hepatic and splenic parenchyma. A diagnosis of malignant mesothelioma of epithelial type with hepatic and splenic metastasis was made.

D-86: EVALUATION OF THE EFFECTS OF EXTENDED FIXATION AND DEMINERALIZATION ON IMMUNOHISTOCHEMICAL PROTEIN EXPRESSION IN CANINE OSTEOSARCOMA

Courtney R Schott, Geoffrey A Wood

Prior to histologic processing osteosarcoma samples frequently require demineralization, presenting a unique challenge. Pre-processing exposure to fixative and demineralization solutions has been reported to alter immunohistochemical expression. The antigens which are susceptible to these effects are not widely known. In order to evaluate the effects on protein expression, tissue samples from the same canine osteosarcoma tumour were subjected to two types of demineralization solution for eight different time periods ranging from 1-48 hours prior to processing. Seven samples from a different osteosarcoma were submerged in formalin for 1-28 days prior to processing. After paraffin embedding, duplicate 1 mm cores for each time point were incorporated into a tissue microarray. Immunohistochemistry was performed using five antibodies, targeting both nuclear and cytoplasmic proteins, including phosphorylated proteins. The immunolabeled slides were digitally scanned for automated image analysis and algorithms were developed for each antigen. Cytoplasmic and/or nuclear H-score was compared between samples for both the fixation and demineralization series. In general, H-score trended downward with increased fixation time for the non-phosphorylated proteins, which was most evident at 14 and 28 days. There was no consistent trend observed for H-score with increased exposure to demineralization solution for non-phosphorylated proteins. For both of the phosphorylated proteins, H-score trended downward with increased exposure to at least one of the demineralization solutions. These results indicate that not all antigens are equally affected by extended formalin fixation nor demineralization when evaluating protein expression by immunohistochemistry.

D-87: URINARY BLADDER UROTHELIAL CARCINOMA WITH PERSISTENT UROABDOMEN AND CONCURRENT LIPOGRANULOMATOUS TUBULOINTERSTITIAL NEPHRITIS IN AN ADULT FEMALE SERVAL (LEPTAILURUS SERVAL)

Nicholas A Crossland, Alissa St. Blanc, Peter M DiGeronimo, Rachel E Cianciolo, Gordon J Pirie, Amy E Thiessen, Nobuko Wakamatsu

Background: A 16-year-old intact female serval (*Leptailurus serval*) presented to the local zoo's veterinary hospital for recurrent hematuria, pollakiuria, intermittent anorexia, and dehydration. Due to continued anorexia, significant weight loss, and lack of response to medical therapy the animal was humanely euthanized.

Methods: A CBC, serum biochemistry, urinalysis, and urine sediment cytology were performed one week prior to euthanasia. A postmortem examination with routine histopathologic evaluation was pursued, complimented with special stains performed on the kidney (Masson's trichrome, PAS, and Jones') and immunohistochemistry performed on the urinary bladder (i.e. cytokeratin 7 and cytokeratin AE1/AE3).

Results: Antemortem ancillary test findings included azotemia, hyperkalemia, hyponatremia, hypochloremia, low USG (1.016), and moderate atypia of urothelial epithelium cells. Gross findings included transmural thickening of 70% of the urinary bladder body by a pale tan-to-red, multinodular, infiltrative neoplasm measuring 3.5 x 3.0 x 0.6cm. Kidneys were within normal limits grossly. Histologically, the urinary bladder mass was consistent with a non-papillary infiltrative urothelial carcinoma. Neoplastic cells had strong cytoplasmic immunoreactivity to cytokeratin AE1/AE3, with no immunoreactivity to cytokeratin 7. Micrometastasis to the lungs was identified. Evaluation of the kidneys revealed mild tubulointerstitial nephritis associated with intraluminal lipid casts and interstitial lipgranulomatous aggregates.

Conclusion: Blood work findings are consistent with a combination or pre-renal, renal, and post-renal disease. The latter was attributed to persistent uroabdomen caused by tumoral compromise of the urinary bladder wall. To the authors' knowledge, urothelial carcinoma of the urinary bladder has not been previously described in a serval.

D-88: ULTRASTRUCTURE OF AIR-CONDUCTING MUCOSA OF DOGS WITH CHRONIC RESPIRATORY DISEASE SUSPECTED OF PRIMARY CILIARY DYSKINESIA

Ileana Miranda, Anibal Armien

Background: The mucociliary clearance is one of the main defense mechanisms of the respiratory tract, which can be inherently impaired in primary ciliary dyskinesia (PCD) or reversibly altered in secondary ciliary dyskinesia (SCD). Candidates for further investigations of PCD should be young individuals, especially those with consanguinity or affected siblings, with chronic oto-sinu-pulmonary disease unresponsive to treatment, history of unexplained neonatal respiratory distress, or those presenting with typical clinical phenotypes such as *situs* abnormalities, hydrocephalus and infertility/subfertility without obvious cause. The lack of clinical diagnostic possibilities of PCD in dogs likely leads to misdiagnosis or underdiagnosis.

Objective: This study aims to evaluate the air-conducting mucosa of dogs with chronic respiratory disease suspected of PCD that were submitted for ciliary ultrastructural evaluation.

Methods: Samples of respiratory mucosa from 15 dogs submitted to the Ultrastructural Pathology Unit at the Minnesota Veterinary Diagnostic Laboratory were evaluated by light and transmission electron microscopy (TEM). Fourteen biopsies and necropsy in one case were performed.

Results: PCD was only confirmed in the necropsied dog, which presented with chronic otitis, bronchopneumonia, hydrocephalus and ultrastructural abnormalities in 84% of the assessed cilia, including absence of dynein arms and microtubular changes. All other 14 cases showed only non-specific alterations in the minority of the evaluated cilia and were classified as SCD.

Conclusion: Ciliary ultrastructural analysis can confirm a PCD diagnosis if specific abnormalities exist. TEM remains an important investigation in Veterinary Medicine, as no other specific test for PCD in dogs has been standardized yet.

D-89: DIAGNOSIS OF COLLAGEN TYPE III GLOMERULOPATHY USING PASH/MASSON'S TRICHROME COMBINATION (PASH/MT) AND PICOSIRIUS RED (PSR) STAINS

Anne Burnum, Brittany McHale, Uriel Blas, Cathy Brown

Collagen type III glomerulopathy (collagenofibrotic glomerulopathy, CG), is a rare familial disease of young dogs that causes proteinuria and progressive loss of renal function. Definitive diagnosis requires transmission electron microscopy (TEM) to confirm expansion of the mesangial matrix and subendothelial space by abnormal type III collagen. However, a presumptive diagnosis can be achieved with light microscopy alone through the use of histochemical stains. CG was diagnosed in a 12 week old male Pug that was one of 3 littermates with acute renal failure. At necropsy, there was generalized subcutaneous edema. Renal sections were evaluated with a panel of stains used for renal biopsies (H&E, periodic acid Schiff-hematoxylin [PASH], Masson's trichrome [MT], Jones methenamine silver, and Congo red [CR]). Glomeruli were globally expanded by deposits of eosinophilic material in the mesangium and capillary wall. The deposits were blue with MT and stained with Jones silver, consistent with collagen. Based on these results, differentials included CG and glomerulosclerosis (GS). Picrosirius red (PSR) and PASH/MT combination stains were used to differentiate type IV collagen from type III. The deposits were blue with PASH/MT and orange-red with green birefringence using PSR, ruling out GS and supporting a diagnosis of CG. For comparison, these stains were repeated on renal tissues from an age-matched control dog and on two dogs with TEM-verified diagnoses of CG or focal segmental GS. Final confirmation of CG was obtained by TEM. This diagnostically challenging case demonstrates the utility of histochemical stains for the diagnosis of glomerular disease.

D-90: PULMONARY ALVEOLAR CAPILLARY DYSPLASIA IN PUPPIES

Katie J Barnes, Csaba Galambos, Kurt Williams

Respiratory distress is a significant cause of morbidity and mortality in neonatal puppies. In spite of the frequency of respiratory disease in young puppies there is little published literature describing non-infectious neonatal pulmonary pathology in veterinary medicine. In contrast, non-infectious interstitial lung disease in human neonates is well documented, with more than 13 entities described. Alveolar capillary dysplasia (ACD) with misalignment of pulmonary veins (MPV) is a rare developmental lung disorder in human infants resulting in severe pulmonary hypertension, hypoxia, and death shortly after birth. Herein we describe the pathology of an interstitial lung disease in five puppies (3 female, 2 male; median age of 4 days) with features similar to ACD/MPV. Lungs from affected puppies were evaluated histologically and using immunohistochemistry for a-smooth muscle actin (SMA), pancytokeratin, and CD31. The lungs had diffusely markedly thickened alveolar septa, medial hypertrophy of pulmonary arteries, and centrally located, dilated thin-walled veins similar to MPV in humans. Pancytokeratin labeling of alveolar epithelial cells highlighted the underdeveloped pulmonary acini. Numerous spindle-shaped SMA-expressing cells were present around small vessels and capillaries, and were often detected beneath the alveolar lining epithelium throughout the thickened interstitium. CD31 labeling identified capillary profiles within the alveolar interstitium; many of the capillaries were dilated and dysplastic, with double capillary layers frequently noted. Based on the current study, we conclude that pulmonary developmental abnormalities with similarities to human pediatric lung diseases are present in the domestic dog, and may be the cause of early life death in puppies.

D-91: FATAL RANAVIRUS INFECTION IN A GROUP OF CAPTIVE MELLER'S CHAMELEONS (TRIOCEROS MELLERI)

Lauren B Peiffer, Nathan Pate, Kathleen Gabrielson, Allan P Pessier, Ellen Bronson, Lisa Mangus

Ranaviruses are well-documented, world-wide pathogens of fish and amphibians that are increasingly recognized as causes of morbidity and mortality in reptiles. This case series documents the first known occurrence of ranavirus-associated disease in chameleons. In November 2015, a group of five Meller's Chameleons from the Maryland Zoo in Baltimore died and were submitted to Johns Hopkins University veterinary pathology service for postmortem examination. All animals presented in thin body condition. The first four animals did not have any other significant gross findings, while the fifth animal exhibited mild transudative coelomic effusion and petechial hemorrhages affecting the tongue and kidneys. Microscopically, there was multifocal necrosis most notably affecting the spleen, liver, kidney, adrenal tissues, and nasal cavity. Moderate to abundant numbers of basophilic intracytoplasmic viral inclusions were present in the liver and nasal mucosa. All animals exhibited varying degrees of necrotizing rhinitis with secondary bacterial and, in one animal, fungal infection. Intravascular bacterial colonies were also observed in the liver of one chameleon, suggestive of intercurrent bacteremia. Samples of liver were sent to the San Diego Zoo Institute for Conservation Research, where samples were positive on Taqman qPCR for a portion of the ranavirus major capsid protein (MCP) gene. Sequencing of portions of the neurofilament-like and MCP genes further identified the virus as a member of the *Frog Virus 3* (FV3) group. This series demonstrates that *Ranavirus* should be considered a differential in lizards that present with sudden death, rhinitis, skin lesions, and splenic / hepatic necrosis.

D-92: DEVELOPMENT OF QUANTITATIVE REAL-TIME PCR ASSAYS TO DETECT MYCOBACTERIUM SPP. IN ZEBRAFISH (DANIO RERIO)

Danielle M Meritet, Donna M Mulrooney, Micheal L Kent, Christiane V Löhr

Background: Infections with *Mycobacterium* spp. are common in zebrafish research colonies. As co-morbidities can influence biological responses to stimuli including experimental treatment and conditions, detection of *Mycobacterium* ssp. infections in zebrafish colonies prove essential.

Methods: We developed quantitative real-time PCR assays to detect the three *Mycobacterium* species most commonly identified in laboratory zebrafish by targeting sequences of the mycobacterial heat shock protein 65 gene. Zebrafish were experimentally infected with *M. marinum*, *M. chelonae* and *M. haemophilum.* The experiment was terminated at 12 weeks post infection.

Results: Simplex PCR assays are both highly specific and sensitive for fresh frozen samples, and highly specific and moderately sensitive for formalin-fixed paraffin-embedded (FFPE) samples. We evaluated two different sampling techniques for FFPE samples for whole, sagittal sectioned zebrafish. DNA quantity and DNA purity was equal for both paraffin cores targeting granulomas containing bacteria and scrolls from the entire fish. The diagnostic sensitivity of cores was superior to scrolls for *M. chelonae* and *M. haemophilum*, but not *M. marinum* likely due to widespread granuloma formation with *M. marinum* versus more localized infections with *M. chelonae* and *M. haemophilum*.

Conclusions: The assays are specific, sensitive, cost effective, and rapid and ideally suited to diagnose common *Mycobacterium* spp. infections in laboratory zebrafish.

D-93: SUBCUTANEOUS PANNICULITIS-LIKE T-CELL LYMPHOMA IN DOGS

Erica Noland, Stefan Keller, Matti Kiupel

Canine nonepitheliotropic cutaneous T-cell lymphomas (NETCL) are poorly characterized and may be confused with other conditions. In humans, a number of distinct subtypes of NETCL have been recognized, including indolent subcutaneous panniculitis-like T-cell lymphoma (SPLTL). We describe 5 dogs with subcutaneous T-cell lymphomas similar to SPLTL. All cases were characterized by proliferations of small to intermediate and sometimes large sized, CD3 positive T-cells that had infiltrated the subcutis in a lace-like pattern and frequently rimmed adipocytes. No epitheliotropism was observed and neoplastic cells were often karyorrhectic and there were regions of extensive necrosis. Heavy infiltrates of histiocytic cells with prominent phagocytosis masked the lymphoid neoplastic cell population in some sections. A clonal T-cell receptor gamma gene rearrangement was determined by PCR in 4 of the 5 cases. The mean age was 8.5 years (5.5 to12 years). No breed or sex predilection was observed. Two dogs presented with an acute onset of multiple skin masses, 2 dogs had solitary masses with subsequent development of multiple smaller masses within 17 to 60 days of diagnosis, and 1 dog had a solitary mass. In at least two dogs, the masses were erythematous and/or ulcerated. Locations included shoulder, neck, and ventral abdomen. While two dogs were euthanized following diagnosis, one dog treated with chemotherapy (CCNU) survived 7 months post diagnosis. While SPLTLs are less aggressive lymphomas in humans, their biological behavior in dogs remains uncertain. SPLTL is a distinct entity in dogs and needs to be accurately diagnosed to better determine clinical behavior.

D-94: COMPARISON OF HISTOLOGIC MARGIN STATUS IN LOW-GRADE CANINE MAST CELL TUMORS EXAMINED BY RADIAL AND TANGENTIAL SECTIONS

Milan Milovancev, Duncan Russell, Camila B Dores

Background: Histologic margin status may predict recurrence in cutaneous malignancies. However, there is a poor understanding of how sectioning technique influences margin outcome.

Objective: The aim of this study was to compare histologic margin status in canine mast cell tumors sectioned by both radial and tangential techniques. The hypotheses were that tangentially sectioned neoplasms are more frequently positive (ie. neoplastic cells at the inked surgical margin), and that complete tangential margins are associated with longer histologic tumor-free margins (HTFMs).

Methods: Eighteen inked surgical margins from canine cutaneous mast cell tumors (Kiupel low grade, Patnaik Grade II) were prospectively collected. All masses were excised with curative-intent, wide surgical excision. Following a complete radial section, tangential sections of the adjacent inked surgical margin were embedded and sectioned for routine histopathology. Tangential margins were graded as negative or positive (clusters with more than 10 mast cells in a 40μm radius).

Results: Tangential sections detected significantly more positive surgical margins than radial sections (8/18 (44%) vs. 2/18 (11%), $p = 0.0256$; Chi-square test). HTFM length was greater in the negative tangential margins (12.6 ± 6.0mm; range 0 – 19mm) compared to the positive tangential margins (5.5 ± 5.2mm; range 0 – 13mm, $p = 0.0176$; unpaired t-test). Tangential section margin categorization was weakly positively correlated to HTFM length ($r2 = 0.304$; $p = 0.0176$; Pearson correlation coefficient).

Conclusion: These data indicate that sectioning technique may influence histologic margin status. Tangential sections could be a more sensitive indicator of incomplete histologic margins.

D-95: ZEBRA ALERT! EASTERN EQUINE ENCEPHALITIS IN PUPPIES IN THE MIDWEST AND NORTHEAST

Caroline Andrews, Jodie Gerdin, Jon Patterson, Scott D. Fitzgerald

Eastern equine encephalitis virus (EEEV) is an alphavirus within the family Togaviridae that is classified as a select agent and is capable of causing mortality in humans and a number of veterinary species. The virus is spread by mosquitoes and has recently been postulated to overwinter in snakes. Herein we describe three cases of EEEV in puppies in Michigan and New York. Two puppies were euthanized following an acute history of seizures and obtundation. A littermate of one of these puppies died two weeks earlier following a history of anorexia and fever. All three puppies lacked significant gross lesions at autopsy and tested negative for rabies virus. In all three puppies, histologic examination revealed a necrotizing, neutrophilic and lymphoplasmacytic meningoencephalitis with strong positive immunohistochemical labeling for EEEV antigen in neurons and glial cells. The diagnosis of EEEV was confirmed by PCR in one puppy and cell culture in the other two dogs. EEEV is rare in dogs and pathologic descriptions have only been reported from puppies in the Southeast. The clinical and pathologic features of the cases described here are similar to those previously reported. The initial clinical signs of EEEV in puppies are typically nonspecific, including anorexia, fever, and diarrhea, but rapidly progress to severe neurologic disease characterized by seizures and recumbency. Although rare, EEEV should be considered as a differential diagnosis for neurologic disease in puppies, especially after more common etiologies, such as canine distemper, rabies, and toxoplasmosis, have been ruled out.

D-96: CLINICOPATHOLOGIC FEATURES OF CANINE LINGUAL T-ZONE LYMPHOMA

Lauren J Harris, Kelly L Hughes, Emily D Rout, Janna A Yoshimoto, Claire M Cannon, Paul R Avery, E.J. Ehrhart, Anne C Avery

Background: Canine T-zone lymphoma (TZL) is a subtype of canine T-cell lymphoma characterized by unique cytomorphologic features and histologic pattern, immunophenotypic loss of CD45 expression, and an indolent clinical course. Dogs with TZL typically present with single to multiple lymphadenopathy and/or lymphocytosis. We report an unusual presentation of this disease involving the tongue.

Objectives: To characterize the clinicopathologic features of a novel extranodal presentation of TZL recently identified in the tongue of dogs.

Results: Twelve dogs presented to the Colorado State University Clinical Immunology Laboratory and Diagnostic Medical Center between 2006 and 2016 for evaluation of multifocal raised erythematous lingual masses. Seven dogs presented with concurrent peripheral lymphadenopathy or lymphocytosis, four presented with both peripheral lymphadenopathy and lymphocytosis, and one presented with neither peripheral lymphadenopathy nor lymphocytosis. All dogs were diagnosed with lingual TZL based on variable combination of immunophenotyping via flow cytometry, cytology, histopathology, and immunohistochemistry. Interestingly, three cases were initially diagnosed as plasma cell tumors based on histopathology suggesting differentiation of these entities might be a diagnostic challenge. With variable treatment protocols six dogs exhibit complete resolution of the lesions.

Conclusions: This case series highlights a unique presentation of canine TZL and presents a new entity that should be considered as a differential for lingual neoplasia. The characteristic indolent behavior of TZL makes differentiation from other, more aggressive, round cell tumors especially important for accurate prognosis and treatment decisions. This report also demonstrates the utility of immunophenotyping via flow cytometry in the diagnosis of extranodal round cell neoplasia.

D-97: GENOMIC DETECTION OF CYTOLOGICALLY CONFIRMED MAST CELL TUMOR LYMPH NODE METASTASIS OTHERWISE NOT DETECTED BY CONVENTIONAL HISTOPATHOLOGY

Gregory A Krane, Barbara Davis, Sarah Beatty, Sarah Boston, Owen Skinner, Julia A Conway

An eight year-old, castrated Boxer presented with a recurrent interdigital forelimb mass 4.5 months after incomplete excision of a Patnaik grade II-Kiupel low grade mast cell tumor (mitotic index of zero). The cytologic diagnosis of the mass and corresponding prescapular lymph node was mast cell tumor with lymph node metastasis. Cytologically, the metastatic mast cells in the lymph node comprised 20-25% of the total cell population and exhibited moderate pleomorphism with increased multi-nucleation and mitoses. There was no detectable splenic, hepatic, or intra-thoracic involvement (abdominal ultrasonography, hepatic and splenic fine needle aspiration with cytology, three-view thoracic radiography). Partial foot amputation and lymphadenectomy followed. The histopathologic diagnosis of the mass was Patnaik grade II-Kiupel high grade mast cell tumor (seven mitoses in ten 400x fields). No metastatic mast cells were present in six histologic step-sections through the entire lymph node (H&E and Toluidine-blue stains). A quantitative nuclease-protection assay performed on mRNA extracted from formalin-fixed, paraffin-embedded sections of the mast cell tumor showed significant expression of Kit (15-fold increase over housekeeping genes), MAP2K1, MAP2K2, Vimentin, and CD11a. The patient's lymph node expressed low levels of Kit, which control lymph node did not express. Cytologic and genomic evaluations of the lymph node indicated metastatic spread of the mast cell tumor despite lack of histologic detection. This case highlights potential limitations of histopathology in evaluation of tumor metastasis, warranting further investigation into the capability of genomic analysis of tumors and lymph nodes to be a sensitive and useful adjunct diagnostic tool in such evaluations.

D-98: CD31 IMMUNOREACTIVITY IN CANINE RENAL CELL CARCINOMAS

Tyler J Peat, Elijah Edmondson, Deidre M DuSold, Margaret A Miller, Jose A Ramos-Vara

CD31 (platelet endothelial cell adhesion molecule 1, PECAM-1) is considered a specific marker of endothelial neoplasms in both humans and dogs. However, CD31 immunoreactivity has been reported in some human non-endothelial neoplasms (carcinomas, sarcomas, mesotheliomas, and lymphomas). Immunohistochemical expression of CD31 in similar canine neoplasms has not been reported. Immunoreactivity for CD31 (mouse monoclonal antibody, clone M0823, Dako, Carpinteria, CA) was detected in 14 of 43 (33%) formalin-fixed, paraffin-embedded canine renal cell carcinomas (RCC): 5 of 16 (31%) papillary, 5 of 13 (38%) solid, and 4 of 14 (29%) tubular); and 2 of 2 (100%) metastatic RCC. The percent of neoplastic epithelial cells with membranous CD31 reactivity was scored as 0 (50%). The intensity of immunoreactivity was scored as 1^+, 2^+, or 3^+. CD31 immunoreactivity was heterogeneous in the 14 positive primary tumors. The mean percent reactivity (MR) and mean intensity (MI) scores were 1.6 and 2.8, respectively, for papillary RCC; 2.6 and 2.6 for solid RCC; 2.2 and 3.0 for tubular RCC. The MR and MI were 1.0 and 2.0, respectively, in metastatic RCC. Patchy membranous and/or cytoplasmic CD31 expression was noted in cortical tubular epithelial cells in 25 (61%) of 41 samples with adjacent non-neoplastic parenchyma. In summary, CD31 reactivity in some canine RCCs expands the expression of this marker, once believed to be specific for endothelial cells.

D-99: COEXPRESSION OF CD3 AND CD20 IN THREE CASES OF CANINE ENTEROPATHY-ASSOCIATED T-CELL LYMPHOMA, LARGE CELL TYPE (TYPE 1)
Erica Noland, Matti Kiupel

The majority of reported primary intestinal lymphomas in dogs are comprised of large T-cells that, according to the human World Health Organization criteria, are subclassified as enteropathy-associated T-cell lymphoma (EATL) type 1. Classically, T-cell lymphomas are thought to only express T-cell markers, i.e. CD3; however, there are growing numbers of reports of human peripheral T-cell lymphomas and cutaneous epitheliotropic T-cell lymphomas (CETCL) that coexpress CD20. Recently, coexpression of CD20 was reported in a CETCL in a dog. The clinical significance of these findings in either species has yet to be determined. We describe three cases of CD3+, CD20+ intestinal T-cell lymphoma in dogs. These lesions are characterized by proliferations of intermediate to large sized, CD3 and CD20 positive lymphocytes within the mucosal lamina propria and focally within the mucosal epithelium. These cells were Pax-5 negative similar to other reports of CD20+ T-cell lymphomas. PCR for rearrangement of the T-cell receptor gamma gene confirmed a monoclonal cell population in all 3 cases. Initial clinical signs included weight loss, inappetence, diarrhea, and/or vomiting. The mean age was 9 years (3 to 12 years). No breed predilection was observed. One dog was euthanized 11 days following diagnosis. One dog treated with CCNU survived 163 days. Interestingly, one untreated dog is alive after 597 days. Given the clinical implications, recognition of CD20+ T-cell lymphomas in order to limit misdiagnosis is important. While the clinical significance and prognostic value of expression of CD20 is unclear, it may allow for treatment with CD20 targeted therapy.

D-100: DEVELOPMENT OF A NOVEL DIAGNOSTIC IMPLANT FOR DETECTION OF SPECIFIC IMMUNE RESPONSES IN MYCOBACTERIUM AVIUM PARATUBERCULOSIS (MAP) INFECTED CATTLE, MAP VACCINATED CATTLE, AND NAÏVE CATTLE
Tracy Lindquist, Saleh Albarrak, Shannon J. Hostetter, Doug E. Jones, Jesse M. Hostetter

Background: Current diagnostics for *Mycobacterium avium subspecies paratuberculosis* (MAP) cannot accurately identify cattle early in the course of infection.

Development of a diagnostic assay detecting cell-mediated immune responses specific to the varying clinical stages of disease observed in MAP infection is needed.

Objective: The aim of this study was to develop a diagnostic tool that consistently identifies cell mediated immune responses specific to MAP infected cattle and MAP vaccinated cattle.

Methods: Using a permeable metallic implant containing MAP antigen emulsified in a collagenous scaffold, serologically positive, serologically negative, vaccinated, and naïve cattle were implanted subcutaneously. After 72-120 hours, the device was removed and the collagen center was collected. A portion of the collagen was formalin-fixed and the inflammatory cell infiltrate was observed histologically using a hematoxylin and eosin stain. The remaining collagen was harvested and cytokines were measured using a quantitative bead-based multi-analysis.

Results: Optimization of implant design as well as implantation technique, implant retrieval, and collagen collection was achieved. Cytokines (TNF-α, IFN-γ, and IL-10) were successfully measured from the collagen and the inflammatory infiltrate was observed within the collagenous scaffold via stained histological sections.

Conclusion: This new diagnostic tool can be used to successfully retrieve and measure specific cell-mediated immune responses occurring within an animal.

D-101: ANGIOSTRONGYLUS CANTONENSIS INDUCED VERMINOUS MENINGOENCEPHALOMYELITIS IN A RED KANGAROO

Sonika Patial, Brooke A Delcambre, Peter DiGeronimo, Rudy Bauer

Angiostrongylus cantonensis is a zoonotic parasitic helminth that normally resides in the pulmonary arteries and the right ventricle of rats, the definitive host, in which it causes little disease. Humans, dogs, opossums, and various zoo animals are “accidental” hosts that acquire infection through the intermediate hosts (gastropods). However, no reports have described an infection in a Red Kangaroo. Here, we report verminous meningoencephalomyelitis caused by Angiostrongylus cantonensis in a 9-month-old male Red Kangaroo (Macropus rufus). This kangaroo first presented as lethargic, recumbent, and hypothermic with severe muscle wasting and diminished withdrawal reflexes in both hind limbs. Complete blood counts, chemistry panel, urine analysis, fecal exams, radiographs, and abdominal ultrasound were all unremarkable. Serology for toxoplasmosis and leptospirosis was negative. The kangaroo progressed to non-ambulatory paraparesis, declined mentally, and died within three weeks. Gross examination at necropsy showed mildly congested vessels and multifocal areas of dark brown discoloration, malacia and cavitation in the brain and the spinal cord. Histopathology revealed the presence of several transverse and longitudinal sections of nematode larvae in focal areas of cerebrum, cerebellum and the spinal cord, which were surrounded by extensive areas of rarefaction characterized by loss of gray or white matter, hemorrhage, spongiosis, neuronal necrosis, and gliosis. Interestingly, the eosinophilic response, commonly observed in human cases, was largely absent and the inflammatory response was minimal. Several nematodes were extracted from the fixed brain and spinal cord via dissection. The structural features and further analysis of these nematode larvae revealed at least one fifth-stage Angiostrongylus larva.

Education Focused Scientific Session

December 6, 2016
8:05 AM – 8:20 AM
Digital Imaging at the AFIP/JPC - 20 Years of Lessons Learned
Bruce H. Williams

Over twenty years ago, the AFIP's Dept. of Veterinary Pathology set up its first server on the nascent Internet and began publishing results from its Wednesday Slide Conference, complete with garish backgrounds and grainy digital images. Sharing staff members with the AFIP's Telepathology Project, the Department became well-known as early visionaries of the digital pathology revolution, publishing results from the WSC, Veterinary Systemic Pathology Online, and being the first to adopt telemedicine consultation (in all its forms: static, real-time and virtual slide). But being early adopters brought with it a number of hard and expensive lessons learned, which are worth remembering (and hopefully not repeated with more advanced systems) as we sit truly poised on a digital revolution in veterinary pathology. This lecture (borrowing liberally from the presenter's six previous ACVP presentations on imaging over the years) will trace the early days of the Internet, digital imaging and pathology, and the factors that held us back as well as brought us to where we are today.

December 6, 2016
8:20 AM – 9:00 AM
RATIONALE AND APPLICATIONS FOR COMPUTER-ASSISTED DIGITAL WHOLE SLIDE IMAGE ANALYSIS IN POSTGRADUATE PATHOLOGY EDUCATION
Mark Simpson, Shelley B Hoover, Munish Puri, Jennifer E Dwyer, Charles H Halsey, Stephen M Hewitt

Whole-slide image (WSI) based digital pathology systems (DPS) are forecast to grow at a compound annual rate of 12.1% from 2016-2021. Drivers include rising cancer prevalence, increasing demand for tele-pathology consultations, as well as utilization for drug discovery and companion diagnostics. In addition to database organization and retrieval of histopathology images, the greatest impacts are anticipated to be due to image analysis and pathology informatics merging into pharmaceutical research, computer-assisted clinical diagnostics, and education today. Veterinary pathologists must create a collective vision to justify the significant investment for integrating DPS into pathology labs to maintain effective roles in health care. Adoption of the technology will be driven by enhanced familiarity and by innovation with the use of DPS, in all facets of visual pattern data streams. Training and consulting in digital pathology will parallel the preceding evolution in digital radiology. Certain segments of pathology are early adopters, and contemporary academic training as well as validation studies in DPS must occur to inform application and clinical deployment. Most notably, the ability to quantify lesion features improves reproducibility and objectivity in pathology assessments and will be increasingly relevant to education for future practice. Multiple methods/applications for automating lesion quantification in WSI, providing analyses superior to subjective assessments obtained using semi-quantitative grading approaches, will demonstrate this fact. Pre-analytic implementation strategies for selecting and establishing computer algorithm parameters include characterization of suitable tissue, stain or immunoassay quality. Validation of DPS in clinical diagnostics will be modeled.

December 6, 2016
9:00 AM – 9:15 AM

HOW WE USE DIGITAL IMAGES FROM SCANNED MICROSCOPIC SLIDES FOR INSTRUCTION OF VETERINARY STUDENTS IN HISTOPATHOLOGY

Joseph S Haynes, Amanda J Fales-Williams

The Department of Veterinary Pathology has used digital images from scanned microscopic slides for teaching histopathology to veterinary students in place of routine microscopy for the past 4 years. We will demonstrate how we accomplish this in our department, along with our view of the benefits and problems associated with using scanned microscopic slides. It is our observation that veterinary students are very comfortable with this technology, which allows them all to view the same image, capture and label photomicrographs, and have access to the microscopic images at any time. Advantages for the instructor include greater confidence that the students are viewing the lesions and are not encumbered by marginal student microscopy skills or microscope quality. IT resources required to utilize digital images from scanned microscopic slides include having access to an appropriate variety of scanned slides, individual computers with appropriate software for each student, and adequate reliable network capacity in the pathology laboratory for a large group of student computers. Anecdotally, with increased incorporation of digital slides into general pathology labs, students appear to rise to the challenge of recognizing histologic features of mast cell tumors in a series of low-grade and high-grade tumors. Similarly, general pathology students more accurately verbalize histologic differences of coagulative and caseous necrosis, referring to specific histologic features. This represents an improvement over rote responses from previous years.

December 6, 2016
9:15 AM – 9:30 AM

DIGITAL PATHOLOGY AND QUANTITATIVE IMAGE ANALYSIS: LEVERAGING THE TECHNOLOGIES

Olulanu H Aina

Digital Pathology involves capture, management, analysis and interpretation of pathology information from a scanned glass slide. The technology uses high-resolution images for computer-based morphologic and quantitative analyses. Scanner systems offer high-throughput automatic loading, high magnification, and autofocus capture of up to 400-slides with scan speeds less than 35 seconds per slide. Practical applications include education/ training, biomedical research and clinical patient care. Advantages include better storage and preservation of original raw data, improved workflow through remote access, rapid image retrieval, synchronized viewing of multiple slides and ability to annotate regions- of- interest. Other advantages include improved ergonomics; elimination and/or reduction of shipping and travel costs, and ease of image capture for presentations. Perhaps the most important advantage of Digital Pathology is Quantitative Image Analysis.

Quantitative Image Analysis (QIA) is the extraction of meaningful information or statistical data from digital images by using mathematical algorithms to generate rapid and objective measurements. Automated algorithms for tissue recognition and segmentation, algorithms for analysis of cellular structures, size and stain intensity are used to generate unbiased, precise and reproducible data which improves the accuracy of interpretation. Stereology, a more sophisticated form of image analysis, is used for 3-dimensional/ volumetric analysis.

Some drawbacks include cost, pre-analytic variables and variability in scanning systems. Regulatory Standardization, Instrument Validation and Data Security are crucial milestones to global acceptance of these technologies.

With advancing technology comes the promise of increased efficiency, higher productivity and rapid turn-around. Adopting new technology can be uncomfortable but the rewards are immeasurable.

December 6, 2016
9:30 AM – 9:45 AM

DIGITAL PATHOLOGY AS INVALUABLE RESOURCE FOR (CONTINUING) EDUCATION: THE EXPERIENCE WITH THE AGP ROUNDS

Alessandra Piersigilli, Paige K Michael, Alexander D Borowsky, Robert D Cardiff

Digital Pathology (DP) is revolutionizing pathology education and services. In particular, DP enhances global training and scientific interchange. The Center for Genomic Pathology has over 15 years of experience with digital imaging.

The quality of digital whole slide images (WSIs) is comparable to optical microscopy but lack depth of field and ultra-high magnification. However, WSIs allow additional operations such as image analysis and annotation on the same images. In addition, archiving of entire cases, whole data sets, and integrated research projects, DP allows rapid retrieval of data and slides for study and research. Most important, the WSI become a permanent record that can be retrieved and used on demand.

DP enhances international Telepathology allowing sharing of WSIs between multiple, remotely distributed users. The dynamic mode enables synchronous visualization of the WSIs at any magnification in any field. The participants, via a live telecommunication link, enjoy active real-time interactions.

The Center for Genomic Pathology has exploited DP potential by establishing a set of e-learning modules for training in mouse pathobiology and a monthly live Telepathology with the global comparative pathology community, the Academy for Genomic Pathology (AGP). During these sessions a moderator presents interesting and controversial cases which are discussed by the attendees. Attendees from North America, Europe and Asia have participated. This web conference has been an effective and useful resource for those pathologists without access to "on-site" experts, qualified second opinion or peer review or who simply want to keep up with continuing education.

December 6, 2016
10:30 AM – 10:45 AM

GOING DIGITAL: ESTABLISHING AN EPATHOLOGY PROGRAM FOR COLLABORATION AND TEACHING AT THE CDC'S INFECTIOUS DISEASES PATHOLOGY BRANCH

Joy M Gary, Yokabed Ermias, Jana M Ritter, Wun-Ju Shieh, Sherif R Zaki

Background: Digital pathology, including whole slide imaging and telepathology, is a rapidly-changing, vital tool that provides a platform for collection, storage, and communication of histopathologic information.

Objective: CDC's Infectious Diseases Pathology Branch (IDPB) is developing a digital pathology program to enhance its mission, which includes identifying and investigating emerging pathogens, conducting and supporting infectious disease research, contributing to public health work, and providing support and training to health departments and other health organizations.

Methods/Results: IDPB is currently utilizing a static telepathology email service (epathology@cdc.gov) that receives images and case inquiries from around the globe and has proved to be valuable for rapid communication with clinicians encountering unusual histologic features and infectious agents. The number of case submissions is expanding. IDPB has also established a real-time, dynamic telepathology system that streams live video feed from a microscope to online viewers. Finally, IDPB is expanding its whole slide imaging to provide capabilities for rapid quality control and case discussions with overseas collaborators, for development of educational courses for visiting residents and scholars, and for archiving rare and unique cases.

Conclusions: Throughout program implementation, digital pathology technologies have been changing rapidly and have provided challenges in the form of information technology security and storage of large files, requiring adaptation and creative solutions. Despite the challenges, digital pathology will allow IDPB pathologists to utilize their unique positions as infectious diseases experts by receiving and sharing valuable information, accomplishing the crucial public health and educational missions of the branch.

December 6, 2016
10:45 AM – 11:00 AM
The Utility and Implementation of Digital Microscopy in Veterinary Pathology Education
Michael K. Keel

The application of digital microscopy in pathology and histology has become much more common in recent years and will likely soon become standard in our industry. Whole-slide images can represent an improvement over the use of glass slides because of more effective archiving of difficult to acquire sections, the ability to share a single tissue section with a large audience, and the ability to share materials with simultaneous users in disparate regions. The finished product also has increased portability, because a microscope is not essential for viewing the slides. Limitations include the cost of scanning equipment, difficulties in streaming extremely large files via the internet, and limited compatibility between different systems due to proprietary technology among industry competitors. Here, we review various platforms available, techniques for implementation of digital microscopy in pathology training, and potential improvements in the near future. The significant advantages of whole-slide images are balanced by multiple challenges in implementation: the large size of files creates difficulty for access, to end users who have limited bandwidth available for streaming images; fine details, such as intracytoplasmic granules may be difficult to see in scanned slides; and, the quality of scanned slides is always limited by the quality of the slides scanned. Sharing digital slides may also be challenging in some settings due to cybersecurity concerns and potential vulnerability in the requisite software. Nonetheless, digital microscopy is becoming increasingly important to pathology training and the tools have improved sufficiently to make the technology accessible to many.

December 6, 2016
11:00 AM – 12:00 PM
APPLICATION OF DIGITAL PATHOLOGY: "WHAT'S YOUR DIAGNOSIS" WORKSHOP
M.K. Keel, K.P. Carmichael, F. Oliveira, D. Rissi

Education Focused Posters

ED-01: CELL-ING IT? THE FLIPPED CLASSROOM MODEL FOR TEACHING THE CELL-BASED MODEL OF COAGULATION IN THE CLINICAL PATHOLOGY COURSE TO SOPHOMORE VETERINARY STUDENTS
Cheryl A Moller, Kenneth D Royal, Jennifer A Neel

Background: The need to streamline the Veterinary Medicine curriculum whilst providing robust current information to students is a challenge faced by academic institutions. This also impacts student knowledge retention. The cell-based model of coagulation is one such example of a relatively recent and complex update.

Objectives: To use an active learning model, the flipped classroom, to teach sophomore veterinary students the cell-based model of coagulation. Further, to evaluate the efficacy of, and student perception of this teaching method.

Methods: Three pre-recorded lectures were posted online, followed by two interactive class sessions. The hemostasis questions in the mid-term and final exams were the same as the previous year. Students completed an anonymous survey immediately following the final exam. Exam scores were compared to the previous year. Survey results were collated and analyzed.

Results: Exam results for the flipped year group were slightly lower than the traditional year group (mid-term mean 88% vs 83% P

Conclusion: Novel teaching methods should be investigated to convey complicated information. The high workload and time constraints of the Veterinary Medicine course necessitate that course content be as time-effective as possible, and minimize the amount of required out-of-class work for students.

ED-02: USE OF AN INTERACTIVE LEARNING PLATFORM (LECTURETOOLS BY Echo360®) TO FACILITATE TEAM-BASED AND CASE-BASED LEARNING IN A LARGE CLINICAL PATHOLOGY COURSE WITH MULTIPLE CAMPUSES.

Shannon J Hostetter, Lisa E Gestrine, Amanda J Fales-Williams

It can be difficult to actively engage students within a large classroom environment, particularly if students are enrolled at multiple campuses. The integration of team-based learning (TBL) and/or case-based learning (CBL) into course design can foster both student participation and active learning in larger classes. One of the obstacles instructors face is finding an interactive learning platform that will facilitate the use of TBL/CBL in a large classroom, particularly if some students are enrolled off campus. LectureTools by Echo360® is an audience response and assessment system with several advantageous features for both TBL and CBL, including real-time polling and slide release and the capacity to incorporate images into student assessment. LectureTools was chosen as the learning platform for a clinical pathology course with an enrollment of 145 second year veterinary students (120 on site and 25 at a remote campus.) The course uses both TBL and CBL throughout the semester. An interactive case bank was established in LectureTools and students were asked to work through these cases by completing histories, generating differentials, interpreting laboratory data (including visual identification of blood smear, urine sediment, and cytologic abnormalities), and selecting additional diagnostic tests. The use of LectureTools cases was highly successful, garnered positive student feedback, and was associated with substantial improvement in final examination scores. This format fostered an interactive learning environment, improved learning outcomes and proved to be a holistic approach to teaching clinical pathology. Additionally, LectureTools was an invaluable tool for the engagement of students at the remote campus.

ED-03: USE OF VOICE RECOGNITION TECHNOLOGY FOR REPORT GENERATION IN AN ANATOMIC PATHOLOGY TRAINING PROGRAM

Teresa L Southard

Background: A high caseload in a pathology training program provides residents with broad exposure to a variety of species and disease processes; however, generating and editing thorough, high-quality pathology reports for all of those cases is a time-consuming task which can negatively affect resident morale. In 2015, the Section of Anatomic Pathology at Cornell University College of Veterinary Medicine began providing voice recognition software (Nuance Dragon NaturallySpeaking) for resident report generation.

Objective: To evaluate the impact of voice recognition technology (VRT) use in a high-volume anatomic pathology training program

Method: Surveys were sent to the residents and faculty members in the Section of Anatomic Pathology.

Results: All but one resident reported using VRT to generate at least some of their reports, and 2/7 used VRT for almost all of their reports. Most residents took more than a week but less than a month to become proficient with the software. Residents reported an average of 5-6 transcriptional errors in a typical biopsy report, and 4/7 faculty members reported a slight or significant increase in time spent editing reports when residents use VRT. 4/7 faculty members reported increased resident productivity and 5/7 reported improved resident morale since the introduction of VRT. None of the residents and only one faculty member was concerned about a negative impact of VRT use on board preparation.

Conclusion: Overall, residents and faculty members reported a positive effect of VRT use on resident productivity and morale; however, most faculty members noted increased editing time on VRT-generated reports.

ED-04: TEACHING THE PATHOMECHANICS OF DEGENERATIVE JOINT DISEASE: A COMPARATIVE CASE STUDY APPROACH

Elizabeth W Uhl, Uriel Blas-Machado, Michelle L Osborn

The most common disease affecting man and animals is degenerative joint disease (DJD, osteoarthritis). Pathomechanical forces induced by how an individual structurally interacts with its environment directly cause the tissue lesions of DJD, but are rarely considered in teaching its pathogenesis. This is unfortunate since therapeutics based upon identification of the sources of mechanical stress are critically needed as treatments focused only on controlling pain and tissue pathology mostly fail to prevent disease progression. Static postural analysis (SPA) is a well-established technique requiring no specialized equipment that can be used to identify the pathomechanical causes of joint pain and damage. Examples of how a pathomechanical perspective places DJD lesions in the context of their mechanical etiology and how SPA analyses of common forms of DJD in horses and dogs, with relevant comparisons to DJD in humans, can be used to augment the teaching of DJD in veterinary pathology courses will be presented. These examples include how 2D models can be used to highlight both species and individual vulnerabilities, and thus illustrate *why* a patient has DJD. This type of functional analysis can be used by pathologists to explain why the lesions form where they do, and by practitioners to educate clients and formulate individualized therapies, both of which will facilitate the transfer of mechanically-based therapeutic approaches between human and veterinary medicine.

ED-05: THE TWEETING PATHOLOGIST – HOW SOCIAL MEDIA CAN ENHANCE YOUR PATHOLOGY PRACTICE AND SHARE YOUR STORY

Jamie L Rothenburger

Background: Twitter and social media are more than a mindless distraction. With over 115 million monthly users, veterinary pathologists can connect to a diverse and global audience.

Objective: The objective of this presentation is to demonstrate how Twitter can enhance pathology practice while fostering collaborative case consultations and research.

Methods: Using examples from personal twitter accounts (@JRothenburger and @Vet_Pathology), as well as those from popular veterinarian and medical doctor (MD) pathologist accounts, I will 1) provide concrete examples of highly impactful tweets and effective use of Twitter; 2) examine diverse uses and ethical issues related to social media.

Results: I will lower the threshold to entering this online community by describing how to navigate Twitter, compose high-quality tweets and use hashtags (#), mentions, videos and pictures to maximize your reach. I will discuss the many benefits of a social media presence which includes 1) the establishment of professional relationships with a diverse array of people including fellow veterinary and MD pathologists, scientists from within and outside of areas of expertise and the general public; 2) informal interactions that offer opportunities for learning, collaboration and informal knowledge transfer; 3) mechanism to keep up with important findings and topics in the broader science community.

Conclusion: As a whole, the veterinary pathology specialty will benefit from an active social media presence by its members since many people don't know exactly what we do or why we do it. Social media offers us the chance to tell our unique story.

December 4, 2016
2:30 PM – 2:45 PM

EVALUATION OF α-SYNUCLEIN (aSyn) SPREAD IN MOUSE BRAIN BY AUTOMATED IMAGE ANALYSIS

Noa Safra, O. Foreman, H. Ngu, H. Lin, A. Robles, B. Brendza, G. Ayalon

Background: α-synuclein (aSyn), a soluble protein localizes to neuronal presynaptic terminals. In Parkinson's disease, aSyn is found in neurons in an insoluble aggregated and phosphorylated form termed Lewy bodies. Evidence from in vivo and cell based studies suggest that fibrillar species of pathological aSyn propagate from cell to cell and spread to adjacent neurons through a nucleation–dependent seeding mechanism.
Objective: Develop a high throughput assay to assess spread of aSyn in mouse brains
Methods: The spreading mechanism is modeled in mice by unilaterally injecting preformed aSyn fibrils into dorsal striatum of PrP.hu.αSyn.A53T transgenic mice expressing the A53T mutant allele of aSyn under the human prion promoter. At study termination, brains are coronally sectioned and stained with anti-aSyn, anti-CD68, anti-Iba1 and anti-NeuN antibodies to assess spread, gliosis and neuronal loss. Subsets of brains are homogenized to measure aSyn by ELISA. Slides are scanned on Leica SCN400 imaging system. Ipsilateral and contralateral regions of interest are manually defined on the digital images and a positive pixel area, optical density and aggregated a-Syn count algorithm are performed using Matlab software. Thresholds for markers are adjusted using control tissues and transfected cell-lines.
Results: Fibrillar aSyn spreads along neuronal circuitry. By day 60, phosphorylated aSyn aggregates are present in the contralateral striatum by immunohistochemistry and ELISA.
Conclusion: This spreading model can be used to study biological mechanisms of aSyn spread, and potentially for drug discovery purposes. Using slide scanning technology and automated quantification algorithms, this method offers non-biased high throughput analysis supporting anti-aSyn efficacy studies.

December 4, 2016
2:45 PM – 3:00 PM

LACK OF NF-KB INDUCING KINASE (NIK) RESULTS IN EOSINOPHILIC ESOPHAGITIS (EOE) AND GASTRIC HYPERPLASIA IN MICE: IMPLICATIONS FOR NONCANONICAL NF-KB SIGNALING IN HUMAN EOE.

Dylan K McDaniel, Bettina Heid, Irving C Allen

The NF-κB pathway is a powerful modulator of inflammation in the gut that can proceed along two distinct arms deemed the canonical and noncanonical pathways. The canonical pathway produces a variety of classic proinflammatory mediators and is well-studied in gastrointestinal disease. However, the noncanonical pathway, a unique signaling cascade that produces a distinct set of chemokines involved in lymphoid stroma management, is undercharacterized. NF-κB inducing kinase (NIK) is a central molecule in noncanonical signaling and is essential for the production of effector molecules. Mice lacking NIK have been previously shown to develop eosinophilic inflammation in major organs such as skin, liver, and lung. However, characterization of the gastrointestinal tract of these mice has been lacking. Here we show that *Nik-/-* mice display a significant eosinophilic esophagitis that has many similar features to human eosinophilic esophagitis (EoE), including intraepithelial eosinophil accumulation and degranulation, microabscess formation, fibrosis, and basal cell hyperplasia. Additionally, these mice display significant gastric hyperplasia at the esophageal junction suggestive of chronic irritation due to reflux. Interestingly, eosinophil infiltration is localized to the esophagus and gastroesophageal junction; the caudal stomach, small intestines, and colons of these mice are unaffected, again similar to the human disease. Esophageal tissue of *Nik-/-* mice contains significantly elevated mRNA levels of IL-1β, as well as a trend towards increased TSLP, a gene associated with EoE. These findings suggest that *Nik-/-* mice may useful as a naturally occurring model of EoE and highlights a novel role for noncanonical NF-κB signaling in eosinophilic gastrointestinal disease.

December 4, 2016
4:30 PM – 4:45 PM

BLOOD-TUMOR BARRIER PERMEABILITY IN THREE EXPERIMENTAL MODELS OF BRAIN METASTASIS FROM BREAST CANCER

L. Tiffany Lyle, Renata Duchnowska, Paul R Lockman, Chris E Adkins, Afroz S Shareef, Emily Hua, Diane Palmieri, David Liewehr, Seth M Steiberg, Wojciech Kloc, Ewa Izycka-Swieszewska, Naema Nayyar, Priscilla K Brastianos, Patrica S Steeg, Brunilde B Gril

Brain metastases of breast cancer are increasing in incidence; chemotherapy is generally ineffective due to poor overall uptake, and heterogeneous permeability. Herein, functional components of the blood-brain barrier (BBB)/blood-tumor barrier (BTB) were characterized in highly permeable and poorly permeable experimental brain metastases from breast cancer. Brain seeking variants of 3 breast cancer cell lines, MDA-MB-231-BR6, JIMT-1-BR3, and SUM190-BR3 were injected into the left cardiac ventricle of mice to produce brain metastases. Mice were injected with 3kd Texas Red dextran and sterilely perfused; highly permeable metastases were identified by exudation of red dye. Immunofluorescence expression of BTB functional components was quantified in uninvolved brain, highly permeable metastases, and poorly permeable metastases. We report a significant increase in endothelial cell and neuroinflammatory components in brain metastases compared to uninvolved brain in all 3 models. There was no trend in these markers in highly permeable versus poorly permeable metastases. Interestingly, increased expression of the desmin+ subpopulation of pericytes was associated with highly permeable metastases compared to poorly permeable metastases in the 231-BR6 ($p=0.0002$), JIMT-1-BR3 ($p=0.004$), and SUM190-BR3 ($p=0.008$) models. This increase was accompanied by a decrease in CD13+ pericytes in the 231-BR6 ($p=0.014$), JIMT-1-BR3 (0.002), and SUM190-BR3 (p=ns) models. Findings for both pericytes populations were corroborated in human craniotomy specimens. A single preliminary functionality experiment demonstrated increased permeability in an *in vitro* model of the BTB with desmin+ pericytes compared to CD13+ pericytes. These alterations provide testable hypotheses for the development of targeted therapeutic strategies to increase chemotherapeutic delivery to brain metastatic lesions.

December 4, 2016
4:45 PM – 5:00 PM

THE UNIQUE PROINFLAMMATORY ROLE OF THE HIPPO EFFECTOR TAZ IN LIVER

Joshua D Webster, Thijs Hagenbeek, Noelyn Kljavin, Anwesha Dey

The Hippo pathway has been implicated in tissue growth and regeneration. Dysregulation of this pathway has also been associated with cancer. YAP and TAZ are effector proteins of this pathway. While YAP and TAZ phosphorylation by LATS 1/2 prevents nuclear translocation, unphosphorylated YAP and TAZ translocate to the nucleus and interact with TEADs to regulate transcription. Currently, there is little understanding of the unique roles of YAP and TAZ. In this study, we utilized a hydrodynamic tail vein system to evaluate the differential roles of constitutively active YAP (YAP-S5A) and TAZ (TAZ-S4A) in cooperation with oncogenic NRas (NRasV12D) in the development of liver cancer. YAP-S5A resulted in consistent oval cell hyperplasia and less frequent cholangiocellular carcinoma with a prolonged median survival time of greater than 50 days. In contrast, mice with TAZ-S4A died within 7 days after hydrodynamic tail vein injection. TAZ-S4A mice had anaplastic cells throughout the liver and marked myeloid infiltration. Giant cells were cytokeratin AE1/AE3 positive and von Willebrand's factor negative, confirming an epithelial rather than megakaryocytic origin. Myeloid cells were primarily CD68 and F4/80 positive macrophages. Markedly elevated serum cytokines, including CCL2 and CXCL1, were also noted in these mice. Lesions in TAZ-S4A mice were abrogated by inhibiting TAZ interactions with TEAD, suggesting a role for TEAD-dependent transcriptional regulation. These results demonstrate unique roles of YAP and TAZ in the liver and link TAZ to proinflammatory signaling, which may play a role in inflammation-driven cancer development.

Experimental Disease Focused Scientific Session II

December 6, 2016
2:30 PM – 2:45 PM

ELUCIDATING THE PATHOGENESIS OF PREECLAMPSIA: REDUCED PLACENTAL EXPRESSION OF REGULATOR OF G-PROTEIN SIGNALING (RGS)-2

Katherine N Gibson-Corley, Katherine J Perschbacher, Jeremy A Sandgren, Sabrina M Scroggins, Donna A Santillan, Eric J Devor, Mark K Santillan, Justin L Grobe

Preeclampsia is a devastating hypertensive disorder of pregnancy, occurring in roughly 5% of all pregnancies and we are currently not able to predict the disorder. It has been shown that placental trophoblasts play a key role in the pathogenesis but it is not known what causes their dysfunction. Multiple hormones, including angiotensin, endothelin and vasopressin have all been implicated. Interestingly, the predominant placental-expressed receptors for these hormones share a final common pathway – the Gαq second-messenger cascade. Regulator of G-Protein (RGS)-2 acts as an endogenous inhibitor of Gαq signaling, and others have recently reported that a loss-of-function mutation in RGS2 is associated with preeclampsia in humans. We therefore hypothesized that disinhibited Gαq signaling within placental trophoblasts (due to reduced RGS2 activity) represents an initiating factor for preeclampsia. RGS2 was strongly expressed in human placenta and was grossly suppressed in placental tissues from preeclamptic patients. Further, RGS2 was variably expressed in layers of placenta from human patients. Immunohistochemical analyses of RGS2 in murine placentas showed cytoplasmic localization within trophoblasts in wild-type animals, consistent with its canonical cytoplasmic GTPase-accelerator protein function. Selective breeding of RGS2-null sires to wildtype dams resulted in heterozygous loss of RGS2 within the feto-placental unit, in the context of a wildtype maternal environment. This breeding resulted in maternal hypertension, but notably this breeding did not cause proteinuria or intrauterine growth restriction. These findings implicate RGS2 – within the placenta – in the pathogenesis of preeclampsia, and highlight the impact of paternal genetics to the development of preeclampsia symptoms.

December 6, 2016
2:45 PM – 3:00 PM

PRELIMINARY PATHOLOGY ASSOCIATED WITH EXPERIMENTAL ZIKA VIRUS INFECTION OF RHESUS MACAQUES (MACACA MULATTA)

Rebekah I. Keesler, Lark L. Coffey, Koen K. A. Van Rompay, Jennifer Watanabe, Eliza Bliss-Moreau, Anne Gibbons, John H. Morrison, Kari Christe, Patricia Pesavento, Anil Singapuri, Rie Watanabe, Eric Delwart, Michael Busch, Graham Simmons, Marion M. Lanteri

Since 2013, Zika virus (ZIKV) has come to the forefront of emerging diseases in the Pacific Islands and Americas. Due to the seemingly innocuous clinical presentation of ZIKV prior to the Pacific Island and South American outbreaks, little research had been conducted since its discovery in 1947. As severe clinical manifestations including microcephaly and Guillain-Barré syndrome associated with recent outbreaks become more widespread, the development of animal models remains essential to elucidate viral pathogenesis, tissue targets, disease progression, and teratogenicity, as well as to test potential drug and vaccine candidates. Due to the biologic and physiologic similarities of non-human primates to humans, rhesus macaques (*Macaca mulatta*) were experimentally inoculated with a 2015 Brazilian strain of ZIKV and necropsied 7 or 14 days post-inoculation (PI). Adult animals had no clinical signs but developed detectable virus and viral RNA levels in plasma for 5 days PI and neutralizing antibody titers beginning 5 days PI. Gross and histologic examinations of the adult animals were also unremarkable. One pregnant animal had *in utero* death of her fetus at 47 days of gestation, 7 days PI. The placenta had multifocal small areas of necrosis, although the fetus did not have histologic lesions. Via quantitative reverse transcription PCR and in situ hybridization for ZIKV, multiple fetal tissues and the placenta were positive. This pilot study demonstrates that rhesus macaques can be infected with a current outbreak strain of ZIKV, and that virus replication can occur in fetal tissues, a factor that may have caused fetal death.

December 6, 2016
3:30 PM – 4:30 PM
Of Animals and Humans: Pathogenic Properties in the Evolving Genus Helicobacter
James G. Fox

December 6, 2016
4:30 PM – 4:45 PM
INHIBITION OF AN AQUATIC RHABDOVIRUS DEMONSTRATES PROMISE OF A BROAD-SPECTRUM ANTIVIRAL FOR USE IN AQUACULTURE
Bethany F Balmer, Rachel L Powers, Benhur Lee, Frederic Vigant, Michael E Jung, Maureen K Purcell, Kevin R Snekvik, Hector C Aguilar

Many enveloped viruses cause devastating disease in aquaculture, resulting in significant economic impact. LJ001 is a broad-spectrum antiviral compound that inhibits enveloped viral infections by specifically targeting phospholipids in the lipid bilayer via production of singlet oxygen ($^1O_2$). This stabilizes positive curvature and decreases membrane fluidity, which inhibits viral-cell membrane fusion during viral entry. Based on previous mammalian studies and the requirement of light for activation of LJ001, we hypothesized that LJ001 may be useful as a preventative and/or therapeutic agent for enveloped viral infections in aquaculture. Here, we report that LJ001 was more stable with a prolonged inhibitory half-life at relevant aquaculture temperatures (15^{0}C) as compared to 37^{0}C in mammalian studies. When LJ001 was pre-incubated with our model virus, infectious hematopoietic necrosis virus (IHNV), infectivity was significantly inhibited *in vitro* (using the EPC fish cell line) and *in vivo* (using rainbow trout fry) in a dose-dependent manner. While horizontal transmission of IHNV in a static co-habitation challenge model was inhibited by LJ001, transmission was not completely blocked at established antiviral doses. Therefore, LJ001 may be better suited as a therapeutic for aquaculture settings that include viral infections with lower viral shedding rates than IHNV or where higher viral titers are required to initiate infection of naïve fish. Importantly, our data also suggests that LJ001-inactivated IHNV elicited an innate immune response in the rainbow trout host, making LJ001 potentially useful for future vaccination approaches.

December 6, 2016
4:45 PM – 5:00 PM
PATHOLOGICAL FINDINGS IN A FERRET MODEL OF NON-ADAPTED FILOVIRUS DISEASE
Hannah S Bender, Bronwyn A Clayton, Sarah Riddell, Jessica Haining, Jean Payne, Jennifer Harper, John Bingham, Deborah Middleton, Glenn A Marsh

The extent of the recent, protracted ebolavirus outbreak in West Africa underscored the significant public health threat of zoonotic filoviruses, galvanizing global efforts to develop new vaccines and therapeutic compounds. Relevant animal models of disease are essential for this work. Dose ranging and time course studies were performed to characterize the pathogenesis of disease caused by wild-type isolates of Zaire ebolavirus (Mayinga), Sudan ebolavirus and Marburg virus in the ferret (*Mustela putorius furo*). Following oronasal challenge, clinical signs and pathology findings were comparable for all three viruses. The onset of viremia was accompanied by fever and rapid clinical progression, culminating in disseminated intravascular coagulation. In all animals, gross lesions and histologic changes were most pronounced within the lungs, which were characterized by severe interstitial pneumonia, with intense neutrophilic and histiocytic inflammation, and perivasculitis. Hepatocellular necrosis and splenic lymphoid necrosis were also observed. Histologic findings are comparable to those reported in a guinea pig model of host-adapted ebolavirus infection, and are distinct from changes in non-human primates and people in several important ways. Nevertheless, the ferret is a reliable model of viral haemorrhagic fever induced by unadapted filoviruses, presenting an attractive adjunct to non-human primate models for preclinical testing of new drugs.

Experimental Disease Posters

E-01: COMPARISON OF PORCINE CIRCOVIRUS TYPE 2(PCV2)-ASSOCIATED LESIONS PRODUCED BY CO-INFECTION OF PCV2 AND PORCINE REPRODUCTIVE AND RESPIRATORY SYNDROME VIRUS(PRRSV)

Jiwoon Jeong, Se Eun Kim, Chanhee Chae

Background: Mixed infection of PCV2 and PRRSV is one of the most common co-infections associated with swine disease in field. However, an in-depth comparison of the virulence and pathogenicity of concurrent infections with the two genotypes of the two viruses has yet to be undertaken.

Objective: The aim of this study was to compare the virulence and pathogenicity of a combination of concurrent infections of two genotypes of PCV2 and two genotypes of PRRSV in terms of PCV2 viremia, and PCV2-associated lesions and antigens in co-infected pigs.

Methods: 90 pigs were divided into 9 groups(10 pigs in each group) at 2 weeks of age: PCV2a/type 1 PRRSV, PCV2a/type 2 PRRSV, PCV2b/type 1 PRRSV, PCV2b/type 2 PRRSV, PCV2a, PCV2b, type 1 PRRSV, type 2 PRRSV, or negative control. At 6 weeks of age, the pigs were inoculated intranasally as follows: 3 ml of each PCV2 genotype inoculum; 3 ml of each PRRSV genotype inoculum. Blood samples were collected at 0, 5, 7, 10, 14, 21, and 28 dpi. Tissues were collected from each pig at necropsy(14 and 28 dpi)

Results: Pigs with PCV2a (or 2b)/type 2 PRRSV had significantly ($P < 0.05$) higher levels of PCV2 viremia, more severe PCV2-associated lesions, and more PCV2 DNA within the lesions compared to pigs with PCV2a (or 2b)/type 1 PRRSV.

Conclusion: The results of this study demonstrate significant differences in the virulence and pathogenicity of type 1 and type 2 PRRSV with respect to the production of PCV2-associated lesions.

E-02: ENGINEERED HIGH AFFINITY SOLUBLE AXL DECOY RECEPTOR THERAPY IMPROVES CLINICAL OUTCOME IN MOUSE MODELS OF ACUTE MYELOID LEUKEMIA

Anh Diep, Yu Miao, Jonathan Nagel, Mihalis Kariolis, Amato Giaccia

Background: Acute myeloid leukemia, AML, kills more than 10,000 people annually in the United States. An improved understanding of the molecular biology of AML has lead to molecularly targeted therapies for AML, however, improvements in overall survival remain elusive due to chemoresistant disease. Axl, a receptor tyrosine kinase, has been implicated a number of malignancies and also plays a role in the pathogenesis of AML. Objective: The aim of this study was to inhibit the Axl pathway using an engineered high affinity soluble Axl decoy receptor in xenograft AML mouse models as a means of reduction of disease burden.

Methods: Subcutaneous xenografts and leukemic engraftments of the MV4-11 and OCI-AML3 human cell lines were created in Nod Scid gamma mice. Mice were randomized into treatment groups to compare the soluble Axl therapy alone or in combination with cytarabine, the standard chemotherapy drug for AML. Survival, tumor burden and toxicity were evaluated throughout the course of the study. Subcutaneous tumors were assessed for apoptosis, proliferation, and activation of Axl.

Results: Inhibition of Axl activation by the soluble Axl receptor resulted in no toxicity and reduced disease burden in both OCI-AML3 and MV4-11. Additionally, inhibition of Axl in combination with cytarabine resulted in a slightly additive effect. Cytarabine alone resulted in high toxicity and minimal tumor control.

Conclusion: High affinity soluble Axl therapy may be a safe and efficacious therapy for AML in combination with cytotoxic chemotherapy in preclinical AML mouse models.

E-03: MICE DEFICIENT IN INTERFERON SIGNALLING DO NOT SUCCUMB TO SYSTEMIC HENDRA VIRUS INFECTION

Emma Croser, Glenn Marsh, Paul Hertzog, Deborah Middleton

Background: In contrast to Hendra virus infection in people, horses, cats, and ferrets, fulminating systemic disease is not encountered in immunologically intact mice. Mice develop self-limiting respiratory tract infection and olfactory route encephalitis.

Objective: Using immune knockout mouse strains we investigated components of the innate immune system that may protect mice from Hendra virus associated systemic disease.

Methods: C57BL6J wild type, Ifnar1-/- and Stat1-/- mice were challenged either intranasally or intraperitoneally with 50,000 50% tissue culture infective dose ($TCID_{50}$) of Hendra virus (Hendra virus/Australia/Horse/2008/Redlands).

Results: All mice that succumbed to infection exhibited acute neurological signs with histopathological lesions of olfactory pathway associated encephalitis. In wild type mice challenged intranasally, replication was confined to the respiratory and olfactory tracts. Virus was not present in the olfactory tract of wild type mice challenged intraperitoneally. In Infnar1-/- and Stat1-/- mice viral antigen was present in the olfactory bulb, lung, liver and spleen regardless of challenge route. Formation of syncytia, vasculitis, severe inflammation and necrosis seen with Hendra virus infection in other animal species was not present.

Conclusion: Inactivation of the interferon signalling pathways facilitates systemic Hendra virus replication as determined by viral load and diversity of infected tissues. However, the systemic vasculitis and severe tissue necrosis seen with Hendra virus infection in other animal species was not observed in mice. We therefore conclude that inactivation of these signalling pathways alone are insufficient to permit the development of severe systemic disease in mice.

E-04: EVALUATION OF α-SYNUCLEIN (aSyn) SPREAD IN MOUSE BRAIN BY AUTOMATED IMAGE ANALYSIS

Oded Foreman

Background: α-synuclein (aSyn), a soluble protein localizes to neuronal presynaptic terminals. In Parkinson's disease, aSyn is found in neurons in an insoluble aggregated and phosphorylated form termed Lewy bodies. Evidence from in vivo and cell based studies suggest that fibrillar species of pathological aSyn propagate from cell to cell and spread to adjacent neurons through a nucleation–dependent seeding mechanism.
Objective: Develop a high throughput assay to assess spread of aSyn in mouse brains
Methods: The spreading mechanism is modeled in mice by unilaterally injecting preformed aSyn fibrils into dorsal striatum of PrP.hu.αSyn.A53T transgenic mice expressing the A53T mutant allele of aSyn under the human prion promoter. At study termination, brains are coronally sectioned and stained with anti-aSyn, anti-CD68, anti-Iba1 and anti-NeuN antibodies to assess spread, gliosis and neuronal loss. Subsets of brains are homogenized to measure aSyn by ELISA. Slides are scanned on Leica SCN400 imaging system. Ipsilateral and contralateral regions of interest are manually defined on the digital images and a positive pixel area, optical density and aggregated a-Syn count algorithm are performed using Matlab software. Thresholds for markers are adjusted using control tissues and transfected cell-lines.
Results: Fibrillar aSyn spreads along neuronal circuitry. By day 60, phosphorylated aSyn aggregates are present in the contralateral striatum by immunohistochemistry and ELISA.
Conclusion: This spreading model can be used to study biological mechanisms of aSyn spread, and potentially for drug discovery purposes. Using slide scanning technology and automated quantification algorithms, this method offers non-biased high throughput analysis supporting anti-aSyn efficacy studies.

E-05: MUTANT IL-7Ralpha AND MUTANT NRAS ARE SUFFICIENT TO INDUCE T-CELL ACUTE LYMPHOBLASTIC LEUKEMIA/LYMPHOMA

Sarah D Cramer, Julie A Hixon, Caroline Andrews, Gisele OL Rodrigues, Xiaolin Wu, Timothy Back, Helen Michael, Maggie Cam, Emilee Senkevitch, Peter D Aplan, Wenqing Li, Scott K Durum

Background: Pediatric leukemia is the most common childhood cancer. While therapy is curative in approximately 80% of children, patients suffer both short and long-term side effects. Targeted therapy could reduce these side effects. Components of the IL-7Ralpha signaling pathways are potential therapeutic targets that are often mutated in pediatric leukemia. However, these mutations typically occur in combination with other mutations. Identifying mutations that collaborate with mutant hIL-7Ralpha to generate leukemia could help to inform development of targeted therapies.

Objective: We sought to determine whether combination of mutant hIL-7Ralpha and mutant NRas is sufficient to generate T-cell leukemia.

Methods: Immature thymocytes were isolated from C57BL/6J mice and transduced with retroviral vectors containing mutant hIL-7Ralpha and mutant NRas. Cells were cultured on the OP9-DL4 bone marrow stromal cell line for 9-10 days and then injected into sublethally irradiated $Rag^{-/-}$ mice. Mice were euthanized at onset of clinical signs, and cells were serially passaged into recipient mice. Lesions were analyzed using histology, immunohistochemistry, and flow cytometry. Clonality was assessed using ligation-mediated PCR.

Results: Mice injected with mutant hIL-7Ralpha-NRas cells developed rapid-onset, highly penetrant disease characterized by lymphoblastic leukocytosis, splenomegaly, and hepatomegaly. Spleen, liver, and lung were infiltrated by sheets of monomorphic T-cells that effaced normal parenchyma. T-cells were a mixed immunophenotype consisting predominantly of $CD4^{+}CD8^{+}$ and $CD8^{+}$ cells. Cells were highly polyclonal, consistent with an origin from many initiating cells. Cells injected into recipient mice generated similar lesions.

Conclusion: Combination of mutant hIL-7Ralpha and mutant NRas generates T-cell leukemia/ lymphoma.

E-06: COMBINED DEFICIENCY OF RGS6 AND OXIDIZABLE CAMKII PRODUCE EMBRYONIC LETHALITY WITH CARDIOVASCULAR AND HEMATOPOIETIC DEFECTS

David K Meyerholz, Bandana Chakravarti, Heather A Flaherty, Rory Fisher

Regulator of G-Protein Signaling 6 (RGS6) is a GTPase activating protein for Gai/o proteins.. Calcium/Calmodulin-dependent protein Kinase II (CaMKII) is a serine/threonine-specific protein kinase. RGS6 and oxi-CaMKII are expressed during different gestational stages of WT embryos. We hypothesized that these two proteins might have functions that overlap during fetal growth and studied homozygous mutant mice deficient in RGS6 and expressing an oxidation resistant mutant of CaMKII.. Mutant mice were macroscopically normal at E9.5. While still viable at E10.5, mutant mice were smaller in size than controls and no embryos were detected alive after E10.5. Evaluation of the viable embryos showed reduced distension and filling of yolk sac vessels by hematopoietic cells. The placental labyrinth had both maternal and fetal vessels that were detectable; however, both vessel types had reduced filling and distension by blood cells. Mutant mice had larger nucleated erythrocytes with larger nuclei and less intense eosinophilic cytoplasmic staining suggestive of reduced hemoglobin production. Lastly, mutant hearts showed reduced free wall thickness and trabeculation by primitive cardiomyocytes. The scope of these lesions paralleled lesions seen in mice with altered Notch1 signaling. Evaluation of Notch1 pathways in several fetal tissues from mutant mice showed dysregulated Notch signaling. These results suggest that RGS6 and oxi-CaMKII have overlapping roles in Notch1 signaling that are essential for proper fetal development.

E-07: PERIPHERAL AND CENTRAL TISSUE RESERVOIRS IN ASYMPTOMATIC CHRONICALLY FELINE IMMUNODEFICIENCY VIRUS-INFECTED CATS

Chrissy Eckstrand, Brian Murphy, Kathryn Pitt, Ellen Sparger

Background Feline immunodeficiency virus (FIV) infection in cats results in life-long persistence and progressive immunopathology. In a cohort of experimentally-infected cats an interesting dichotomy occurred where peripheral blood CD4+ leukocytes progressively declined over time in spite of apparent viral latency. In this cohort, we recently found evidence that lymph-node tissue harbors foci of ongoing viral replication suggesting that tissue reservoirs are important in the immunopathogenesis during the asymptomatic phase.

Objective We sought to further characterize cellular and tissue reservoirs of viral replication during the asymptomatic phase by examining spleen, mesenteric lymph node (MLN), and intestine from FIV-infected and uninfected cats.

Methods/Results Survival surgery was performed to collect spleen, MLN, and intestine. Blood collected at surgery demonstrated severe CD4+ depletion, undetectable plasma viral-*gag*-RNA and rarely detectable peripheral blood mononuclear cell-associated viral-*gag*-RNA by real-time PCR. In spite of viral transcriptional inactivity in blood, FIV-*gag*-RNA was detectable in all three tissue sites from 3/4 FIV-infected cats. A novel *in situ* hybridization assay identified follicular domains as foci of ongoing replication, and had superior sensitivity over real-time PCR, as all 4 FIV-infected cats were positive. Additionally, we demonstrated that CD4+ depletion occurred in tissues, and that CD4+ and CD21+ cells are important reservoirs of replication, while CD8+ contain provirus but are less productive.

Conclusions These findings support the hypothesis that tissue reservoirs harbor foci of ongoing viral replication in the late asymptomatic phase, in spite of a relatively inactive viral replication signature in blood. Future lentiviral eradication strategies should take tissue reservoirs into consideration.

E-08: CONDITIONAL LIVER SPECIFIC LOSS OF TRISTETRAPROLIN FAMILY OF RNA BINDING PROTEINS RESULTS IN HEPATITIS IN MICE

Sonika Patial, Gordon P. Flake, Perry J. Blackshear

Background: Tristetraprolin (TTP) family of RNA binding proteins, including, TTP, ZFP36L1, and ZFP36L2 affect mRNA levels by binding to AU-rich elements on the 3'-untranslated regions of specific mRNAs and promoting their rapid turnover. Despite similar biochemical functions, whole-body deletion of the three proteins in mice results in three unrelated phenotypes. While the deletion of TTP results in polyarticular arthritis, dermatitis, and myeloid hyperplasia; deletion of ZFP36L1 is embryonic lethal; and the deletion of ZFP36L2 results in defective hematopoiesis and early post-natal death.

Methods: In order to test the role of the three family members in a single tissue/cell type, here, we utilized Cre-LoxP system to conditionally delete the three proteins in mouse liver. We generated three independent/single knockout mice (KO) strains, a triple KO mice strain, and three strains that contained deletion of all but one allele of each family member.

Results: While the independent liver specific deletion of the three proteins did not result in any apparent phenotype in mice, simultaneous deletion resulted in defects in bile acid metabolism and chronic active hepatitis. These mice progressed to developing hepatocellular carcinomas later in life. Interestingly, keeping one wildtype allele of either ZFP36L1 or ZFP36L2 but not that of TTP, intact, protected mice from exhibiting this phenotype.

Conclusions: Our data suggests that TTP family of RNA binding proteins play an essential role in maintaining liver immune homeostasis. Moreover, these proteins act in both specific and redundant manner with the role of both ZFP36L1 and ZFP36L2 as indispensable in mouse liver.

E-09: TRANSCRANIAL APPLICATION OF A2A RECEPTOR AGONISTS DECREASES NEOCORTICAL CELL DEATH FOLLOWING MILD TRAUMATIC BRAIN INJURY

Kara N Corps, Kenneth A Jacobson, Dorian B McGavern

Background: Traumatic brain injury (TBI) represents a complex group of brain injuries that vary in severity, pathology and clinical outcome. Mild TBI (mTBI) is the most frequently diagnosed type. Previously we found that early pathological changes in our model are mediated by production of reactive oxygen species (ROS), damage to the glia limitans, and purinergic signaling that drives changes in microglia. Adenosine signaling has been shown to modulate a variety of disease and injury states including spinal cord injury.

Objective: We hypothesized that transcranial modification of adenosine signaling would decrease cell death after mTBI.

Methods: Our lab has developed a focal, compressive, closed-skull murine model of mTBI. We applied agonists and antagonists of the four adenosine receptors transcranially to anesthetized mice subjected to mTBI. We subsequently used two-photon laser scanning microscopy to investigate the effect of A2A agonism on the glia limitans, ROS production and changes in microglial morphology *in vivo.*

Results: Agonizing the A2A receptor, but not the A1, A2B, or A3 receptors, significantly decreased cell death in the neocortical parenchyma eight hours following mTBI in two genetically diverse strains of inbred mice. Co-application of an A2A agonist and an A2A antagonist eliminates this protective effect, while co-application of an A2A agonist with antagonists to other adenosine receptors does not. We tested the FDA-approved A2A agonist, Regadenoson, and recapitulated these results. Compared to controls, A2A agonism produced no difference in glia limitans leakage, ROS production, or microglial morphology.

Conclusion: Agonizing A2A receptors protects the neocortical parenchyma following mTBI.

E-10: EXPRESSION KINETICS OF RANTES AND MCP-1 IN THE BRAIN OF DEER MICE (Peromyscus maniculatus) INFECTED WITH VESICULAR STOMATITIS NEW JERSEY VIRUS

Leonardo P Mesquita, Fábio RP Bruhn, Paulo César C Maiorka, Elizabeth W Howerth

Background: The vesicular stomatitis virus is a RNA virus that causes encephalitis in deer mice. Chemokines, such as RANTES and MCP-1, are important for chemotaxis and activation of inflammatory cells in central nervous system (CNS) during the course of VSV encephalitis. However, the role of CNS resident cells regarding chemokine expression is poorly characterized.

Methods: Samples ofCNS of deer mice inoculated intranasally with 10^6 $TCID_{50}$/50μL of vesicular stomatitis New Jersey virus (VSNJV) were collected in 1 to 7 days post inoculation. Paraffin embedded tissues were submitted to histological evaluation; immunohistochemistry for RANTES (CCL-5) and MCP-1 (CCL-2) and double-immunofluorescence to evaluate chemokine expression within neurons, astrocytes and microglia using antibodies against NeuN, GFAP and Iba-1 respectively.

Results: RANTES and MCP-1 were expressed only in the olfactory bulb (OB), where the virus was restricted. This chemokine expression was followed by the influx of inflammatory cells to the OB later in the course of acute disease. Neurons, astrocytes and microglia expressed RANTES, whereas MCP-1 was expressed by neurons and astrocytes. VSNJV antigens were localized predominantly within neurons in the OB.

Conclusion: Although astrocytes and microglia responded to VSNJV infection by expressing chemokines, neurons were the predominantly infected cell type. Therefore, VSNJV-infected neurons may have a critical role in initiating an immune response in the OB. Strikingly, the signaling between infected neurons and other CNS resident cells might be critical for activation of astrocytes and microglia during infection by neurotropic RNA viruses.

E-11: DELIVERY OF ALX-0171 BY INHALATION GREATLY REDUCES DISEASE BURDEN IN A NEONATAL LAMB RSV INFECTION MODEL

Alejandro Larios Mora, Jack M Gallup, Albert Van Geelen, Linde Duprez, Thomas Stohr, Laurent Detalle, Mark R Ackermann

RATIONALE: A trivalent Nanobody, ALX-0171, targeting the RSV F-protein was developed and its therapeutic potential evaluated in a neonatal lamb model following daily nebulization for 3 or 5 consecutive days.

METHODS: Colostrum-deprived newborn lambs were nebulized with RSV-M37 (~3.5 x 10^7 FFU/lamb) on day 0, and were subsequently treated by daily nebulization with either ALX-0171 or placebo and examining 2 different treatment regimens (day 1-5 or day 3-5 post-infection).

RESULTS: On day 3 post-infection, corresponding to the day of peak viral loads, the general well-being of the vehicle-treated lambs declined as was reflected by inactivity, lethargy, weakness, drooping of ears, and lack of appetite. Both viral RNA and cultivatable virus was consistently present in all lung lobes of these lambs on day 6. In addition, gross lung examination revealed extensive viral lesions involving ~40% of all lung lobes which were correlated to viral antigen expression in the bronchioli/alveoli. Histologically, the lesions were consistent with those described previously for lambs and infants after RSV infection. In contrast, the ALX-0171-treated lambs had a >10,000-folddecline in cultivatable virus, markedly reduced lung viral antigen expression, reduced lung viral lesions and reduced histological changes. ALX-0171 treatment exerted a positive effect on clinical parameters (as assessed by respiratory rates, wheeze, expiratory efforts, temperature, malaise and body weights) of the lambs and was well-tolerated.

CONCLUSION: ALX-0171 was well-tolerated in RSV infected neonatal lambs and exerted a positive effect on RSV-induced lung lesions and inflammation and markedly reduced symptoms of illness.

E-12: BOVINE HERPESVIRUS-4 AS A DELIVERY PLATFORM FOR THEILERIA PARVA ANTIGENS.

Laura B.A. Williams, Lindsay M. Fry, Ivan I. Morrison, Giulia Tebaldi, Donald P. Knowles, Gaetano Donofrio

Background: East Coast Fever (ECF) is an often fatal disease of cattle in Africa caused by the tick-borne protozoan *Theileria parva*. The development of a sustainable vaccine is critical to the long term control of ECF. A BoHV-4 vaccine vector could establish a persistent infection to facilitate the long-term expression of *T. parva* antigens. BoHV-4 has shown success in experimental vaccines against a variety of infectious diseases by producing robust humoral immune responses. The aim of thisstudy is to test the BoHV-4 platform in the effectiveness of delivering *T. parva* antigens to incite an antigen-specific immune response.

Methods: Recombinant BoHV-4s expressing the *T. parva* antigens Tp2 and Tp9 were constructed from an apathogenic strain of BoHV-4 genome cloned as a bacterial artificial chromosome.The *in vitro* stability of the viral constructs was determined by nested PCR across the inserted genes, and cattle were immunized with the recombinant viruses. Nested PCR detected BoHV-4 in peripheral blood of immunized cattle. The humoral immune response to BoHV-4 was detected serologically using a commercially available indirect ELISA and by immunofluorescent antibody detection (IFA) using a monoclonal antibody. Antibody to Tp2/Tp9 was detected by indirect ELISA.

Results: The recombinant viruses were developed and verified. The constructs maintained the antigen insertion *in vitro*. Cattle immunized with the recombinant viruses had short-term, sporadically detectable BoHV-4 DNA, and developed sustained serologic responses to BoHV-4. Antibody to Tp2/Tp9 was detected.

Conclusion: Preliminary data suggest that BoHV-4 can be used as a novel viral vector for use in cattle.

E-13: AEROSOLIZED REFLUX TO PIG VOCAL FOLDS: AN IN VIVO STUDY

Abigail C Durkes, Preeti Sivasankar

Objective: The objective of this study was to investigate epithelial changes in response to repeated, aerosolized acidified-pepsin exposures in an *in vivo* porcine model.

Study Design: Prospective, *in vivo* study

Methods: Twelve pigs inhaled acidified-pepsin (simulated reflux) or saline (sham) through a nose cone attached to a nebulizer. The pigs were challenged for a total of 60 exposures (3 times per day, 5 days per week, for 4 weeks). Vocal fold, nasal mucosa, trachea and lung tissue morphology were evaluated histologically. Vocal fold ultrastructural alterations, epithelial intercellular space diameter, and microridge height were examined via transmission electron microscopy (TEM). Complementary DNA microarray analysis of vocal fold epithelium was conducted and followed up with real-time polymerase chain reaction investigating the gene expression of E-cadherin (Ecad), zona occludin-1 (ZO-1), cystic fibrosis transmembrane conductance regulator (CFTR), and epithelial sodium channel (SCNN1α).

Results: All animals tolerated the repeated, aerosolized acidified-pepsin challenges. Even after sixty simulated reflux exposures, vocal fold epithelium had preserved morphology and transcriptional and translational levels of epithelial proteins.

Conclusions: This is the first report of an *in vivo* simulated aerosolized reflux challenge in a pig. This study adds to the growing body of data suggesting that the healthy, uninjured larynx is robust and resilient to multiple simulated reflux exposures.

E-14: Non-human primate model of Type 2 diabetes mellitus: cerebral expression of glucose transporters Glut1 and Glut3 and alterations associated with insulin resistance and cognitive impairment

Laura Cavicchioli, Shana Elkind, Karen Dalecki Boisvert, Faramarz Taheri, Sheila Macri, James Rowlett, Susan Westmoreland

It is well known that cognitive impairment could be associated with Type 2 diabetes mellitus (T2DM). As there is a significant role for insulin in normal cognitive functioning, T2DM-related insulin dysregulation may contribute to cognitive impairment. For this study, 15 female Rhesus macaques were investigated to evaluate aspects of the insulin resistance hypothesis using a three-pronged approach: confirm body composition changes with physiological testing; gauge cognitive capability using a Novel Object Recognition (NOR) task; and measure expression levels of two glucose transporters, GLUT1 and GLUT3, in the brain regions associated with NOR-related learning and memory.

A strong correlation has been observed between abdominal fat, body weight and insulin resistance, characterized by hyperinsulinemia and euglycemia. In particular, omental fat weight correlated positively with Intravenous Glucose Tolerance Test mean insulin levels, indicating that animals with more omental fat have higher insulin levels. Within physiological parameters correlated positively with performance at NOR task, an altered glucose metabolism, in particular pronounced insulin resistance, was linked to poorer cognition.

Even if NOR performance did not correlate significantly with the expression of glucose transporters, both positive and negative correlations among physiological parameters and the expression levels of the glucose transporters were found. Clear associations are between GLUT1 and GLUT3 immunohistochemical expression levels in brain regions of interest and increased body fat and insulin resistance.

Overall, results provide some support for the "insulin resistance hypothesis" of T2DM-mediated cognitive deficits, although in this phase of disease progression deficits do not appear to be mediated through glucose transporter-regulated mechanisms.

E-15: XENOGENIC GRAFT VERSUS HOST DISEASE IN CANINE PBL-SCID MICE: A POTENTIAL MODEL OF CANINE IMMUNE MEDIATED HEMOLYTIC ANEMIA

Jeremy B Foote, Emily C Graff, Russel C Cattley, Richard C Bird

BACKGROUND: Xenotransplantation of human peripheral blood mononuclear cells (PBMCs) intoNOD SCID common IL2γ-/- mice (NSG) provides a model to study donor specific immunity, however, engrafted mice develop xenogeneic graft versus host disease (xGVHD). Objective: determine whether NSG mice are permissive for xenotransplantation of canine PBMCs and to characterize xGVHD.

METHODS: 8 week-old female NSG mice were irradiated with 100cGy of γ irradiation and engrafted with 2 x 10^7 canine PBMCs. Mice were monitored for engraftment of canine lymphocytes and xGVHD by histology and flow cytometry. Bone marrow, blood smears, and CBC analyses were performed to determine whether mice developed bone marrow aplasia and anemia.

RESULTS: Post engraftment, canine T and B-lymphocytes expand and recirculate between the bone marrow, spleen, and peritoneal cavity. At 21 days post engraftment canine $CD8^+$ T and Ig^+ $CD21^+$ B cells predominate. Engrafted mice produced significant quantities of canine immunoglobulins beginning at 14 days post engraftment. Mice eventually develop multi-organ inflammation (meninges, CNS, liver, spleen, stomach, omentum, pancreas, and uterus) and immune mediated hemolytic anemia beginning at 28 days post engraftment. By flow cytometry we demonstrated that agglutinated erythrocytes from anemic mice are coated with canine antibodies.

CONCLUSIONS: NSG mice are permissive for engraftment of canine PBMCs but eventually develop multi-organ inflammation. Engrafted mice develop an immune mediated hemolytic anemia coinciding with canine B cell expansion and antibody production. These findings demonstrate that xGVHD will complicate studies focused on donor specific immunity, however, canine PBL-SCID mice may provide a novel model to study IMHA.

E-16: SURVIVAL TIME AND STABILITY PROPERTIES OF DISEASE-ASSOCIATED PRION PROTEIN IN CHRONIC WASTING DISEASE OF ELK

Jo Moore, Catherine E. Vrentas, Heather H. West Greenlee, Qingzhong Kong, Justin J Greenlee

Background: The Rocky Mountain elk (*Cervus elaphus nelsoni*) prion protein gene exhibits amino acid polymorphism at codon 132, with 132L (leucine) and 132M (methionine) allelic variants present in the population. We have previously shown that following experimental oral challenge with chronic wasting disease (CWD) the incubation times of 132LL elk are prolonged, while incubation times for 132LM elk are intermediate, and incubation times for 132MM elk are short.

Objective: Investigate potential mechanisms for variations in incubation time in elk of different prion protein genotypes.

Methods: The conformational stability of disease-associated prion protein (PrP^{CWD}) from 132MM, 132LM, and 132LL CWD-positive elk from a naturally infected game farm herd was assessed by treatment with increasing concentrations of the denaturant guanidine hydrochloride. Elk brain homogenate was bioassayed in mice expressing the 132MM elk prion protein and the resulting stability of PrP^{CWD} was determined.

Results: The stability of PrP^{CWD} from 132MM and 132LM elk were similar to each other and less stable than that from 132LL elk. On first passage and second passage, mice challenged with 132LL elk PrP^{CWD} had prolonged survival times and more stable PrP^{CWD} compared to mice challenged with 132MM PrP^{CWD}. On first passage, survival times and stability profiles for 132LM-challenged mice were similar to 132MM-challenged mice, while on second passage, they were similar to 132LL-challenged mice.

Conclusions: In CWD of elk, genotype-associated variations in survival time appear to be associated with differences in conformational stability of PrP^{CWD} in both the natural host and in transgenic mice.

E-17: MODELING HUMAN ENTERIC DYSBIOSIS AND ROTAVIRUS IMMUNITY IN GNOTOBIOTIC PIGS

Erica Twitchell, Christine Tin, Ke Wen, Husen Zhang, Sylvia Becker-Dreps, M. Andrea Azcarate-Peril, Samuel Vilchez, Guohua Li, Ashwin Ramesh, Mariah Weiss, Shaohua Lei, Tammy Bui, Xingdong Yang, Stacey Schultz-Cherry, Lijuan L. Yuan

Background: Intestinal microbiota imbalance (enteric dysbiosis) is thought to contribute to reduced oral vaccine efficacy in infants from low-income countries. An animal model will help to determine how microbiota influences vaccine efficacy in a controlled environment.

Objective: Our aim was to develop a gnotobiotic (Gn) pig model of enteric dysbiosis to study effects of human gut microbiota (HGM) on immune responses to oral rotavirus vaccination, and the effects of rotavirus challenge on gut microbiota.

Methods: Gn pigs were colonized with HGM from Nicaraguan infants with high enteropathy scores and no seroconversion after RotaTeq® vaccination or the converse to represent unhealthy HGM (UHGM) and healthy HGM (HHGM), respectively. Immune responses and gut microbiota were evaluated in the pigs after rotavirus vaccination and after challenge.

Results: Compared to UHGM pigs, HHGM pigs had more rotavirus-specific IFN-γ+ T cells after vaccination and more rotavirus-specific antibodies in intestinal contents, suggesting HHGM induces stronger cell-mediated immunity and mucosal immunity. Significant correlations between multiple Operational Taxonomic Units and T cell frequencies were observed. Based on viral shedding and diarrhea postchallenge, the vaccine was more efficacious in HHGM pigs. Differences in microbiome composition were detected between the groups both before and after rotavirus challenge.

Conclusion: Stronger T cell and mucosal immunity, and better vaccine efficacy in HHGM pigs, along with microbiome alterations and correlations with T cell response warrant further investigation. Our results support the use of HGM transplanted Gn pigs as a model of human dysbiosis and oral vaccine responses.

E-18: TARGETING MATRIX METALLOPROTEINASES IN A RABBIT MODEL OF CAVITARY TUBERCULOSIS

Elizabeth A Ihms, Michael Urbanowski, Mariah Klunk, Shichun Lun, William R Bishai

Background:

Tuberculosis (TB) is an enduring global health issue, with a worldwide death toll rivaling that of HIV/AIDS (WHO 2015). A subset of TB patients will progress to develop cavitating pulmonary disease, which is associated with increased rates of bacterial drug resistance and transmission. Despite its clinical significance, the pathogenesis of cavity formation remains poorly understood. Previous work has demonstrated that matrix metalloproteinase (MMP) expression is upregulated in cavitary lesions. As MMPs are implicated in both physiologic and pathologic matrix destruction and tissue remodeling, we hypothesized that inhibition of MMPs during infection would prevent lung destruction and reduce the rates of cavitation.

Methods:

In the current study, we infected rabbits with *Mycobacterium tuberculosis* to model cavity development. Rabbits were then treated with Trocade - a novel MMP inhibitor with activity against MMPs 1, 8 and 13 - to assess the effect of MMP inhibition on cavity progression. Cavitation was assessed by serial computed tomography, followed by gross and histopathology.

Results & Conclusions:

Trocade successfully inhibits MMP-1 in vitro, and shows activity in rabbit plasma. However, treated animals show progressive pulmonary cavitation in the face of MMP inhibition. Contrary to our hypothesis, we found that MMP inhibition in this model worsens outcomes by increasing both the frequency and size of pulmonary cavities.

E-19: TURTLE CLAWS AND EXPERIMENTALLY INDUCED SELENOSIS
David L Haskins, Elizabeth W Howerth, Tracey D Tuberville

Sliders are turtles common throughout the US that can inhabit contaminated bodies of water. In this study, the effect of different selenium levels was evaluated in the yellow-bellied slider (*Trachemys scripta scripta*). Juvenile sliders were dosed with 0 mg/kg, 15 mg/kg or 30 mg/kg seleno-L-methionine (Se; a naturally occurring form of selenium with high bioavailability) via oral gavage, weekly for 5 weeks and then euthanized; some mortality occurred in the higher dose group. Liver, kidney, and muscle were collected to measure selenium concentrations, with a small portion fixed in 10% buffered formalin for histopathology for a subset of animals. Because sloughing of claws was also noted, claws from the same subset of animals were also fixed in formalin. Histologically, animals in Se treated groups had tubular degeneration and regeneration in the kidney, which has also been reported in birds with Se toxicosis. More interesting were the changes in the claws seen in both Se treatment groups, ranging from epidermal hyperplasia with disorganization and intercellular edema to necrosis and ulceration, with accumulation of seroheterophilic exudate between the epidermis and cornified layer. Changes in skin, hair and nails are common in Se toxicosis in all species, but the histology of these changes is poorly reported. However, the changes in the claws of these turtle is similar to those reported in birds with Se toxicosis. Sliders may be a good model to study the pathogenesis of selenosis, particularly in regards to its effects on cornified epithelium.

E-20: ROLE OF EXOSOMES IN THE CONCEPTUS-ENDOMETRIAL CROSS TALK: POTENTIAL IMPLICATIONS IN THE PATHOPHYSIOLOGY OF EARLY EMBRYONIC MORTALITY IN PIGS

Mallikarjun C Bidarimath, Kasra Khalaj, Rami T Kridli, Chandrakant Tayade

Background: Exosomes are nanovesicles (30-120 nm) released from cells and can contain miRNAs and proteins that affect cells at distant sites. Recently microvesicles containing miRNA were identified in the uterine microenvironment in pigs, suggesting their participation in embryo-endometrial cross talk. However, little is known about the role of conceptus-derived exosomes in placental angiogenesis, a vital process that determines its fate.

Objective: To determine the role of exosomes in pathophysiology of early embryonic mortality; and to establish the effects of porcine trophectoderm-cell derived exosomes (PTr2-Exo) on the proliferation of maternal endothelial cells (PAOEC).

Methods: PTr2-Exo and PAOEC-derived exosomes (PAOEC-Exo) were analyzed using transmission immune-electron microscopy. The exosomes, placenta and PTr2 cells were evaluated for the expression of CD63, exosomal marker, by westernblot and Immunofluorescence/Immunocytochemistry. Using RT-PCR, 14 selected exosomal miRNAs were analyzed. *In-vitro* cell culture model was established to demonstrate bidirectional exosome shuttling between conceptus and endometrium. Exosomal proteins were investigated by mass-spectrometry. Finally, Endothelial cell proliferation was evaluated by WST-1 assay.

Results: PTr2-Exo and PAOEC-Exo ranged from 26 to 150nm in diameter. CD63 positive exosomes contain miRNAs and proteins that are crucial for placental angiogenesis. miR-126, miR-296-5P, miR-16, and miR-17-5P were among the most abundant angiogenic miRNAs. Successful delivery of PTr2-Exo promoted endothelial cell proliferation, while their absence did not have an effect.

Conclusion: This data suggests that conceptus-derived exosomes may contribute to placental angiogenesis, while their absence may be detrimental to growing conceptus. Additional controlled studies are merited to decipher exosomal miRNA cargo as markers for early embryonic mortality.

E-21: EXPERIMENTAL INFECTION OF GOATS WITH A NEWLY ISOLATED STRAIN OF AKABANE VIRUS CAUSING ENCEPHALOMYELITIS

Myeon-Sik Yang

Background: Akabane virus (AKAV), a mosquito-transmitted virus, is the major cause of congenital abnormalities and encephalomyelitis in ruminants. In 2010, there was a large-scale outbreak of bovine encephalomyelitis in Korea and AKAV strain (AKAV-7) was newly isolated.

Objective: To identify the neuropathogenicity of AKAV-7 in adult goats.

Methods: Twenty five female goats were used in this study and the goats were divided into five groups: intracerebral (IC) and intrasubarachnoid (IS) each with eight animals, intravenous (IV) and vaccinated before intravenous each with four animals and a negative control animal.

Results: All animals inoculated with AKAV-7 had AKAV-neutralizing antibodies at 6-8 dpi and there was no clinical sign in infected animals. In IC group, five of eight goats had nonsuppurative encephalomyelitis in cerebrum. Viral antigens were detected in nearly all brain parts. In IS group, three of eight goats had encephalomyelitis in cerebrum, cerebellum and spinal cord. At 7 and 21 dpi, viral antigens were measured mostly in spinal cord, especially around injection part (L5-6). The neutralizating antibody titers of serum in vaccine treated group showed early onset and increase in antibody titer on 4 dpi compared to IV group. There were no obvious lesions in central nervous tissues in vaccinated group, whereas one of four goats in IV group showed encephalomyelitis in parietal lobe.

Conclusion: Newly isolated AKAV-7 can cause encephalomyelitis in goats after experimental injection. Futhermore, attenuated AKAV vaccine currently used in Korea may provide protective immunity against AKAV-7 infection in goats

E-22: BRAIN-DERIVED NEUROTROPHIC FACTOR (BDNF) IS PREFERENTIALLY UP-REGULATED IN GOLDEN RETRIEVER MUSCULAR DYSTROPHY HEART

Sarah M Schneider, Cynthia J Balog, Peter P Nghiem, Garett T Sansom, Candice L Brinkmeyer-Langford, Joe N Kornegay

Background: Golden retriever muscular dystrophy (GRMD) is a genetic and phenotypic homologue for Duchenne muscular dystrophy (DMD). Both develop late-onset dilated cardiomyopathy. Prior studies suggested brain-derived neurotrophic factor (BDNF) is increased in GRMD hearts and may be protective in DMD patients.

Methods: Left (LV) and right (RV) ventricle from 41 dogs (7 normal, 6 carrier, 28 affected) were analyzed by quantitative real-time PCR for BDNF and 4 genes (utrophin, matrix metalloproteinase 9 [MMP9], a disintegrin and metallopeptidase domain 12 [ADAM12], and osteopontin [SPP1]), known to contribute to DMD pathogenesis. Raw threshold cycle (Ct) values were normalized with endogenous HPRT expression and a $2^{-\Delta Ct}$ transformation. ANOVA measured effects of genotype, LV versus RV, and gender on gene expression. Age and gene expression in GRMD were correlated using Pearson R^2 coefficients. Significance was set at < 0.05.

Results: Only BDNF in the LV differed among the three genotypes, showing a 10-fold increase (p = 0.013) in GRMD versus normal dogs. GRMD BDNF also was increased 3-fold (p = 0.005) in the LV versus RV. GRMD Utrophin and BDNF levels showed opposite trends compared to age, with utrophin increasing (r =0.304, p=0.027) and BDNF decreasing (r =-0.279, p=0.041). Most gene expression values, particularly those for SPP1, had high standard deviations, which likely precluded detection of significance.

Conclusion: Our results extend earlier work suggesting BDNF is preferentially increased in the GRMD left ventricle. Increased BDNF levels might contribute to early cardiac sparing. Declining BDNF with age could leave the heart more susceptible to disease.

E-23: Pigs with naturally-occurring severe combined immunodeficiency (SCID) fail to develop Porcine Reproductive and Respiratory Syndrome (PRRS)-associated pulmonary disease

Ada G Cino-Ozuna, Catherine L. Ewen, Jack C.M. Dekkers, Maureen C. Kerrigan, Carol R. Wyatt, Raymond R.R. Rowland

Infection of alveolar macrophages with porcine reproductive and respiratory syndrome virus (PRRSV) is followed by interstitial pneumonia with the infiltration of lymphocytes and macrophages. However, the mechanisms leading to interstitial pneumonia are largely unknown. In this study, pigs with naturally-occurring severe combined immunodeficiency (SCID) and wild type littermates were infected with PRRSV. At 10 days after infection, wild type pigs showed pulmonary pathology consistent with acute PRRS, including interstitial pneumonia, perivascular edema, and increased CD3+ (T cell) staining surrounding pulmonary blood vessels. PRRS-associated lesions were absent in the infected SCID littermates. Flow cytometry phenotypic analysis showed similar expression of CD172, SLA-II, and CD163 expression on wild-type and SCID porcine alveolar macrophages (PAMs). PRRSV PCR showed no difference in the presence of virus in alveolar macrophages, bronchial alveolar lavage fluid (BALF) and serum. The results suggest that the productive infection of macrophages is not the principal source for the pulmonary pathology observed during acute PRRSV infection and that T and/or B lymphocytes play a role in the pathogenesis PRRS-related lung pathology.

E-24: INVOLVEMENT OF FLT3 SIGNALING-DEPENDENT DENDRITIC CELLS IN HELICOBACTER FELIS INDUCED GASTRITIS

Du-Min Go, Jun-Hee Choi, Su-Hyung Lee, Jee-Yong Eun, Jae-Hoon Choi, Dae-Yong Kim

Dendritic cells (DCs) are known for important immune mediators in the host response to *Helicobacter pylori* (Hp) infection. Several studies reported functions of DCs in Hp immune escape, but the precise mechanisms remain unclear. To investigate the roles of DCs in *Helicobacter*-induced gastritis, we used Flt3 Knock-out (KO) mice which has reduced numbers of classical DCs and *Helicobacter felis* (Hf) commonly used in a murine gastritis model. Flt3 KO mice displayed decreased gastric DCs compared to WT mice, and injection of flt3 ligand-expressing B16 melanoma cell to WT mice induced increased gastric DCs, implying that these gastric DCs are flt3 signaling-dependent. Mice were inoculated with Hf and then sacrificed respectively 2, 4 and 8 weeks after last inoculation. Gastric tissues from wild-type (WT) mice and Flt3 KO mice were examined by histopathology, FACS analysis, immunohistochemistry and quantitative real-time RT-PCR. We observed that gastric DCs subset, including $CD103^{+}CD11b^{-}$ DCs and $CD103^{-}CD11b^{+}$ DCs, expanded during Hf infection. Histopathologically, Hf-infected Flt3 KO mice showed significantly increased gastritis compared to WT mice. In addition, we found that $CD8^+$ T cell infiltration was considerably increased in Flt3 KO mice compared to WT mice, accompanied by increased mRNA expression levels of inflammatory cytokines. Meanwhile, $CD4^+$ T cell and $Foxp3^+$ Treg cell infiltration were not different. In association with increased gastritis, Hf bacterial load was decreased in Flt3 KO mice stomach. Overall, our data indicate that Flt3 signaling-dependent gastric DCs engage in Hf-induced $CD8^+$ T cell immunity and have an impact on Hf survival.

E-25: VASCULAR-RELATED PULMONARY PATHOLOGY IN INDIAN RHESUS MACAQUES EXPERIMENTALLY CO-INFECTED WITH MYCOBACTERIUM TUBERCULOSIS AND SIMIAN IMMUNODEFICIENCY VIRUS

Denae N LoBato, Taylor W Foreman, Peter J Didier, Deepak Kaushal

Background: Mycobacterium tuberculosis (Mtb), the causative agent of Tuberculosis (TB), and Human Immunodeficiency Virus (HIV) are the leading infectious causes of disease and mortality in people worldwide. HIV infection increases rates of reactivation from latent TB to active disease. Nonhuman primates completely recapitulate human TB, including the ability to be co-infected with Simian Immunodeficiency Virus (SIV). This co-infection model provides a unique opportunity to study the synergy of dual infection with these pathogens. There is little information regarding histopathologic differences—and no previous descriptions of vascular-related pulmonary lesions—in co-infected animals. In this study, we developed a scoring system to evaluate pulmonary pathology and vascular-related lesions in Rhesus macaques experimentally co-infected with SIV and Mtb.

Methods: Latently Mtb-infected macaques were co-infected with SIV and subsequently classified as either reactivators (R) or non-reactivators (NR) based on clinical parameters. A scoring system was developed to characterize severity of ancillary changes including septal thickness, septal cellularity, consolidation, type II pneumocyte hyperplasia, increased alveolar macrophages, perivasculitis, and lymphangitis and/or vasculitis. Findings were statistically compared between groups.

Results: Animals that progressed to active TB demonstrated greater total pulmonary pathology, including number and extent of granulomas. Reactivating animals also had greater perivasculitis, vasculitis, and lymphangitis. Vascular-related changes were significantly lower in animals infected with either SIV or Mtb alone.

Conclusions: Vascular-related pathology is significantly greater in co-infected animals that progress to active TB. Higher vascular-associated inflammation in co-infected animals may contribute to the synergy of these two diseases, and may facilitate disease progression and dissemination.

E-26: COMPARATIVE IMMUNOPATHOLOGICAL STUDY BETWEEN MYCOBACTERIUM LEPRAE AND MYCOBACTERIUM LEPROMATOSIS

Ahmad A Saied, Deanna A Hagge, Rahul Sharma, Maria T. Pena, Nashone A. Ray, Ramanuj Lahiri, Richard W. Truman, Linda B. Adams

Background: Hansen's Disease, or Leprosy, is a chronic disfiguring neuropathological and dermatological disease and a major cause of non-traumatic neuropathy worldwide. *Mycobacterium leprae* was thought to be the sole causative agent of human leprosy. However, a new species of mycobacterium, *M. lepromatosis*, has been reported to be associated with diffuse lepromatous leprosy and Lucio's phenomenon (acute necrotic skin reactions), suggestive of possible greater virulence than *M. leprae*. In this study, we characterize the host immunopathological response to *M. lepromatosis* in comparison with *M. leprae* in gene knockout mouse models that have been examined in experimental leprosy: $IFNg^{-/-}$ which show a susceptible TH2 response, $IL\text{-}10/NOS2^{-/-}$ which display a highly inflammatory TH1 response, along with C57/BL6 control mice.

Methods: Four months after inoculation of $IFNg^{-/-}$, $IL\text{-}10/NOS2^{-/-}$, and C57/BL6 control mice with either *M. leprae* or *M. lepromatosis* into both hind foot pads (FP), FP granulomas were harvested and evaluated for inflammatory cell populations, cytokine expression, and *M. leprae* antigen responsiveness.

Results: In general, minor differences were seen between infiltrating cell populations responding to *M. leprae* or *M. lepromatosis* within each mouse group; however, the differences were more pronounced between the different mouse strains.

Conclusion: These results suggest that host immunity rather than leprosy species variance plays a greater role in the type of clinical disease presentation manifested.

E-27: FROM NERVES TO BRAIN: A TIME-BASED STUDY OF PSITTACIFORM 1 BORNAVIRUS (PaBV) PATHOGENESIS IN COCKATIELS (Nymphicus hollandicus)

Jeann Leal de Araujo, Raquel Rech, Ian Tizard, Jill Heatley, Judith Ball, Jianhua Guo, Aline Rodrigues-Hoffmann

Background: Psittaciform 1 bornavirus (PaBV) is the causative agent of a fatal disease in psittacine birds called proventricular dilatation disease (PDD). Although the viral etiology of PDD has been well documented, its pathogenesis remains to be clarified.

Objective: The aim of this study is to evaluate progression of PDD disease and establish course of infection of cells and tissues in cockatiels experimentally inoculated with PaBV.

Methods: Cockatiels were inoculated with PaBV (28 cockatiels) or mock inoculated (6 cockatiels) via intramuscular route. Birds were euthanized and necropsied at 5, 10, 20, 25, 30, 35, 40, 60, 80, 100, and 114 days post-inoculation (dpi). Tissue samples were collected and processed for histopathology, immunohistochemistry targeting the PaBV

N protein and reverse transcriptase polymerase chain reaction (RT-PCR) targeting the PaBV M protein gene.

Results: Clinical signs of dyspnea, lethargy and regurgitation were observed in 2 birds. Macroscopic lesions of crop and/or proventricular dilatation were observed in 4 birds. After 35 dpi, all infected birds had meningoencephalomyelitis and/or ganglioneuritis. Lymphoplasmacytic encephalitis was the most frequent histologic finding (13/19), followed by ganglioneuritis (11/19), also involving the pericardial ganglions (10/19), myelitis (9/19) and adrenalitis (9/19). PaBV-1 RNA was first detected by RT-PCR at 20 dpi at the site of inoculation (13/19), and within the CNS (12/19) in cockatiels from 35 to 114 dpi.

Conclusions: PDD was successfully reproduced in cockatiels. Lesions and virus distribution throughout the peripheral and central nervous system detected at early infection are allowing us to better understand the pathogenesis of PaBV.

E-28: PROTECTION AGAINST SHIGELLA OCULAR CHALLENGE IN PREVIOUSLY INFECTED GUINEA PIGS: POTENTIAL MODEL FOR DECIPHERING IMMUNE CORRELATES OF PROTECTION

Mark A Smith, Kevin L Hinton, Brianna N McKinney, Kristen A Clarkson, Carly R Strelez, Robert W Kaminski

Background: Both humans and macaques develop protective immunity against *Shigella* following re-infection with a homologous strain. Extension of these findings to the guinea pig keratoconjunctivitis (Sereny) model could aid in determining immune correlates of protection.

Objective: To determine whether guinea pigs challenged with *Shigella* develop protective immunity against re-challenge using the Sereny model.

Methods: Two groups of guinea pigs were infected with *Shigella* in one eye. One group was infected in the opposite naive eye 5 weeks later and the other was infected in both eyes 13 weeks later. A control group was infected in one eye only. Ocular washes were performed for IgA quantification. Postmortem examination was performed 5 days after re-infection; eyes were evaluated histologically and assigned a histologic severity score (0-5).

Results: Control animals developed severe suppurative keratoconjunctivitis with corneal ulcers and edema, and conjunctival erosions. Naive eyes from the 5 and 13 week guinea pigs demonstrated only mild to moderate keratoconjunctivitis. There were no significant differences in the specific IgA titer between naive and previously infected eyes in any group.

Conclusions: These results demonstrate that prior infection with *Shigella* elicits a protective immune response in the guinea pig Sereny model, as has been reported in macaque models and humans. The difference in disease severity between naive and previously challenged eyes upon re-challenge is likely not solely due to differences in local IgA, but perhaps also to chronic inflammatory changes to the eye that require further investigation.

E-31: HISTOPATHOLOGIC AND IMMUNOHISTOCHEMICAL CHARACTERIZATION OF VARIOLA VIRUS INFECTION IN HUMANIZED MICE

Jana M. Ritter, Christina L. Hutson, Ashley V. Kondas, Zachary D. Reed, Sharon M. Dietz-Ostergaard, Johanna S. Salzer, Nadia F. Gallardo-Romero, Heather M. Hayes, Sherif R. Zaki, Victoria A. Olson

Background: Smallpox was eradicated in 1980; however, research involving variola virus (VARV) and other poxviruses affecting humans and animals remains important for developing antivirals, improving vaccines, and maintaining biosecurity vigilance. Conventional rodent models are not permissive to VARV infection. A variety of humanized mice are available, comprising immunodeficient strains with engrafted autologous human cells and/or tissues in an effort to recapitulate the human immune system.

Objectives: We characterized the histopathology and poxviral antigen distribution in necropsy tissues from three types of humanized mice (PBMC, NSG, and BLT; Jackson Labs) up to 21 days after intranasal VARV inoculation.

Methods: Mice were housed in a CDC BSL4 laboratory (VARV is a select agent subject to select agent regulations (42 CFR part 73)), and tissues were formalin-inactivated and fixed before removal from the BSL4 lab.Fixed tissues were processed routinely and stained by H&E and immunohistochemistry using a rabbit polyclonal antibody against VARV.

Results: NSG and BLT mice had high mortality and widespread tissue necrosis with paucicellular inflammation and IHC labeling of poxviral antigen in liver, adrenal gland, and reproductive and lymphoid tissues, and bacterial sepsis in all challenged animals. Grafted human tissues in BLT mice showed abundant IHC labeling. PBMC mice had delayed mortality and overall similar findings, with less liver and bone marrow and more lung involvement, slightly more prominent inflammation, and absence of bacterial sepsis.

Conclusions: Humanized mice may be useful as models of acute, systemic smallpox infection and for the study of next generation medical countermeasures.

E-32: TEMPOROSPATIAL DISTRIBUTION OF MICROGLIAL ACTIVATION IN A MURINE MODEL OF SCRAPIE

Alyona V. Michael, Tyler A. Harm, Justin J. Greenlee, M. Heather West Greenlee, Jodi D. Smith

Mouse models of prion disease offer the advantages of genetic homogeneity and short incubation times while retaining the disease phenotype of natural mammalian hosts. Intracranial (IC) inoculation of C57BL/6 mice with a mouse-adapted scrapie strain

(RML) yields uniform incubation periods with a rapid manifestation of histopathologic changes including spongiform change, abnormal PrP accumulation, and astrogliosis. The current study aimed to further characterize neuropathologic markers of prion disease by examining temporospatial patterns of microglial activation. Weaned C57BL/6 mice were inoculated with RML brain homogenate and killed at 30-day intervals until the onset of clinical signs at 155 days post-inoculation. Immunoreactivity of ionized calcium-binding adapter molecule 1 (Iba1), a cytoplasmic microglial protein upregulated in response to neuroinflammation, was quantified across 17 brain regions within 6 standard transverse sections and compared to mock-inoculated controls. Iba-1 immunoreactivity demonstrated an increased number of microglia and a transition from resting (ramified) to active (phagocytic) morphology at 90 days post-infection that progressively increased through the terminal stages of disease. The microglial response to infection was most robust in the midbrain and thalamus, correlating to previously reported patterns of astrocyte activation in this model. Systematic characterization of microgliosis in this model suggests a role for microglial activation early in disease progression.

E-34: A CANINE MODEL OF CHRONIC GRAFT-VS.-HOST DISEASE: HISTOPATHOLOGICAL ANALYSIS

Smitha Pankajavally Somanathan Pillai

Graft versus host disease (GVHD) is a common complication of allogeneic hematopoietic stem cell transplantation (HCT). The existing murine models have been poor mimickers of human GVHD, the chronic form of which has been stubbornly refractive to treatments. The canine model of allogeneic HCT has been very productive in preclinical studies. Following the observation of pathological and clinical evidence of chronic GVHD in transplanted dogs on various protocols, a canine model of chronic GVHD was developed at Fred Hutchinson cancer research center in Seattle, WA. Clinical findings included skin ulcerations, keratoconjunctivitis sicca, rhinitis, gingivitis, elevated liver enzymes, vomiting and/or diarrhea. The most commonly affected organ was the skin with erythema progressing to dry scabs especially on the trunk and abdomen. Lichenoid lymphoplasmacytic interface dermatitis with folliculitis, sebaceous adenitis, and multifocal ulcerations was common, involving the nose, ear, paw, trunk. A Schirmer's tear test showed decreased tear production (~9mm/min compared to 19mm in control dogs). Histopathologically this correlated with lymphocytic adenitis of the lacrimal gland and conjunctivitis. Bronchiolitis obliterans, a feature highly suggestive of chronic GVHD, was present in 2 dogs. Lymphoplasmacytic infiltrates within liver portal triads were present in 5/8 dogs. Most dogs had evidence of severe esophagitis with lichenoid lymphoplasmacytic infiltrates along the mucosal-submucosal junction, adenitis, dochitis and ulceration. Multifocally within the colon, there was cryptitis and crypt loss. Salivary gland involvement ranged from unremarkable to chronic inflammation with duct obliteration and interstitial fibrosis. The lesions of chronic GVHD in the canine model recapitulate those seen in humans.

E-35: TRANSCRIPTOME ANALYSIS OF HUMAN OSTEOSARCOMA TO IDENTIFY DRIVERS OF METASTASIS IN A PATIENT-DERIVED XENOGRAFT MODEL SYSTEM

Amanda L Koehne, Leanne C Sayles, Alex G Lee, Stanley G Leung, Marcus R Breese, Aviv Spillinger, Alejandro Sweet-Cordero

Background: Osteosarcoma (OS) is the most common human primary bone tumor with approximately 400 cases diagnosed annually in the United States. Early metastatic dissemination to the lungs is a hallmark feature of OS, and complications related to metastatic disease remain the most common cause of cancer-related deaths. Despite the prevalence of metastasis in OS, the pathogenesis is poorly understood.

Objective: The aim of the study was to identify drivers of OS metastasis by comparing the transcriptome of primary site OS to OS lung metastasis in a patient-derived xenograft (PDX) model system.

Methods: 16 PDXs, derived from 10 primary site tumors and 6 metastasis, were submitted for RNA sequencing. Several candidate genes were validated in independent OS samples by qRT-PCR and immunohistochemistry. Small hairpin RNA (shRNA), CRISPR, and overexpression vectors were used to knockdown, silence, and overexpress candidate genes in cell lines and PDXs. The contribution of these candidate genes to the metastatic propensity of OS was characterized using *in vitro* assays as well as intravenous injection and orthotopic mouse models.

Results: Gene expression analysis identified a significant number of genes differentially expressed between metastatic and non-metastatic-derived PDXs. One candidate gene that was highly expressed in metastasis-derived PDXs, ENPP1, is responsible for mineralization of the ECM and has been implicated in breast cancer metastasis to bone. By manipulating ENPP1 expression in cell lines and PDXs, we have altered the metastatic phenotype of OS in mouse models.

Conclusion: Our results demonstrate that ENPP1 may be a novel driver of OS metastasis.

E-36: Hepatic lesion development in a murine model of Leishmania donovani infection

Waldo Luis L Garcia-Jimenez, Isadora Lima, Jacobo Carrisoza, Karin Seifert, Francisco Javier J Salguero

Visceral leishmaniosis (VL) is considered a potentially fatal human disease caused by the intracellular protozoan parasites *Leishmania donovani* and *Leishmania infantum* (*chagasi*). In this study, we aim to provide a better understanding of development of the typical granulomatous lesions induced in the liver together with the local hepatic immune response, at different disease stages, using a time course experimental infection in BALB/c mice. Animals were infected intravenously with a dose of 2 x 10^7 amastigotes and sacrificed at 15, 35 and 63 dpi. Histopathology and immunohistochemical (IHC) techniques for the detection of cell types (CD3, CD45, F4/80) and iNOS were used in this study.

Granulomatous lesions were identified as soon as 15 dpi in all infected animals. Three categories were used to classify granulomas (immature, mature and clear). Clear granulomas were exclusively detected from 35 dpi.

IHC for F4/80 was used to localise macrophages, including Kupffer cells, which were predominant in immature granulomas regardless of the time point (TP) analyzed. iNOS was observed mainly in the cytoplasm of fused Kupffer cells and compared with F4/80$^+$ cells, the highest expression was observed in TP2. T cells (CD3) and B cells (CD45) were predominant in more advances granuloma stages as well as the latest TP, indicating the establishment of acquired immunity.

Our results reflect the role of macrophages during the early susceptibility stage and the necessity for the establishment of a cellular response to control the disease in more advanced stages.

E-37: PANCREATIC REMODELING AND ISLET CELL MORPHOLOGY IN A FERRET MODEL OF CYSTIC FIBROSIS-RELATED DIABETES

Katherine N Gibson-Corley, Weiliang Xie, Ananta Poudel, Manami Hara, John E Engelhardt

Pancreas disease in cystic fibrosis (CF) is a significant cause of morbidity with 50% of patients developing diabetes by 30. Little is known about the pathogenesis of CF-related diabetes due to a lack of animal models that recapitulate this disease. Using a ferret model of CF, we recently demonstrated that an early phase of exocrine-mediated pancreatic inflammation significantly impairs glucose tolerance and that this “crisis” is followed by a phase of functional recovery of the endocrine pancreas. Here we have characterized in more detail morphologic changes to CF islets during these phases of altering glycemia: Phase 0 – newborn; Phase I – 9-19 days (normal glycemia); Phase II – 26-53 days (abnormal glycemia); Phase III – 68-120 days (recovery to normal glycemia). We quantified each endocrine cell type and the total islet area normalized to the pancreas area. There was no significant change in endocrine cell area in the wild-type (WT) from Phase I to III, however in the CF, there was a significant increase in the percent endocrine cell area from Phase II to III. We quantitatively analyzed islet size/shape distribution, cellular composition, circularity and diameter. While WT ferret islets showed a gradual increase in number and size of islets, an increase of large islets/clusters was prominent at Phase II and III in CF islets. We also found the ratio of beta- to alpha-cells changed in large islets during Phase I markedly in CF animals. Importantly, we found similar pancreatic pathology occurs in both humans and ferrets with CF.

E-38: TRANSMISSION OF CHRONIC WASTING DISEASE OF WHITE-TAILED DEER TO SUFFOLK SHEEP FOLLOWING INTRACRANIAL INOCULATION

Justin J Greenlee, Leisa Z Mandell, Robert A Kunkle

Background: Interspecies transmission studies are an opportunity to better understand the potential host ranges of prion diseases. Chronic wasting disease (CWD) of cervids and scrapie of sheep and goats have a similar tissue distribution of abnormal prion protein (PrP^{Sc}) and prion disease exposure across species could occur in pasture or range situations. We previously demonstrated that white-tailed deer are readily susceptible to scrapie from sheep after intracranial or oronasal inoculation, but less is known about the potential for sheep to develop CWD.

Objective: To determine the susceptibility of sheep to CWDfrom white-tailed deer after intracranial inoculation.

Methods: Suffolk lambs (n=15) were inoculated intracranially with brain homogenate from CWD-infected white-tailed deer. After inoculation, sheep were observed daily for the occurrence of clinical signs. Immunohistochemistry (IHC), enzyme linked immunosorbent assay (ELISA), and western blot were performed on brain and lymphoid tissues to assess for PrP^{Sc} accumulation.

Results: Two sheep developed clinical signs and were euthanized and necropsied at 26 and 36 months post-inoculation. PrP^{Sc} was demonstrated by ELISA and IHC in the brainstems of these sheep and one additional sheep that did not develop clinical signs. The majority of sheep were negative by IHC, ELISA and western blot of brain. PrP^{Sc} was not present in lymphoid tissues of any animal in this study.

Conclusions: CWDfrom white-tailed deer is transmissible to sheep, but with a limited attack rate and detection of abnormal prion protein in the brain only.

E-39: SYNONYMOUS CODON CHANGES INDUCE GENE SPECIFIC CHANGES IN EXPRESSION OF MEASLES (HMV) AND CANINE DISTEMPER (CDV) PROTEINS

Elizabeth W Uhl, Michelle L Osborn, Frank J Michel, Robert J Hogan

To determine the effects of codon usage bias (CUB) on morbilliviral gene expression, HMV and CDV gene sequences optimized and suboptimized to human and canine CUB were constructed and placed in plasmid expression vectors. All of the codon changes made were synonymous in that the amino acid coded for was unchanged. Protein expression from human and canine CUB optimized and sub-optimized constructs of the N, F, M, H and P/C proteins was assessed in transfected human (293 HEK) and canine (A-72) cells using fluorescence, Cellomic and, for proteins for which antibodies were available, western analysis. Sequences were also assessed for CpG/UpA dinucleotide frequencies. Results: 1) codon optimization increased, and suboptimization decreased, protein expression of HMV and CDV N and F proteins in both cell lines; 2) synonymous codon changes did not affect expression of the HMV and CDV P and C and the HMV M proteins, but both optimization and sub-optimization to canine CUB increased expression of the CDV M protein in human cells; 3) synonymous codon changes increased expression of the HMV H protein in both cell lines, but did not significantly affect CDV H protein expression; 4) although CpG/UpA frequencies were increased in some of the suboptimized sequences having low protein expression, treatment with the protein kinase inhibitor C16 did not affect protein expression. These results indicate that synonymous codon changes can alter protein expression in a variety of ways, some of which are likely to impact viral pathogenicity.

E-40: DEVELOPMENT OF A SYNGENEIC IMMUNOCOMPETENT MOUSE MODEL SYSTEM TO EVALUATE NOVEL ONCOLYTIC HERPES SIMPLEX VIRUS TYPE-1 (HSV-1) FOR THE TREATMENT OF MELANOMA

Natalie W Fowlkes, Kelli A Clemons, Ramesh Subramanian, Brent Stanfield, Konstantin G Kousoulas

Melanoma is the sixth most common cancer in the U.S. Since metastatic melanoma is often resistant to traditional chemotherapy, recent advances have aimed to harness the immune system's natural ability to eliminate cancer cells and have included immunomodulatory oncolytic viral therapy. Currently, the B16F10 syngeneic mouse model is the most widely used model for preclinical testing of melanoma therapeutics. In our laboratory, *in vitro* infection of B16F10 with our oncolytic herpesvirus OSV confirmed that B16F10 cells are at least partially resistant to HSV-1 infection, due to lacking an important entry receptor, nectin-1. Previous reports indicate that propagation of HSV-1 in baby hamster kidney (BHK) cells enhance herpesvirus entry into resistant cell types. Therefore, we performed flow cytometric analysis of OSV infected B16F10 cells and demonstrated that OSV infection was enhanced more than 5-fold when virus was propagated in BHK cells rather than Vero cells. Subsequently, a mouse melanoma model with uniform tumor growth and consistent, spontaneous lymph node metastasis was developed for *in vivo* testing. In this model, approximately 2 x 10^6 B16F10 cells are injected subcutaneously over the flank of C57BL/6 mice, harvested at 2 weeks, and processed into suspension. Then approximately 200,000 harvested cells are engrafted into pinna of additional C57BL/6 mice. Using this methodology 50% (2/4) animals necropsied at 3 weeks and 100% (6/6) animals necropsied at 4-4.5 weeks had gross evidence of metastasis in sentinel nodes. Our methods allow our current testing of oncolytic herpesviruses in a spontaneously and predictably metastasizing, immunocompetent, syngeneic mouse model.

E-41: CHARACTERIZATION OF THE SWITCH TO IRREVERSIBILITY IN MURINE MODELS OF MAMMARY GLAND INVOLUTION

Katherine Hughes, Christine J Watson

Background: Rodent models of mammary gland involution are used to study cell death in a physiological context in the adult animal and are critical to interrogating the pathogenesis of involution-associated breast cancer. Mammary gland involution has an initial reversible phase of cell death during which lactation may be restored. A subsequent transition to irreversibility, traditionally viewed as 48 hours after induction of forced involution, heralds further cell death, remodelling, and the inability to lactate.

Objective: Mice with a mammary-specific deletion of the transcription factor Stat3 have an extended period during which involution is reversible. Comparison of these mice with controls offers the opportunity to study factors controlling the switch to irreversible involution.

Methods: *In vivo* experiments were used to characterize the timing of the transition to irreversibility. Histopathology and transcriptional analysis at specific time points identified a panel of 'switch marker genes'.

Results: Commitment to irreversible involution occurs earlier than previously suggested. 'Switch marker genes' include genes associated with immune cell influx and chronic inflammation.

Conclusions: We demonstrate that there is an intimate association between the transition to irreversible involution and mammary remodelling and inflammation. Interestingly, however, other putative switch markers such as cleaved caspase 7, are actually upregulated in the absence of Stat3, suggesting that this cleavage is not critical to initiation of the second wave of cell death in the irreversible phase. We suggest that the timing of the involution 'switch point' is earlier than previously thought, which may have implications for future studies using murine involution models.

E-42: EARLY PRION TISSUE ACCUMULATION IN DEER EXPOSED TO CHRONIC WASTING DISEASE

Clare E Hoover, Kristen D Davenport, Nathaniel D Denkers, Davin M Henderson, Candace K Mathiason, Mark D Zabel, Edward A Hoover

Natural infection with infectious, chronic wasting disease (CWD) prions (PrP^{CWD}) is believed to occur through environmental exposure by mucosal aerosol or oral routes. Previous studies in white-tailed deer mimicking these exposure routes with relatively large challenge doses detected PrP^{CWD} in the retropharyngeal lymph node at 1.5 months and tonsil at 3 months using the immuodetection methods of western blotting and immunohistochemistry. However, PrP^{CWD} distribution and the chronological progression of disease during early infection remains unknown. To study the early disease pathogenesis, we challenged white-tailed deer with CWD prions by mucosal routes and serially sacrificed animals between 1 and 4 months post-exposure (MPE) to assess the tissue distribution of PrP^{CWD}. To enhance PrP^{CWD} detection at early infection time points, tissues were assayed using several amplification techniques: real-time quaking induced conversion assay (RT-QuIC) and tyramide signal amplification immunohistochemistry (TSA-IHC). Although we were unable to detect PrP^{CWD} in the immediate period following mucosal exposure, we detected RT-QuIC CWD prion seeding activity in the oropharyngeal lymphoid tissues (tonsil, retropharyngeal, mandibular, and parotid lymph nodes) at 1 and 2 MPE. At 3 and 4 MPE, PrP^{CWD} was detected in systemic lymphoid tissues and was not yet evident in central nervous system tissues. These results indicate that the earliest trans-mucosal entry by CWD prions is in the upper alimentary tract with replication in the oropharyngeal lymphoid tissues followed by rapid dissemination to systemic lymphoid tissues prior to neuroinvasion.

E-43: LACK OF NF-KB INDUCING KINASE (NIK) RESULTS IN EOSINOPHILIC ESOPHAGITIS (EOE) AND GASTRIC HYPERPLASIA IN MICE: IMPLICATIONS FOR NONCANONICAL NF-KB SIGNALING IN HUMAN EOE.
Dylan K McDaniel, Bettina Heid, Irving C Allen

The NF-κB pathway is a powerful modulator of inflammation in the gut that can proceed along two distinct arms deemed the canonical and noncanonical pathways. The canonical pathway produces a variety of classic proinflammatory mediators and is well-studied in gastrointestinal disease. However, the noncanonical pathway, a unique signaling cascade that produces a distinct set of chemokines involved in lymphoid stroma management, is undercharacterized. NF-κB inducing kinase (NIK) is a central molecule in noncanonical signaling and is essential for the production of effector molecules. Mice lacking NIK have been previously shown to develop eosinophilic inflammation in major organs such as skin, liver, and lung. However, characterization of the gastrointestinal tract of these mice has been lacking. Here we show that *Nik-/-* mice display a significant eosinophilic esophagitis that has many similar features to human eosinophilic esophagitis (EoE), including intraepithelial eosinophil accumulation and degranulation, microabscess formation, fibrosis, and basal cell hyperplasia. Additionally, these mice display significant gastric hyperplasia at the esophageal junction suggestive of chronic irritation due to reflux. Interestingly, eosinophil infiltration is localized to the esophagus and gastroesophageal junction; the caudal stomach, small intestines, and colons of these mice are unaffected, again similar to the human disease. Esophageal tissue of *Nik-/-* mice contains significantly elevated mRNA levels of IL-1β, as well as a trend towards increased TSLP, a gene associated with EoE. These findings suggest that *Nik-/-* mice may useful as a naturally occurring model of EoE and highlights a novel role for noncanonical NF-κB signaling in eosinophilic gastrointestinal disease.

E-44: IL-27 DIRECTLY ACTS ON COLONIC EPITHELIUM, INDIRECTLY ON SMALL INTESTINE VIA IMMUNE CELLS
Caroline Andrews, Scott K Durum

Background: The cytokine interleukin (IL)-27 is a promising potential therapy for inflammatory bowel disease, having shown efficacy in three models of murine colitis; however, its mechanism of action has yet to be fully characterized. Mucosally-delivered IL-27 induces phosphorylation and nuclear translocation of STAT1 (pSTAT1) in the colonic epithelium of colitic mice, suggesting that IL-27 may act on intestinal epithelial cells.

Objective: The objective of this study was to determine if IL-27 directly acts on intestinal epithelial cells, or if immune cells mediate these effects.

Methods: Small and large intestinal crypts from C57BL/6 or interferon-gamma receptor knockout mice (Ifn-gammaR$^{-/-}$) were cultured in vitro to produce intestinal organoids. Organoids were stimulated with IL-27 and/or inflammatory cytokines and endotoxin or supernatants from T cells, macrophages, or neutrophils stimulated with IL-27 and/or inflammatory cytokines and endotoxin. Immunohistochemistry or capillary western blot for pSTAT1 was used to assess organoid activation.

Results: IL-27 alone induced pSTAT1 in the cytoplasm of colonic epithelial cells after 12 hours and within the nucleus after 60 hours. In small intestinal epithelial cells, IL-27 did not induce pSTAT1 in3, 12, or 60 hours. Activated T cell supernatants induced nuclear pSTAT1 in small intestinal epithelial cells, and preliminary data suggested that IL-27 augmented this effect. Activated T cell supernatants did not induce pSTAT1 in small intestinal organoids from Ifn-gammaR$^{-/-}$ mice.

Conclusions: IL-27 directly acts on large, but not small, intestinal epithelial cells. Potential effects of IL-27 in the small intestine may be mediated by T cell production of interferon-gamma.

E-45: CHALLENGES IN DEVELOPING A GLANDERS VACCINE AND POSSIBLE ROLE OF MYELOID-DERIVED SUPPRESSOR CELLS IN A MURINE MODEL OF CHRONIC INFECTION

Tomislav Jelesijevic, Jeremy Dyke, Shawn Zimmerman, Frank Michel, Eric Lafintaine, Robert Hogan

Burkholderia mallei is a Gram- negative bacterium that causes the life-threatening zoonosis glanders. The disease is endemic in soliped populations of Asia, Africa, the Middle East and South America. Due to high mortality, history of use as a bioweapon, lack of vaccines, and paucity of antibiotic treatment options, *B. mallei* is classified as a Tier 1 Select Agent and developing countermeasures is a priority. The pathogenesis of glanders is complex and involves the extracellular/intracellular replication of *B. mallei* and dissemination to target tissues (lungs, spleen, liver, lymph nodes) where it forms chronic lesions that are difficult to treat. Efforts during the last decade to develop glanders vaccines have yielded a range of candidates including inactivated bacteria, subunit vaccines, and live attenuated strains (LAS). So far LAS provide the best protection in acute lethal infection studies, but fail to provide sterile immunity. Survivors develop pyogranulomatous lesions in target tissues and ultimately succumb to chronic infection. Efforts to understand mechanism of persistent infection have led to our discovery that myeloid-derived suppressor cells (MDSCs) are present in target organs of chronically-infected mice. Preliminary data also demonstrate a direct correlation between the extent of bacterial burden in tissues and the relative numbers of MDSCs. Given their ability to suppress innate and adaptive immune responses, we propose that MDSCs accumulate in target organs of mice surviving acute lethal infection and suppress the immune response against *B.mallei*. This, in turn, may preclude elimination of the infection and result in the development of hallmark chronic glanders lesions.

E-46: EMBOLIC PNEUMONIA WITH INTRALESIONAL BACILLI IN TWO MARMOSETS INTRANASALLY INFECTED WITH BURKHOLDERIA MALLEI

Tomislav Jelesijevic, Shawn Zimmerman, Frank Michel, Robert Hogan, Eric Lafontaine

Burkholderia mallei, a host-adapted Gram negative immotile bacterium, is the causative agent of the highly fatal zoonotic disease glanders, which primarily affects equine species. The organism is also classified as a Tier 1 Select Agent due to concerns regarding its use as a bioweapon. The most common routes of infection are respiratory and percutaneous. The clinical presentation may range from latent to acute, and bacteremia. Glanders is difficult to diagnose and if left untreated, the mortality rate is 95%. The respiratory route of infection is of particular concern with respect to the use of *B. mallei* as a biothreat agent. With this in mind, we established a marmoset model of intranasal glanders and demonstrated parallels with clinical and pathological manifestations of the disease observed in humans and equines. Here, we present gross and microscopic findings of two marmosets infected with 25,000 bacteria. Both animals developed clinical signs of disease, reached humane end points 96h post infection, and had similar histopathological presentations. Pyogranulomatous rhinitis and osteomyelitis with scattered multinucleated giant cells and intravascular cellular and bacterial emboli were found in the nasal cavity. Examination of the lungs revealed rare scattered foci of angiocentric pneumonia, while microabscesses found in both spleen and liver. All examined tissues contained intralesional bacilli that were observed in HE, modified Gram and immunostained tissue sections. Taken together, the date indicated role of bacteriemia and embolic spread of glanders in intranasally infected animals. Taken together, the data indicate bacteremia and embolic spread of *B.mallei* in intranasally infected animals.

E-47: IMMUNE RESPONSE CELLS OF THE HEART: LESSONS FROM RHESUS MACAQUES (MACACA MULATTA) ABOUT THE DISTRIBUTION OF T AND B CELLS AS WELL AS IDENTIFICATION OF SHORT- AND LONG-LIVED MACROPHAGES

Daniel Petkov, Peter Didier, Xianhong Liu, Carolina Allers, Elizabeth Didier, Andrew Lackner, Marcelo Kuroda

Background: Myocarditis is one cause of morbidity in HIV-infected individuals, but mechanisms of pathogenesis are not well understood. We previously reported that SIV affects short- and long-lived macrophages differently during progression to AIDS in macaques.

Objective: Here we analyzed the distribution of myocytes, macrophages, and lymphocytes in heart and skeletal muscle of non-SIV-infected macaques.

Methods: (Immuno)histochemistry was used to examine heart from rhesus macaques. Macrophage distribution was assessed in cardiac muscle that was deemed normal by histopathology. Tissues from animals at 1st trimester to 24 years of age were evaluated. Cell populations were detected with antibodies to CD163, CD206, HAM56, CD3, and CD20. BrdU and dextran were injected into monkeys to identify short- and long-lived macrophages, respectively.

Results: Higher cellularity was observed in right atrium and ventricles of macaques < 2 years old compared to animals > 2 years of age. Also, more cells were counted in the right atrium with a range of 164 (+ 30) to 448 (+ 58) nuclei per field (NPF) compared to ventricles with a range of 130 (+ 18) to 368 (+ 50) NPF. T and B cells were rarely found throughout heart muscle, whereas myocytes and macrophages were evenly distributed. Among macrophages, 79% were CD163+, 51% of HAM56+ cells were CD163+, and only 3% of CD163+ cells expressed CD206. More than 85% of cardiac macrophages incorporated dextran and were considered long-lived.

Conclusions: These results establish a foundation for phenotypic comparison of immune cells in heart during SIV infection to study pathogenesis of myocarditis.

E-48: Role of Toll-like receptor 3 (TLR3) in the pathogenesis of genital-tract pathology induced by Chlamydia infection in mice

Sebastian E Carrasco, Sishun Hu, Denise Imai, George E Sandusky, Frank X Yang, Wilbert A Derbigny

Chlamydia trachomatis is the leading cause of bacterial sexually transmitted infections in the United States and can progress to severe upper reproductive tract pathology and infertility in women. Understanding how host innate immune sensors interact with *Chlamydia* are important in the development of new therapeutic strategies to control *Chlamydia* infections in humans. Our early work demonstrated that *Chlamydia muridarum* (*Cm*) induces IFN-β in oviduct epithelial cells (OE) in a TLR3-dependent manner, and that cytokine and chemokine expression is diminished in *Cm*-challenge OE from TLR3$^{-/-}$ mice. We hypothesized that TLR3 signaling has a protective role against *Cm*-induced genital tract pathology in mice. *Using the Cm* model of chlamydial genital tract infection, we have examined the impact of TLR3-deficiency *in vivo*, by comparing cytokine and chemokine production, *Chlamydia* shedding and burden, CD4/CD8 profiles, and reproductive pathology between wild-type (WT) and TLR3$^{-/-}$ mice during infection. Our results showed that TLR3-deficiency results in diminished IFN-β, IL-1β, and IL-6, but enhanced IL-10, TNF-α, and IFN-γ secretion during early genital infection. TLR3$^{-/-}$ mice had increased *Cm* shedding during early and mid-genital infection. Using flow cytometry we found that TLR3$^{-/-}$ mice had a delayed recruitment of CD4^{+} T cells into the reproductive tracts during early *Cm* infection. Histopathological, immunohistochemical, qPCR studies are underway to thoroughly assess histopathological lesions, leukocyte infiltrates, bacterial burden respectively, in the genital tracts of TLR3$^{-/-}$ mice during *Cm* infection. Based on these findings, TLR3 signaling is more influential in early infection, is involved in delaying pathogen colonization and recruiting the cellular inflammatory response.

E-49: ASSESSMENT OF A NOVEL RABBIT ATHEROMA MODEL UTILIZING COMBINED DIETARY MODIFICATION, BALLOON INJURY AND ADMINISTRATION OF PRO-CALCIFIC AGENTS

John Keating, Brett Zani, Peter Markham

Background: Preclinical evaluation of endovascular therapies in calcified lesions is limited by the cost of large animal models and inadequacy of many rabbit models (e.g., induction of non-calcified atheromas).

Objective: Generate an improved rabbit model of calcified atherosclerosis using a combination of dietary modification, balloon injury and administration of pro-calcific agents.

Methods: Six New Zealand White rabbits underwent high fat/ high cholesterol diet commencing on Day -21, and application of transdermal nicotine patch through the end of study. Balloon-mediated aortic-iliac injury was performed on Day 0, with commencement of administration of calcium carbonate three times weekly and vitamins A and D once weekly. On Day 146 radiography, orbital atherectomy, euthanasia and necropsy were performed. Arterial sections were stained with H&E, Movat's pentachrome and von Kossa stains.

Results: Large arteries including aorta, iliacs and carotids were grossly enlarged/ thickened and pale, as were liver and adrenal. There were prominent histologic arterial changes, with notable luminal occlusion by marked foam cell accumulation in neointima and media. Cholesterol clefts were minimal; calcification absent to minimal; neointimal myocytes negligible; necrosis usually absent; and neointimal, medial or adventitial fibrosis not noted. Affected vessels were fragile and procedural was injury common at the time of pre-terminal atherectomy.

Conclusions: The model was successful in generating marked intimal and medial atheromatous change/ foam cell accumulation but limited by the absence of features such as calcification, necrosis and fibrosis, and vessels were prone to procedural trauma.

E-50: A PROPOSAL FOR DRONE-BASED MONITORING PROJECT TO CONTROL ENVIRONMENT FACTORS AND PRESERVING HUMAN AND ANIMAL HEALTH.

Leonardo Leonardi, Giovanni Di Guardo, Federica Piro

Background: In order to contribute to the goal of producing and using chemicals in such a way as to contain adverse effects relevant to human health and to the environment, the data collected through a drone-based monitoring system will be made available, comparable and interoperable, in order to improve their use for human health and environmental protection.

Methods: Setting and testing of territorial drone-based investigation models and methods working on "animal sentinels" for the evaluation of environmental pollutants present in selected regions. Retrieval of data for the assessment of the main environmental risk factors responsible for different diseases, with domestic and wild animals selected in the area under investigation acting as "sentinels" for the potential exposure and accumulation of environmental pollutants.

Results and conclusions: Expected results are related to the development and testing of a drone-based monitoring system working on animals as biological indicators of the various ecosystems in relation to human health, with special emphasis on the concentration of chemicals in the environment. Development and dissemination of data on the presence of specific chemicals in the areas investigated, in relation to the qualitative and quantitative assessment of the same contaminants in animal tissues and, finally, to the epidemiology of related human and animal diseases. Establishment of a territorial epidemiological map for comparatively investigating the main human and animal diseases in relation to chemicals that will be identified and quantified by means of toxicological and biomolecular investigation methods both in the environment and in biological tissues.

E-51: THERMOELASTIC STRESS STUDIES ON TRIDIMENSIONAL (3D-PRINTING) PROSTHETIC CANINE PELVIC BONE MODEL.

Leonardo Leonardi, Federica Piro, Enrico Bellezza, Roberto Marsili, Luca Mechelli, Marco Cibeca, Gianluca Rossi

L. Leonardi, F. Piro, E. Bellezza, R. Marsili, L. Mechelli, M. Cibeca, G. Rossi.

Background: 3-D printing (first named sterolithography) also call additive manufacturing is a process, described first by C.W. Hull in 1986, to produce three-dimensional objects and represent one of the most important way to produce viable tissues and bioscaffolds for reconstructive/reparative medicine and surgical procedures. Thermoelasticity stress studies are based on thermoelastic effects in solid that changes its temperature consequently to application of external forces.

Methods: In collaboration with Department of Engineering in Perugia we created a 3-D metallic lumbosacral prosthesis, for a six months male German sheperd dog with hip dysplasia and right sacral facet joint OCD.

Results: The prosthesis made with 3D print adapts perfectly to the part of bone that needs replacing because it's built for the specific patient. Other features are prosthesis lightness and engraftment ability due to its trabecular geometry. Stress distribution on the prosthesis we made has been measured by the thermoelasticity non contact measurement technique obtaining a spatial distribution of temperature changes recorded with thermocamera to study mechanical resistance, especially when it replaces bone parts stressed. The related temperature fluctuations obtianed in our study were referred to specific signal of reference, compared with the loading cycle and mechanical component.

Conclusion: The results obtained show an interesting distribution of the stress concentration of the prosthesis to conclude that this measurement methodology represent an innovative system for evaluation of points least resistance to physical stress and interface failure.

December 4, 2016
9:00 AM – 9:15 AM

GENOMIC MAPPING IN OUTBRED MICE INDICATES OVERLAP IN GENETIC SUSCEPTIBILITY FOR HZE ION AND γ-RAY INDUCED TUMORS

Elijah Edmondson, Michael M. Weil, Christina Fallgren, Debra Kamstock, Dan Gatti

Cancer risk from galactic cosmic radiation exposure is considered a potential "showstopper" for a manned mission to Mars. Calculating the actual risks that will be confronted by spaceflight crews is complicated by our limited understanding of the carcinogenic effects of high charge, high energy (HZE) ions, a radiation type for which there are no human epidemiological data. Here, we examine some of the assumptions underpinning the current NASA model used to assess space radiation cancer risk. Through a genetics approach using carcinogenesis data from heterogeneous stock (HS) mice, we found that the spectrum of tumors induced by accelerator produced HZE ions was similar to the spectra of spontaneous and γ ray-induced tumors. We mapped 51 quantitative trait loci (QTL) for 11 neoplasms with an average 95% confidence interval of 3.5 Mb. The genetic architecture for tumor phenotypes was complex, with multiple QTL explaining a small proportion of the total variance. QTL controlling susceptibilities to spontaneous, γ ray-induced, and HZE ion-induced tumors were analyzed to determine coincident loci using clustering procedures. Results indicate shared susceptibility loci for specific tumor histotypes rather than clustering based on radiation exposures, suggesting tumorigenesis mechanisms for the two radiation qualities predominantly overlap. These findings support the assumptions underlying the current model used by NASA to estimate cancer risks from space radiation exposures. In addition, because sufficiently powered lifetime carcinogenesis studies have not been previously undertaken in highly recombinant outbred mouse populations, many of the discovered QTL presented here are novel and publicly available in our dataset.

December 4, 2016
9:15 AM – 9:30 AM

OVERCOMING ERLOTINIB RESISTANCE IN HEAD AND NECK SQUAMOUS CELL CARCINOMAS BY BLOCKING INTERLEUKIN-1 SIGNALLING

Katherine N Gibson-Corley, Aditya Stanam, Nnamdi Ihejirika, Andrean L Simons

Tumor resistance to epidermal growth factor tyrosine kinase inhibitors (EGFR TKIs) such as erlotinib is a major obstacle in the success of targeted therapy in head and neck squamous cell carcinoma (HNSCC) patients. To determine the mechanism(s) of erlotinib resistance, we developed four erlotinib-resistant (ER) cell lines from erlotinib-sensitive (ES) HNSCC cell lines and compared their gene expression profiles. Enrichment analysis of microarray data revealed deregulation of the IL-1 signaling pathway in ER-HNSCC cells compared to ES-cells. In ER- cells, IL-1 pathway gene expression was upregulated by > 2 fold, and IL-1RA was significantly reduced in ER-cells compared to ES-cells. Xenografts of ER and ES-cells were performed using athymic *nu/nu* mice with endpoints including tumor morphology, immunohistochemistry, and serum cytokines (IL-1alpha, IL-1beta, IL-6, and IL-8). Blockade of IL-1 signaling using a recombinant IL-1R antagonist inhibited growth of ER-xenografts but not ES-xenografts. Also, IL-1R antagonist ± erlotinib treated ER-xenografts were less vascularized than controls. Mice carrying ER-xenografts had significantly less circulating G-CSF and IL-1N when treated with IL-1R antagonist ± erlotinib as compared to those treated with water. We then looked at levels of *IL1A* or *IL1RAP* mRNA in HNSCC patients and found that patients with high expression of both *IL-1A* and *IL1RAP* were associated with shortened survival as compared to those with low expression. Altogether, IL-1 signaling may be upregulated in ER-HNSCC cells and, therefore, its blockade overcame erlotinib resistance in HNSCC xenografts. This may represent a novel strategy to overcome EGFR inhibitor resistance for treatment of HNSCC patients.

December 4, 2016
9:30 AM – 9:45 AM

HEPATOTOXICITY ASSOCIATED WITH LOW-DOSE AZOXYMETHANE IN A MOUSE MODEL OF COLON CANCER.

Kathryn A Eaton, Sara A Poe, Chriss Vowles, Natalie Anderson, Trisha J Denike

Background: Azoxymethane (AOM) is commonly used to model colon carcinogenesis. In high doses, AOM is hepatotoxic, but colon cancer models generally use low chronic doses (10 mg/kg, presumed to be non-toxic) in association with either chemically-induced or infectious colitis. The current study was designed to study colon carcinogenesis in a model of colitis in gnotobiotic IL-10 knockout (KO) mice.

Methods: Germ-free IL-10 KO mice were monocolonized with *E. coli* followed by 6 weekly injections with 10mg/kg AOM. Mice were necropsied when they became moribund.

Results: Early deaths occurred within 4 days of an AOM injection. Mice that were affected early (after the first or second AOM injection) had severe, acute centrilobular - massive hepatocellular necrosis. Mice that were affected later in the study had both chronic and acute lesions including hepatocellular necrosis, regeneration, and disorganization of lobules. Sporadic lesions included hepatic lipidosis or multifocal neutrophilic or granulomatous hepatitis.

Conclusions: AOM is a known hepatotoxin, but acute toxicity has not been described at doses less than 20mg/kg. In this study, both acute and chronic toxicity were present in mice given only 10mg/kg. The pathogenesis of increased toxicity is not yet known, but may be attributable to mouse strain, germ-free conditions, or AOM lot. Important findings of this study are 1) that severe liver damage may occur in some mice given low dose AOM, and 2) that liver damage may occur in mouse models of colon carcinogenesis, possibly leading to unexpected experimental outcomes.

December 4, 2016
9:45 AM – 10:00 AM

SILYMARIN MAY PREVENT DOXORUBICIN-INDUCED CHRONIC CARDIOTOXICITY IN BALB/C MICE VIA ALTERING TOPOISOMERASE-II BETA GENE EXPRESSION

Emrah Ipek

Conclusion: Silymarine may prevent doxorubicin-induced chronic cardiotoxicity in balb/c mice via decreasing Top2β expression.

December 4, 2016
11:30 AM – 12:00 PM

NEONATAL AND JUVENILE OCULAR DEVELOPMENT IN SPRAGUE-DAWLEY RATS: A HISTOMORPHOLOGICAL STUDY

Vanessa Vrolyk, Ancuta Apreutese, Cedric Gordon, Roy Foster, Andrew Graham, Bernard Palate, Julius Haruna, Marie-Odile Benoit-Biancamano

Background: While the normal histomorphological features of the eye in mature rats have been described, the physiological developmental changes occurring in the eye of post-natal rats remain incompletely characterized.

Objective: To characterize the histomorphological developmental features of the eye in Sprague-Dawley rats from birth to postnatal day (PND)30, focusing on cell proliferation and apoptosis.

Methods: Both eyes were collected from 51 rats divided in 13 time points from PND1 to PND30. Hematoxylin and eosin staining was performed for histological description, while immunohistochemistry staining was performed to highlight apoptosis (caspase-3) and cell proliferation (Ki-67).

Results: During postnatal development, dramatic remodeling changes occurred within most structures of the eye, where apoptosis and cell proliferation coexist. At birth, the retina was characterised by a large neuroblastic layer presenting a prominent band of Ki-67 positive mitotic cells at the posterior aspect of this layer. A complete outer plexiform layer was present by the end of the first week of life. The formation of the inner and outer nuclear layers was accompanied by apoptotic capsase-3 positive cells. Before the opening of the eyelids, the corneal epithelium was composed of one to two cell layers which transformed into stratified squamous epithelium at PND14. The lens continuously developed by addition of new lens fibers from the lens germinal zone of the equator.

Conclusion: During the first month of life, rat's eyes undergo a rapid phase of remodeling and growth, reaching maturity by PND30. These observations will serve as a historical database useful in pediatric ocular drug development.

December 6, 2016
2:30 PM – 3:00 PM

UNRAVELING RETINOPATHY IN ALBINO HANNOVER-WISTART RATS ORALLY ADMINISTERED AFQ056: A NEGATIVE ALLOSTERIC MODULATOR of mGluR5

I.M. Pruimboom-Brees, A. Cordier, E. Tritto, V. Dubost, K. Darribat, A. Doelemeyer, D. Thiel, H. Schadt, F. Pognan, M. Schneider, S. Stefanski, T. Fawler, A. Hartmann, F. Gasparini

The retina of albino rats is very sensitive to light. Toxic retinopathy resulting from alterations in the phototransduction cascade resembles that of spontaneous light-induced cases except for its earlier onset or higher incidence or severity in chronic studies. This has been described with many neurologically active drugs and thus, establishing the mode of action is critical for human risk assessment. AFQ056 is a negative allosteric modulator of mGluR5 developed for the treatment of various neurological conditions. When orally administered to albino rats for 17 days, AFQ056 induced retinopathy in both genders at high dose but with greater severity in females. The finding was not observed in subsequent 4 and 26 weeks toxicology studies. In a 2-year carcinogenicity study, AFQ056 exacerbated the onset, incidence and severity of spontaneous retinopathy in high dose females, which had higher systemic exposures than high dose males. High margins of safety were established relative to clinical exposures. Retinopathy was not observed in dogs or mice. Time-course studies were initiated in albino rats, under various lighting conditions (darkness, normal and high light intensity) and using pharmacological and suprapharmacological doses. These studies established the kinetics of immunohistochemical (whole retinal mount), biochemical (retinal neurotransmitters), electrophysiological (ERG) and genomics events leading to photoreceptor degeneration. The data indicated that AFQ056-mediated retinopathy results from non-pharmacological Müller cell toxicity, which is only possible during the light ON phase of the circadian cycle during which nocturnal rodent species are particularly vulnerable and can be monitored by ERG (b-wave).

December 6, 2016
3:30 PM – 4:30 PM

Microbiota and Reproducibility of Animals Models

Craig L. Franklin

December 6, 2016
4:30 PM – 5:00 PM

MORPHINE TREATMENT POTENTIATES CITROBACTER RODENTIUM VIRULENCE, SYSTEMIC DISSEMINATION AND EXACERBATES GUT DYSBIOSIS IN MICE

Fuyuan Wang, Jingjing Wang, Li Zhang, Sabita Roy

Background: Opioid analgesics are frequently prescribed in the United States and worldwide. However, severe side effects such as immunosuppression and gastrointestinal symptoms limit their use. It is unclear how opioids modulate bacterial virulence and gut homeostasis. *Citrobacter rodentium* is a natural mouse pathogen that models intestinal infection by enteropathogenic Escherichia coli (EPEC) and enterohemorrhagic E. coli (EHEC) in humans.

Objective: Here, by using a mouse-model of *C. rodentium* infection, we determined effects of morphine on gut homeostasis and host resistance against bacterial infection.

Methods: C57/BL6J wild-type mice were implanted with 25mg morphine or placebo pellet subcutaneously. Mice were infected with *C. rodentium* through oral gavage at day 1 post morphine treatment. The *C. rodentium* virulence factors, systemic dissemination, gut microbiome and host immune response were detected at day 5 post infection.

Results: In a mouse model of *Citrobacter rodentium* infection, morphine treatment resulted in 1) the promotion of *C. rodentium* systemic dissemination, 2) increase in virulence factors expression with *C. rodentium* colonization in intestinal contents, 3) altered gut microbiome, 4) damaged integrity of gut epithelial barrier function, 5) inhibition of *C. rodentium*-induced increase of goblet cells, and 6) dysregulated IL-17A immune response.

Conclusion: This is the first study to demonstrate that morphine promotes *C. rodentium* infection, indicating morphine modulates virulence factor-mediated adhesion of pathogenic bacteria and induces gut dysbiosis and disruption of host defense. This study reveals over-prescription of opioids may increase the risk of infections in the emergence of pathogenic strains and should be used cautiously.

Industrial and Toxicologic Pathology Posters

T-01: QUANTITATION OF BIOMARKERS USING FLOW CYTOMETRY IN CLINICAL PATHOLOGY: COMPARATIVE SENSITIVITY AND RELIABILITY OF MULTI-ANALYTES PLATFORM AND MULTIPLEX CYTOMETRY IN THE QUANTITATION OF CIRCULATING CYTOKINES

Alain Stricker-Krongrad, Catherine Shoemake, Miao Zhong, Jason Liu, Guy Bouchard

Background: Circulating Th1/Th2/Th17 cytokines IL-2, IFN-g, TNF-a, IL-4, IL-6, IL-10, and IL-17a become elevated in response to inflammation and are key regulators of immune responses. Measuring the expression profiles of cytokines is important in monitoring polarization of the immune response; results should be independent of the quantitation methods if they are to be accepted as validated clinical pathology biomarkers.

Objective: The aim of this study was to evaluate the effect of quantitation methods on the detection of biomarkers of inflammation.

Methods: Female C57BL6 mice were treated orally with vehicle or dexamethasone, then challenged with lipopolysaccharide (LPS) IV 1-1.5 hours later. At 0.5, 1, 2, 4, and 6 hours after LPS challenge, blood samples were collected and plasma was analyzed with the Rodent MAP V3.0 Antigens assay or the Mouse Cytokine Panels A & B assay (Myriad-RBM; Luminex platform), and the Mouse Th1/Th2/Th17 cytokines (Becton-Dickinson: BD CBA) assay on a BD Accuri C6 flow cytometer platform. Results were compared between assays.

Results: Reproducible quantitation of circulating TNF-a and IL-6 levels were obtained with assays from both Myriad-RBM and BD Biosciences. The BD CBA cytokine assay was not as sensitive as the Myriad-RBM assays in detecting and quantitating circulating IL-2 and IL-4 and IL-17A levels, but was more sensitive and reliable in measuring circulating IFN-g levels. Reliable circulating IL-4 measurements were not achieved by either assay.

Conclusions: Quantitation of circulating biomarkers of inflammation can be achieved using multiplexed flow cytometry, but careful considerations have to be made for the validation of assays.

T-02: DISCOVERY OF ETHYL UREA DERIVATIVES AS INHIBITORS OF HUMAN ISLET AMYLOID POLYPEPTIDE FIBRILLIZATION AND CYTOTOXICITY

Jessica S. Fortin, Marie-Odile B.-Biancamano, René C.-Gaudreault

The pathological process amyloidosis, in which misfolded proteins form toxic oligomers and fibrils, occurs in more than 25 diseases, including type 2 diabetes. Islet amyloid is made up of human islet amyloid polypeptide (hIAPP) and occurs in almost all patients with type 2 diabetes. Accumulation of islet amyloid deposits develops over time and contributes significantly to the progression of the disease. A number of naturally occurring products bearing polyphenol moieties have been shown to inhibit hIAPP fibrillization. Generally, those molecules lack selectivity and exhibit poor bioavailability, thus limiting their clinical application. Further discovery of new molecules is needed to create a diverse set of anti-amyloid compounds with appropriate pharmaceutic properties. We have recently designed and synthesized *N*-phenyl-*N'*-(2-ethyl) ureas (EUs) that are non-cytotoxic small molecules to evaluate the role of the aryl-substituted moiety on the inhibition of hIAPP fibrillization. Thirty EUs were tested *in vitro* for their anti-amyloidogenic activity using the fluorometric ThT assay. Transmission electron microscopy (TEM) was used to visualise amyloid fibrils and qualitatively compare between potent molecules and controls. Cell survival assays in pancreatic MIN-6 cells were also conducted. The findings of the present study demonstrate that compound EU-362 is able to significantly inhibit the formation of hIAPP fibrils and protect cells from amyloid cytotoxic effects. Our results suggest that increasing the nucleophilic potency of the aryl moiety, as demonstrated in EU-362, significantly enhances the anti-amyloidogenic activity of the molecules. This structure-activity relationship study of EU may present a new opportunity to develop inhibitors of pancreatic amyloidosis.

T-03: TARGETED DELIVERY OF SIRNA THERAPEUTICS: ASGPR TISSUE EXPRESSION EVALUATION IN RAT, MOUSE AND NON-HUMAN PRIMATE

Brenda A Carito, Carole E Harbison, Maja M Janas, Natalie D Keirstead

Background: The asialoglycoprotein receptor (ASGPR) mediates cellular uptake of terminal galactose- or*N*-acetylgalactosamine (GalNAc)-containing glycoproteins. Physiologic roles for the ASGPR include hepatic clearance of apoptotic cell remnants and a wide variety of desialylated proteins from circulation, such as immunoglobulins and transferrin. In addition to its physiologic functions, the relative hepatic specificity of ASGPR expression has been exploited for delivery of GalNAc conjugated oligonucleotide therapeutics to the liver. In species commonly used for preclinical drug safety testing, ASGPR expression in non-hepatic tissues has been inconsistently reported, which may have implications for systemic distribution of GalNAc-conjugated oligonucleotide therapies.

Methods: The goals of this study were to analyze systemic ASGPR protein expression in healthy control CD-1 mice, Sprague Dawley rats, and cynomolgus macaques by immunohistochemistry (IHC), and to determine the age related onset of expression in the rat fetus.

Results:ASGPR1 IHC labeling was observed in hepatocytes and in scattered interstitial cells, most likely of macrophage/dendritic cell origin, in all organs tested. Expression was detected in the rat fetal liver after gestation Day 16, and this was comparable to the adult. No labeling was observed in the parenchyma of the kidneys, thyroid, brain, salivary glands, or testis in any species. qPCR was performed to confirm ASGPR expression patterns and identified high levels of ASGPR mRNA in the liver, but low levels in other organs, suggesting the ASGPR1 antibody may have cross-reactivity with other closely related lectins. This study supports the use of ASGPR for highly specific targeting of GalNAc-siRNA conjugates to the liver.

T-04: NEONATAL AND JUVENILE OCULAR DEVELOPMENT IN SPRAGUE-DAWLEY RATS: A HISTOMORPHOLOGICAL STUDY

Vanessa Vrolyk, Ancuta Apreutese, Cedric Gordon, Roy Foster, Andrew Graham, Bernard Palate, Julius Haruna, Marie-Odile Benoit-Biancamano

Background: While the normal histomorphological features of the eye in mature rats have been described, the physiological developmental changes occurring in the eye of post-natal rats remain incompletely characterized.

Objective: To characterize the histomorphological developmental features of the eye in Sprague-Dawley rats from birth to postnatal day (PND)30, focusing on cell proliferation and apoptosis.

Methods: Both eyes were collected from 51 rats divided in 13 time points from PND1 to PND30. Hematoxylin and eosin staining was performed for histological description, while immunohistochemistry staining was performed to highlight apoptosis (caspase-3) and cell proliferation (Ki-67).

Results: During postnatal development, dramatic remodeling changes occurred within most structures of the eye, where apoptosis and cell proliferation coexist. At birth, the retina was characterised by a large neuroblastic layer presenting a prominent band of Ki-67 positive mitotic cells at the posterior aspect of this layer. A complete outer plexiform layer was present by the end of the first week of life. The formation of the inner and outer nuclear layers was accompanied by apoptotic capsase-3 positive cells. Before the opening of the eyelids, the corneal epithelium was composed of one to two cell layers which transformed into stratified squamous epithelium at PND14. The lens continuously developed by addition of new lens fibers from the lens germinal zone of the equator.

Conclusion: During the first month of life, rat's eyes undergo a rapid phase of remodeling and growth, reaching maturity by PND30. These observations will serve as a historical database useful in pediatric ocular drug development.

T-05: ZEBRAFISH AS A PRECLINICAL TOXICOLOGY MODEL
Steve Cassar, Bruce E LeRoy

Zebrafish are a vertebrate model with established and emerging applications for toxicology and drug discovery. Their fecundity, small size, rapid development, and transparency through organogenesis allow for rapid in vivo, non-invasive assessment of experimental molecules' effects on multiple organs using very little test substance. At Abbvie, a dedicated zebrafish toxicology laboratory was recently established and currently delivers several fit-for-purpose assays. Testing is informed by prospective target safety assessments or early toxicity findings. Decisions to use fish are made when translatable biology can be predicted and toxicity endpoints can be assessed more efficiently compared with mammalian models. Differences between zebrafish and human will undoubtedly result in a lower number of translatable findings across all physiology, but by interrogating appropriate systems, we can increase the probability of translation. Here we describe five toxicity assays using larval zebrafish at different times between 3 and 7 days post fertilization. Our novel gastrointestinal assay measures transit rate of fluorescently-labeled food using a plate spectrophotometer. Heart rates are measured from videos of beating hearts in un-anesthetized fish. A behavioral tracking system is used to quantify movement under disparate conditions to assess potential seizure liability, ocular toxicity, or locomotor effects. Based on performances of reference drugs in these assays, zebrafish positive predictive values range from 70% to 100%. Further assay metrics and details are provided.

T-06: CARDIO-RESPIRATORY DEVELOPMENT IN JUVENILE SPRAGUE-DAWLEY RATS: A HISTOMORPHOLOGIC AND IMMUNOHISTOCHEMICAL STUDY

Vanessa Vrolyk, Ancuta Apreutese, Cedric Gordon, Roy Foster, Andrew Graham, Bernard Palate, Julius Haruna, Marie-Odile Benoit-Biancamano

Background: To adequately identify drug-related lesions affecting the cardio-respiratory system in juvenile toxicity studies, it is essential to recognize correctly the post-natal physiological developmental remodeling events.

Objective: To describe the histomorphological events occurring in the cardio-respiratory system of Sprague-Dawley rats from birth to postnatal day (PND) 30, focusing on cell proliferation and apoptosis.

Methods: 51 Sprague-Dawley rat from 6 time-mated rat dams Crl:CD(SD) were divided into 13 time points from PND1 to PND30. Body and heart weights were recorded, and the aorta, heart, trachea and lungs were fixed with 10% buffered formalin. Hematoxylin and eosin was used for histological evaluation, while immunohistochemistry staining for vimentin, caspase-3, and Ki-67 or PHH3 was performed to identify mesenchymal cells, apoptosis and cell proliferation, respectively.

Results: At birth, the aortic wall was hypercellular with rare elastic fibers compared to PND30, where thick elastic fibers developed. Perinatally, the myocardium was hypercellular containing Ki-67-positive primitive cardiomyocytes characterized by poorly developed myofibrils and rare cross-striations. Marked expression of caspase-3 was seen in immature endothelial cells of the endocardium. The trachea of neonatal rats was lined by an immature columnar epithelium with no goblet cells. Furthermore, the lung parenchyma was composed of "primary saccules" delineated by a double capillary network, while at PND10, alveolar septa were markedly thickened by proliferating vimentin-positive interstitial cells. Throughout the first 30 PNDs, alveolar septa remodeling involved both cell proliferation and apoptosis.

Conclusion: The cardio-respiratory system of Sprague-Dawley rats is partially developed at birth but develops rapidly during the first month of age.

T-07: URO-GENITAL DEVELOPMENT IN JUVENILE SPRAGUE-DAWLEY RATS: A HISTOMORPHOLOGIC AND IMMUNOHISTOCHEMICAL STUDY

Vanessa Vrolyk, Ancuta Apreutese, Cedric Gordon, Roy Foster, Andrew Graham, Bernard Palate, Julius Haruna, Marie-Odile Benoit-Biancamano

Background: Preclinical juvenile toxicity studies are crucial and require a better understanding of the remodeling changes occurring in the uro-genital system of post-natal rats.

Objective: To describe the histomorphological events occurring in the uro-genital system of Sprague-Dawley rats from birth to postnatal day (PND) 30, focusing on cell proliferation and apoptosis.

Methods: 51 Sprague-Dawley rat pups from 6 time-mated rat dams Crl:CD(SD) were divided into 13 time points from PND1 to PND30. Body, testes and kidneys weights were recorded, and kidneys, ovaries and testes were fixed in 10% buffered formalin. Hematoxylin and eosin staining was used for the histological evaluation and immunohistochemistry staining was performed to highlight apoptosis (caspase-3) and cell proliferation (Ki-67).

Results: The kidney grew at a maximal rate in the first 10 PNDs. At birth, the renal subcapsular nephrogenic zone included different stages of developing nephrons, with abundant Ki-67-positive cells. Developing renal tubular epithelial cells in the cortex had intranuclear expression of caspase-3, and frequent caspase-3-positive apoptotic bodies were present within the kidney papilla. Perinatally,the ovarian parenchyma included numerous oocyte nests, characterized by clusters of germ cells, and a clear cortico-medullary demarcation at PND26. At birth, seminiferous tubular epithelium was composed of one to two layers of spermatogonia and Sertoli cells, including frequent Ki-67 positive mitotic cells. The first round spermatids were identified at PND26.

Conclusion: The uro-genital system of Sprague-Dawley rats is partially developed at birth but develops rapidly in the first month of age through active cell proliferation and cell death.

T-08: EXPERIMENTAL KIDNEY INJURY IN CATS INDUCED BY A HIGH DOSAGE REGIMEN OF MELOXICAM: A WELL CONTROLLED-RANDOMIZED PROSPECTIVE STUDY.

Pablo E Piñeyro, Nicolas F Villarino

The administration of non-steroidal anti-inflammatory drugs (NSAIDs) in cats causes acute kidney injury (AKI) within a very short period of time by mechanisms that remain poorly understood. This severely limits the possibility to provide optimal treatment for a variety of conditions that require long-term administration of NSAIDs, such as cancer. An *in-vivo* model of AKI in cats will help unveil the mechanism underlying their predisposition to NSAID-induced AKI and open the way to substantial advances in the treatment of inflammation and pain in these animals. The *objective* of this study was to induce AKI in catsusing a high dose regimen of meloxicam.*Methods:*In order to achieve this objective, eight female cats were allocated randomly to two experimental groups: control (n=4;

saline) and meloxicam (n=4).

Cats in the meloxicam group were treated with 0.3 mg meloxicam/kg subcutaneously every 24 hrs until cats showed biochemical changes (serum creatinine, specific urine gravity, urine protein to creatinine ratio, etc.) associated with AKI. All animals were euthanized within 24hs after the last treatment administration.Samples from both kidneys were collected for histopathological evaluation. *Results:*Cats in the meloxicam group had abnormal biochemical markers of kidney function and integrity within 18 days of treatment. In contrast to cats in the control group, all cats in the meloxicam group had severe kidney changes associated with acute injury, including glomerular and tubular damage and interstitial lymphocytic infiltration. *Conclusion:*The administration of meloxicam could be used to develop an experimental model of NSAID-induced kidney injury in cats.

T-09: LIGHT-INDUCED RETINAL DEGENERATION IN SPRAGUE-DAWLEY RATS IN A DRUG SAFETY STUDY AT A NONCLINICAL RESEARCH LABORATORY

Laura Zwick, Joshua Bartoe, Dale Cooper, Jay Albretsen, Daniel Patrick

Retinal degeneration from intense or prolonged exposure to light can occur in albino rats and has the potential to confound drug safety evaluation. During a 26-week oral toxicity study of a pharmaceutical test article in Sprague-Dawley rats, ophthalmoscopic examination prior to termination revealed bilateral retinal atrophy in four rats treated with the test article and two with the water vehicle. All affected rats were housed in the uppermost cages. Histopathologic examination of the eyes was limited to rats in the control and high dose groups. Microscopic lesions consistent with retinal phototoxicity were present in the eyes of nine rats in the control group and two rats in the high dose group, including the three rats in these groups that had ophthalmoscopic lesions. All but one of the affected animals was housed in the top row of cages indicating a relationship of the retinal lesions to the intensity of the light exposure. Retinal degeneration was of minimal to moderate severity, typically bilateral, and involved the central retina bordering the optic disc. Affected portions of the retina were hypocellular and exhibited variable loss of rod and cone photoreceptor processes, disorganization of the inner and outer nuclear layers, and narrowing or absence of the plexiform layers. Light-induced retinal degeneration is uncommon at this facility and was most likely due to the maximal light intensity exceeding the recommended range. Additional in-house studies are currently investigating the pathophysiology of light-induced retinal lesions in rats and evaluating the efficacy of multiple preventative measures.

Natural Disease Focused Scientific Session I

December 4, 2016
8:00 AM – 8:15 AM

MORBILLIVIRUS ASSOCIATED LIPID PNEUMONIA IN ARCTIC FOXES (VULPES LAGOPUS)

David Rotstein, Raphaela Stimmelmayr, Susan Sanchez, Greta Krafsur, Brian Person

Complete necropsies and sampling was done for 24 arctic foxes (Vulpes lagopus) (2012) as part of an ongoing long-term North Slope Borough Department of Wildlife Management arctic fox health assessment project initiated in 2008. Histopathologic findings included lipid pneumonia in 33% (n = 8) of the foxes of which lymphoid depletion was observed in half. Rare suspect syncytial cells were observed within the lung and lymph node of two foxes. GMS, AFB, and Gram stained sections of lung were negative. CDV immunohistochemical staining was conducted on sections from 7 arctic foxes including two cases with no lipid pneumonia or lymphoid depletion. Immunohistochemistry yielded positive staining in the lung, lymph node, spleen, adipose, and renal pelvic urothelial cells. In arctic foxes, morbillivirus infection has not been well-described except for a serosurvey in Norway. In the current study, there were no observable clinical signs and death was not illness-associated. Morbillivirus in this wild population may be subclinical with more severely affected animals not observed. Additional epidemiological, pathological, and molecular analyses are needed to better understand the sylvatic cycle.

December 4, 2016
8:15 AM – 8:30 AM

FATAL INTERSTITIAL PNEUMONIA IN CAPTIVE TIGERS AND LIONS NATURALLY INFECTED WITH CANINE DISTEMPER VIRUS – FROM A CASE REPORT TO AN OUTBREAK

Mainity Batista Linhares, Karen A. Terio, Herbert E. Whiteley, Jonathan P. Samuelson, Debbie L. Cassout, Fred Froderman, Shih-Hsuan Hsiao, Patrick J. Roady, Adam W. Stern, Ian T. Sprandel, David A. Coleman

Results: Affected felids had an interstitial pneumonia with type II pneumocyte hyperplasia, syncytia, and both intranuclear and intracytoplasmic viral inclusion bodies. One tiger had concurrent histoplasmosis and a second had toxoplasmosis. Inclusion bodies were evident in multiple organs, but related inflammation was primarily in the lungs. Nervous system involvement was rare and minimal. Hemagglutinin gene sequence was most similar (>96%) to those of CDV strains reported in wildlife from the American Midwest.

Conclusion: Results of this investigation reconfirm the susceptibility of felids to Canine Distemper Virus. Lesions in this current outbreak were primarily pulmonary, in contrast to some previous outbreaks in felids which had concurrent or primary neurologic lesions and signs.

December 4, 2016
8:30 AM – 8:45 AM

EQUID HERPESVIRUS TYPE 1-INDUCED PLATELET ACTIVATION IS INHIBITED BY LOW MOLECULAR WEIGHT HEPARIN EX VIVO

Priscila B da Silva Serpa, Tracy Stokol

Background: Abortion and myeloencephalopathy in horses infected with equid herpesvirus type 1 (EHV-1) has been partly attributed to ischemia from vascular thrombosis. We have found that EHV-1 activates platelets through virus-associated tissue factor triggering thrombin generation. Inhibition of platelet activation may reduce thrombosis in infected horses. We hypothesized that low molecular weight heparin (LMWH) would inhibit EHV-1-induced platelet activation. To test this, we evaluated the *ex vivo* effect of LMWH on P-selectin upregulation on platelets exposed to EHV-1.

Methods: Platelet-rich plasma was derived from citrated venous blood of healthy horses by low speed centrifugation. Platelets (1 x 10^6 cells/mL) were exposed to abortigenic (RacL11) and neuropathogenic (Ab4) EHV-1 strains at 1plaque forming unit/cell for 10 minutes at 37°C after *ex vivo* treatment with LMWH at 0.1, 0.5, 1, 2.5 and 5 µg/mL.

Controls included thrombin (0.15 U/mL) and heparin vehicle. Platelet activation was assessed by flow cytometric detection of surface P-selectin. Median percentage of positive cells was compared between vehicle- and LMWH-treated cells with a Wilcoxon sign rank test and Bonferroni adjustment. Anti-factor Xa activity was measured in platelet-poor plasma spiked with LMWH to confirm an inhibitory effect.

Results: LMWH inhibited platelet activation by thrombin, RacL11 and Ab4 in a dose-dependent manner, with complete suppression occurring at a dose of 2.5 µg/mL ($p<0.01$). LMWH yielded detectable anti-factor Xa activities (0.2-0.3 U/mL) at 5 µg/mL.

Conclusion: LMWH inhibits platelet activation by EHV-1 and holds promise as a potential adjunctive treatment to limit thrombosis in EHV-1-infected horses.

December 4, 2016
8:45 AM – 9:00 AM

CHARACTERIZATION OF NATURALLY ACQUIRED CANINE HERPESVIRUS-ASSOCIATED ENCEPHALITIS

Mason C. Jager, Erica A. Sloma, Morgan T. Shelton, Andrew D. Miller

Canine herpesvirus-1 (CHV) is an α-herpesvirus of canids with a global distribution that causes significant mortality in neonatal puppies. While histologic lesions of brains in experimentally infected puppies were described in the 1970s, characterization of naturally acquired infections is limited. The aim of this study was to describe the histologic, immunohistochemical, and in-situ hybridization features of canine herpesvirus-1 encephalitis. Six female and eleven male puppies ranging in age from stillborn to 57 days old were studied. Histologically, lesions in (14/17) cases were characterized by multifocal glial nodules, neuronal satellitosis, and cerebellar cortical necrosis; however, significant inflammation was not a primary feature in any of the cases. Immunohistochemistry for Iba1, CD3, and CD20 was performed and in situ hybridization for canine herpesvirus-1 was performed to determine localization of viral antigen. In all cases, the glial nodules were immunoreactive for Iba1 consistent with reactive microglia with cases having immunoreactivity for CD3 (13/17) and CD20 (3/17). In situ hybridization for canine herpesvirus revealed extensive viral antigenicity in the granular neurons of the cerebellar folia (8/9) and multifocally within endothelial cells in the meninges (11/14). Other tissues with positive hybridization included neurons and endothelial cells in the cerebrum (5/15), brainstem (4/9), hippocampus (1/2), and thalamus (2/5). These results further clarify the cellular tropism of canine herpesviral infection and suggest that developing cerebellar granular neurons are an important site of viral replication. Extensive recruitment of inflammatory cells is not a prominent feature of the disease, but rather the lesions are defined by proliferating microglia.

December 4, 2016
9:00 AM – 9:15 AM

AN UNUSUAL LARGE CETACEAN STRANDING EVENT ALONG THE PACIFIC COAST OF BRITISH COLUMBIA, CANADA

Heindrich N. Snyman, Paul Cottrell, Kathleen A. Burek-Huntington, Kate Savage, Kathi Lefebvre, Lisa Spavin, Deborah Fauquier, Stephen Raverty

From April 2015 to May 2016 there was a sudden unprecedented increase in large whale strandings along the northeastern Pacific. There were 17 dead whales, including seven humpback whales (*Megaptera novaeangliae*); five fin whales (*Balaenoptera physalus*) (three simultaneously stranded on a beach); three grey whales (*Escherichtius robustus*); one sperm whale (*Physeter macrocephalus*); and a minke whale (*Balaenoptera acutorostrata*). These stranding coincided with an anomalous warming of water (blob) with harmful algal blooms (HAB's) in the region. An Unusual Mortality Event (UME) was simultaneously declared by the US National Marine Fisheries Service following similar reports in Alaska. Necropsies were performed on eight of the affected whales with tissues collected for histopathology and ancillary diagnostic testing. For some whales, carcasses were inaccessible, presented in advanced autolysis or field conditions were unsafe and precluded necropsy. For some of these cases, sampling was restricted to photo documentation and external morphometric evaluation with limited tissue collection. Three whales presented with evidence of ship strike and microscopically, fat emboli were identified in the pulmonary microvasculature along with degenerative changes in sections of the skeletal muscle. Domoic acid (ng/ml) and Saxitoxin (ng/g) were detected by ELISA within samples of colonic contents of six whales, suggesting a potential role for HAB's in this unusual stranding event. At present, there is little known regarding the impact of detected levels of domoic acid or paralytic shellfish poisoning on cetaceans and investigations are ongoing to better characterize the impact of HAB's on the health of these iconic marine mammals.

December 4, 2016
9:15 AM – 9:30 AM

NANNIZZIOPSIACEAE SPP. INFECTION IN AQUATIC CHELONIANS

Daniel B Woodburn, Matthew C Allender, Carol W Maddox, Andrew N Miller, Katherine Haman, Jennifer N Langan, Caryn P Poll, Stefanie Bergh, Drury R Reavill, Karen A. Terio

Background: The family Nannizziopsiaceae comprises significant reptile pathogens including *Ophidiomyces ophiodiicola* and *Nannizziopsis vriesii*. While these fungi are well-documented pathogens of lizards, snakes, and crocodilians, there are no known reports of disease in turtles. However, a novel presentation of shell disease has recently been recognized in free-ranging and captive aquatic chelonians associated with fungi that are morphologically and genetically consistent with Nannizziopsiaceae spp.

Objective: The aim of this study was to describe the histologic lesions of Nannizziopsiaceae spp. shell infections and characterize the fungal organism.

Methods: A retrospective review of chelonian cases submitted to the Zoological Pathology Program identified 38 individuals with shell lesions and histologic sections available for review. Of these, 18 individuals had similar lesions and intralesional fungi compatible with Nannizziopsiaceae spp. Infection was confirmed via PCR and culture in a subset of these cases. Cultured isolates were characterized by morphology and sequence analysis.

Results: Histologic lesions associated with Nannizziopsiaceae spp. infection were unique and were similar across the 18 affected individuals. The lesions consisted of necrotizing dermatitis and osteomyelitis with dermal inclusion cysts lined by keratinized squamous epithelium and containing morphologically compatible fungal hyphae. Sequence data obtained from frozen tissue samples and phylogenetic analysis of the fungal ITS region indicated that fungi infecting chelonians comprise a monophyletic clade within the Nannizziopsiaceae distinct from species infecting other reptiles.

Conclusion: Nannizziopsiaceae spp. infection results in characteristic shell lesions in aquatic chelonians, and the causative agent is distinct from other related reptile pathogens. Additional morphologic and molecular characterization is ongoing.

December 4, 2016
9:30 AM – 9:45 AM

BARTONELLA INFECTION AS A POSSIBLE CAUSE OR COFACTOR OF FELINE ENDOMYOCARDITIS

Taryn A Donovan, Nandhakumar Balakrishnan, Philip R Fox, Iago C Barbosa, Taylor McCoy, Ed B Breitschwerdt

Background: Feline endomyocarditis (FEMC) is an important idiopathic cause of acute death in domestic cats that is most often diagnosed at postmortem examination. *Bartonella* spp. have been sporadically reported in humans with myocarditis and sudden death. These Gram negative, intracellular, vascuolotropic, alpha proteobacteria are a well-recognized cause of culture-negative endocarditis in humans and animals.

Methods: Archived formalin fixed, paraffin embedded (FFPE) feline heart tissues from 66 cats were divided into three groups on the basis of histopathologic findings. Group 1 (n=41) had a histologic diagnosis of endomyocarditis, Group 2 (n=12) had a gross and histologic diagnosis of hypertrophic cardiomyopathy, and Group 3 (n=13) had grossly and histologically normal hearts. DNA extracted from FFPE tissues were tested with *Bartonella* genus, *B. koehlerae* species-specific PCR and *Bartonella vinsonii berkhoffii*> subspecies-specific PCR using multiple primer sets targeting *Bartonella* 16-23S ITS elements. Special precautions were taken to minimize cross contamination.

Results: In group 1, twenty cats (49%) were positive for *Bartonella spp*. DNA sequencing revealed that 10 cats were infected with 1 *Bartonella spp* and the remaining 10 were infected with more than one *Bartonella spp*. Within group 2, only one cat (8%) was positive for *Bartonella koehlerae*. In group 3, none of the 13 cats were PCR positive for either *Bartonella* genus or species-specific 16-23 ITS elements PCR.

Conclusions: These results suggest that *Bartonella* spp. is a possible cause or cofactor of FEMC. More work is needed to clarify this relationship and support causation.

December 4, 2016
9:45 AM – 10:00 AM

MULTI-CENTRIC ROUND CELL NEOPLASMS OF WILD TURKEYS (MELEAGRIS GALLOPAVO)

Kevin D Niedringhaus, Heather MA Fenton, Justin D Brown

Multiple potentially oncogenic viruses including lymphoproliferative disease virus (LPDV) and reticuloendotheliosis virus (REV) have been detected in wild turkeys (*Meleagris gallopavo*). However, infection with these viruses appears to be more common than the presence of lymphoproliferative infiltrates. No studies have directly addressed the incidence of multi-centric neoplasms in wild turkeys. Clinical case records from wild turkeys submitted to the Southeastern Cooperative Wildlife Disease Study from 1975-2015 were reviewed for cases diagnosed with neoplastic lesions. Neoplasia was reported in 35 out of 751 wild turkey cases submitted (4.7%), which is approximately five times what has been observed in other wild bird species. Retrovirus testing resulted in no viruses being detected (n=4), LPDV detection (n=20), REV detection (n=3), and both LPDV and REV detection (n=1). Cases originated from fourteen states, and no sex or age bias was observed. The most common gross lesions were poor physical condition (n=16), multifocal nodules in the skin (n=13), nodules in the liver and/or spleen (n=14), and splenomegaly (n=8). The neoplastic cellular infiltrates were predominately pleomorphic round cells with large eccentric nuclei and prominent nucleoli resembling lymphocytes or lymphoblasts (n=34), except for one case that appeared to be of myeloid cell origin. This study suggests that neoplastic diseases are a rare cause of morbidity and mortality in wild turkeys, emphasizes the importance of histology and ancillary testing in suspect cases, and highlights the need for further studies to better understand the significance of retrovirus infection in neoplasms of wild turkeys.

December 4, 2016
10:30 AM – 11:30 AM

Update on the Epidemiology of High Pathogenicity Avian Influenza in Wild Birds

Thomas DeLiberto

December 4, 2016
11:30 AM – 11:45 AM

PERIVENOUS LIPID DEPOSITION AND PHLEBITIS PRECEDE PECTORAL MYOFIBER DEGENERATION IN WOODEN BREAST MYOPATHY DEVELOPMENT IN BROILER CHICKENS

Erin M. Brannick, Michael P. Babak, Marie F. Mutryn, Behnam Abasht

Wooden Breast Disease, a myopathy emerging in commercial broiler chickens worldwide, is characterized by development of palpable firmness and irregular angularity within pectoral muscle. Previous descriptions of Wooden Breast lesions relied on tissues from birds at market age (5-7 weeks old). By this advanced stage, myofiber degeneration and necrosis, myositis, phlebitis, and edema are present alongside reparative lesions of myofiber regeneration and fibrosis. Thus, it is not possible to determine the order of lesion development to identify a potential cause or progression (pathogenesis) for the disease. To determine the early pathogenesis of Wooden Breast, 350 male broiler chickens from a lineage known to develop the myopathy were raised under commercial conditions with weekly biopsies and necropsies performed on subsets of the population from 1 to 7 weeks of age. Pectoral muscle specimens were evaluated by gross and histologic analysis, electron microscopy, and RNA sequencing for Wooden Breast lesions. Histologic lesions were present at all time points with gross lesions detectable by 3 weeks of age. At 1 week, phlebitis and perivenous lipid accumulation were first observed with related venous congestion and edema. Myofiber degeneration, necrosis, and myositis began at 2 weeks, myofiber regeneration at 3 weeks, and fibrosis at 4 weeks of age, with progression of lesion severity over time. In conclusion, venous disturbances and/or impaired lipid metabolism in pectoral muscle may trigger subsequent development of myopathic lesions leading to clinically detectable Wooden Breast Disease. Meat quality may also be affected at a cellular level even in clinically normal animals.

December 4, 2016
11:45 AM – 12:00 PM

ASSESSMENT OF THE VALUE OF GANGLION CYTOLOGY FOR THE RAPID POST MORTEM DIAGNOSIS OF EQUINE GRASS SICKNESS

Chiara Piccinelli, R.C. Jago, E.M. Milne

Focus: Natural Disease

Background: Equine grass sickness (EGS or equine dysautonomia) is a frequently fatal disease associated with neuronal degeneration in the autonomic nervous system. It affects grazing horses; the etiology is unknown, but a soil-borne ingested agent is suspected. Gold standard for diagnosis is histopathological examination of hematoxylin-eosin (H&E) sections of autonomic ganglia but processing takes 2-3 days. As space-time clustering of cases occurs, rapid diagnosis is essential, even in fatal cases, to allow early implementation of management changes for other at-risk horses.

Objective: The purpose of the study was to evaluate the accuracy of rapid cytological examination of cranial cervical ganglion scrapings obtained post-mortem for diagnosis of EGS.

Methods: Three cranial cervical ganglion smears were obtained during post-mortem examination from each case (20 controls and 16 EGS cases). For each, the slides were stained with May-Grünwald Giemsa (MGG), H&E and Cresyl Fast Violet (CFV). The slides were examined blind and assigned to EGS or control category; H&E-stained sections were the gold standard.

Results: Considering all 3 stains together, a correct diagnosis was reached for every case (sensitivity and specificity: 100%). However, a small number of individual smears

(4/107, 3.7%) were non-diagnostic due to low cellularity. Also, for a small number of individual smears the final diagnosis was correct but not certain (5/36 CVF, 2/35 H&E and 4/36 MGG) due to low cellularity or suboptimal morphology, the latter especially for CFV stain.

Conclusion: Cranial cervical ganglion cytology is reliable for rapid post mortem diagnosis of EGS, particularly using MGG and H&E stains.

Natural Disease Focused Scientific Session II

December 4, 2016
1:30 PM – 1:45 PM

MOLAR APICAL ELONGATION IN CAPTIVE MICROTUS VOLES: DISEASE CHARACTERISTICS, RISK FACTORS AND MANAGEMENT IMPLICATIONS FOR AN ENDANGERED SPECIES (MICROTUS CALIFORNICUS SCIRPENSIS)

Denise M Imai, Risa Pesapane, Chris Conroy, Christina N Alarcón, Nora Allan, Jennifer Fung, Brian G Murphy, Janet Foley, Frank JM Verstraete

Since the establishment of a captive breeding colony of endangered Amargosa voles (*Microtus californicus scirpensis*), molar apical elongation has been the leading cause for euthanasia or death. Severity may vary but all molars are affected. In extreme cases, the overgrown molar apex can deform surrounding bone and protrude into the nasal passages, the calvarium and through the ventral margin of the mandible. Histologically, the lesion is characterized by apical odontogenic hyperplasia and dysplasia. Overall prevalence is 63% (92/146 voles) and increases to 77% in aged voles (>1yo). No sex predisposition was observed. Clinical signs include ocular discharge, abnormal mastication, dyspnea, abnormal mentation, weight loss, and death. Mean age of onset is 16 weeks (7-41 weeks). Progression to extreme severity occurs over 4-12 weeks with at least 1 incidence of regression. Evaluation of *Microtus* museum specimens revealed molar apical elongation in 2 other *M. californicus* subspecies and 3 other *Microtus* species. In the museum specimens, half the adults were affected and age was the only significant risk factor. A probable genetic influence was identified both in the colony of *M. c. scirpensis* and in the museum collection of *M. c. californicus.* The condition has also been documented in wild *Microtus spp.* The etiopathogenesis of molar apical elongation is likely multifactorial, due to 1) continuous odontogenic proliferation in elodonts, 2) inadequate attrition and 3) probable heritable disease susceptibility. Thus, in captivity, dietary management of occlusal attrition to prevent or delay molar apical elongation is a fundamental concern.

December 4, 2016
1:45 PM – 2:00 PM

CLINICAL AND PATHOLOGICAL OUTCOMES OF EXPERIMENTAL INFECTION WITH TWO DIFFERENT BLUETONGUE VIRUS SEROTYPE 1 FIELD ISOLATES.

Giantonella Puggioni, Davide Pintus, Giorgio Meloni, Anna Maria M. Rocchigiani, Daniela Manunta, Eleonora Melzi, Giovanni Savini, Massimo Palmarini, Maria Dattena, Annalisa Oggiano, Ciriaco Ligios

Background: Europe is recently being the scenario of several Bluetongue virus (BTV) incursions, causing from unapparent to fatal clinical manifestations in ruminants.

Objective: We experimentally studied whether BTV serotype 1, which was responsible for 2006 and 2013 outbreaks in Italy, shapes different clinical and pathological outcomes of the infection.

Methods: Two groups of rams were inoculated with 2 Italian different field isolates of BTV serotype 1 (BTV1-2006, BTV1-2013) and then sequentially euthanized. At necropsy tissues were collected for BTV serotype 1 RNA detection, histopathology as well as for viral VP7 and NS2 protein immunohistochemistry.

Results: Rams showed a notably different clinical course depending on the BTV isolates, with the BTV1-2006 being the most virulent and resulting in a fatal *exitus*. Interestingly, in all the rams the clinical disease was characterized by the involvement of the genital tract, with edema, hyperthermia or cyanosis of the scrotum.

Histologically, although the severity of clinical signs and lesions was different between the 2 groups of rams, all displayed the degeneration of the tubular germinative epithelium and, in the rams inoculated with BTV1-2006, even the loss of the Sertoli's cells. By immunohistochemistry, BTV was found in the endothelial cells of the testicular, epididymal and scrotal skin capillaries from 5 days post-infection, both in the BTV1-2006 and in the BTV1-2013 infected rams.

Conclusion: We found that BTV serotype 1 infection involves the reproductive tract of rams leading to testicular degeneration and that variability of its virulence is powerfully shaped by the field isolate.

December 4, 2016
2:00 PM – 2:15 PM

VERMINOUS ARTERITIS DUE TO CRASSICAUDA SP. IN CUVIER'S BEAKED WHALES (ZIPHIUS CAVIROSTRIS)

Josué Díaz-Delgado, Yara Bernaldo de Quirós, Manuel Arbelo, Aina Xuriach, Eva Sierra, Blanca Mompeo, Lilian Pérez, Marisa Andrada, Juliana Marigo, Jose Luis Catão-Dias, Kátia Regina Groch, John F. Edwards, Antonio Fernández

Background: The vascular system of Cuvier's beaked whales (CBW) (*Ziphius cavirostris*; family Ziphiidae), an extremely deep, prolonged-diving cetacean, is increasingly receiving anatomic and physiologic study due to possible anthropogenic interactions; however, vascular pathology rarely has been reported in this species.

Methods: Thirteen CBW stranded in the Canary Islands from June 2008 to June 2014 were autopsied. A careful dissection of the thoracic and abdominal vasculature was performed on these animals.

Results: All had moderate to severe and extensive chronic fibrosing arteritis with aneurysms, hemorrhages, and thrombosis primarily involving the mesenteric and gastroepiploic arteries and the thoracic and abdominal aorta. Microscopically, the lesions varied from subacute subintimal hemorrhages and severe neutrophilic, eosinophilic, and histiocytic dissecting arteritis with intralesional nematode larvae to marked, chronic, fibrosing arteritis with thickening and distortion of the vascular wall with calcification and occasional cartilage metaplasia. In addition, adult nematodes in renal arteries and veins, renal parenchyma and/or ureter were identified morphologically as *Crassicauda* sp. Nucleic acid sequenced from renal nematodes from 2 animals yielded closest nucleotide identity to *C. magna*. The pathogenesis is proposed to involve a host response to larval migration from the intestine to the kidney through the mesenteric arteries, abdominal aorta, and renal arteries.

Conclusion: Severe consequences for such lesions are possible and could vary from reduced vascular compliance to chronic renal disease and predisposition to the development of disseminated intravascular coagulation and multiorgan failure. Severe chronic arteritis in CBW is associated with renal parasitism by *Crassicauda* spp.

December 4, 2016
2:15 PM – 2:30 PM

PIGEON CIRCOVIRUS IN A FLOCK OF RACING PIGEONS WITH CONCURRENT VIRAL, BACTERIAL, AND MYCOTIC INFECTIONS

Tiffany A Peterson, Atsushi Kawabata, Ingrid Cornax, Christine Higbie, Mustajab Mirza, Rudy Bauer, Nobuko Wakamatsu

Background: Immunodeficiency within a flock of birds becomes apparent through multifactorial infections. A flock of racing pigeons infected with pigeon circovirus had concurrent viral, bacterial and mycotic infections. Pigeon circovirus replicates in lymphoid tissues, causing destruction of lymphocytes and macrophages, leading to immunodeficiency and subsequent secondary infections.

Methods: Six juvenile racing pigeons from the same flock presented to LADDL for postmortem examination. Pigeons were unvaccinated; however, parents were vaccinated for poxvirus, paramyxovirus-1, and *Salmonella*. Deworming was performed using moxidectin and praziquental. Birds were housed in two large holding pens. Clinical signs included cutaneous nodules of the beak, cere, and eyelids, anorexia, weight loss, and lethargy. Due to disease progression one pigeon was euthanized and subsequent pigeons succumbed to disease naturally. Postmortem examinations were performed with ancillary diagnostic testing including virology, bacteriology and mycology.

Results: Cutaneous nodules were confirmed histopathologically as avian pox with secondary bacterial and/or mycotic infections. In addition, multiple histopathological changes indicated other infectious diseases, including basophilic intracytoplasmic botryoid inclusions in the bursa of Fabricius (PCR confirmed pigeon circovirus), eosinophilic intranuclear inclusions in multiple organs (PCR confirmed columbid herpesvirus-1), bacterial blepharitis (mixed bacteria) and mycotic pneumonia and airsacculitis (aspergillosis) and ingluvitis (candidiasis).

Conclusions: Immunosuppression due to primary infection with pigeon circovirus is believed to have led to secondary viral, bacterial, and mycotic infections, ultimately resulting in death. Infection of circovirus in the rearing loft has been suggested as the major route of transmission in young pigeons; therefore, improved hygiene in rearing lofts would aid in preventing outbreaks.

December 4, 2016
2:30 PM – 2:45 PM

GROSS AND HISTOLOGIC LESIONS IN 93 BALD EAGLES WITH SPONTANEOUS LEAD INTOXICATION

Leah K Manning, Arno Wuenschmann, Anibal Armien, Michelle Willette, Kathleen MacAulay, Jeff Bender, John P Buchweitz, Mike Murphy, Patrick T Redig

Background: This retrospective review describes and documents the frequency of lesions associated with lead toxicity in bald eagles.

Method: The review included bald eagles with blood lead concentration greater than 1.0 ppm that subsequently underwent necropsy between 2004 and 2015.

Results: Of the 93 identified cases 79 were euthanized. Gross cardiac changes were documented in 51 (54.8%) cases and included multifocal, pale tan areas within the myocardium, cardiomegaly with rounding of the cardiac apex and/or hydropericardium. Histologically, cardiac abnormalities consisted of cardiomyocyte degeneration and/or necrosis (66 cases; 80.0%), fibrinoid vascular necrosis (40 cases; 43.0%) and/or myocardial fibrosis (38 cases; 40.9%). Gross changes were infrequent within the brain (12 cases; 12.9%) and included hemorrhages in brainstem and cerebellum or less commonly yellowish discoloration in the brainstem. Histologic changes observed in the brain (55 cases; 59.1%) included hemorrhage, fibrinoid vascular necrosis, edema, pan-necrosis and/or infarction, predominately within the brainstem, mesencephalon or cerebellum. Fibrinoid vascular necrosis occurred within the choroid of the eyes (24 out of 84 cases; 28.6%), and was occasionally present in other tissues.

Conclusion: Heart lesions but particularly brain and eye lesions attributable to lead toxicity are more common in bald eagles than previously reported and include a spectrum of acute to chronic changes in the heart and acute lesions in the brain and eye. Fibrinoid vascular necrosis with parenchymal degeneration, necrosis and/or hemorrhage and subsequent fibrosis may represent the principal underlying lesion and may be a key to understanding the pathogenesis of lead toxicity in bald eagles.

December 4, 2016
2:45 PM – 3:00 PM

BLASTOMYCES ANTIBODY AND ANTIGEN LEVELS OVER ONE YEAR IN A CANINE OUTBREAK OF BLASTOMYCOSIS

Janelle S. Renschler, Michelle M. Durkin, Lawrence J. Wheat

Background: An outbreak of blastomycosis occurred in 2015 in a Tennessee kennel of ~57 hound dogs.

Objective: Track *Blastomyces dermatitidis* (Bd) antibody and antigen levels in an exposed population over time.

Methods: Serum was tested in the recombinant BAD-1 repeat antibody and Bd antigen assays at 0, 2, 4 and 12 months.

Results: Seven clinical cases occurred, with cytological evidence of Bd yeasts in 2 dogs. Initial antibody results were moderate-high positive (>20 enzyme units [EU]) in 12 dogs (including all clinical cases), low positive in 7 dogs and negative (<10 EU) in 38 dogs. Mean antibody levels decreased over time from 14.2 to 7.6 EU ($p<0.05$). Of the initial antibody negative dogs, 21.1% developed detectable antibody at a later time point. Highest antibody value at any time point was higher in clinical cases compared to non-cases (mean 62.4 versus 11.4 EU; $p<0.0001$). Serum Bd antigen was positive in 91.7%, 33.3% and 29.7% of dogs in the high positive, low positive and negative antibody groups, respectively. Highest antigen value at any time point was higher in clinical cases (mean 4.8 ng/mL) compared to non-cases (mean 0.49 ng/mL; $p<0.0001$).

Conclusion: Higher rBAD-1 antibody and serum antigen levels were associated with clinical blastomycosis and positive antibody was not retained long-term in exposed dogs. The rBAD-1 antibody assay may serve as a marker of active infection or recent exposure to blastomycosis.

December 4, 2016
3:30 PM – 3:45 PM

CORRELATION BETWEEN KIT EXPRESSION AND C-KIT MUTATION IN CANINE ORAL MELANOCYTIC NEOPLASMS

Rebecca C Smedley, Tuddow Thaiwong, Matti Kiupel

Background: The oncogene *c-Kit* plays a central role in differentiation and proliferation of melanocytes. *c-Kit* mutations have been identified in 15-40% of certain human melanoma subtypes, including those histologically similar to canine oral melanocytic neoplasms (COMNs). A response to tyrosine kinase inhibitors (TKIs) has been demonstrated in those human patients.

Objective: The aim was to determine differences in KIT expression and *c-Kit* mutation status between subtypes of COMNs.

Methods: Twenty-six COMNs were characterized histologically and immunohistochemically using Ki67 and KIT. Neoplasms were classified as malignant melanomas (MMs) or histologically well-differentiated oral melanocytic neoplasms (HWDMs). Next Generation Sequencing (NGS) of *c-Kit* coding regions was performed using the Ion Torrent™ System.

Results: Fourteen neoplasms were classified as MMs; 12 as HWDMs. Intraepithelial nests of neoplastic melanocytes expressed KIT in 9/10 (90%) MMs and 5/7 (71%) HWDMs. Three of 11 (27%) MMs, but no HWDMs expressed KIT in at least 10% of submucosal neoplastic melanocytes. NGS identified 65 variants in *c-Kit*: 26 within introns; 17 poylmorphisms; 22 missense mutations. Eight of the 22 missense mutations (1 each in 6 MMs, 2 in 1 HWDM), resulted in amino acid changes that were predicted to affect protein structure by PROVEAN analysis.

Conclusion: Relevant *c-Kit* mutations were more common in MMs (43% versus 8%). Two MMs with KIT expression in submucosal melanocytes had a relevant mutation. There was no significant correlation between *c-Kit* mutations and KIT expression. Integrating genetic features of COMNs with morphologic phenotyping may lead to more accurate diagnosis and selection of targeted therapy.

December 4, 2016
3:45 PM – 4:00 PM

DISCOVERY AND PROFILING OF FELINE MICRORNAS BY HIGH-THROUGHPUT SEQUENCING USING THE SOFTWARE PIPELINE MIRWOODS

Christiane V. Löhr, Jummy Bell, David Sampson, David A Hendrix

Background: MicroRNAs (miRs) are a class of endogenous, eukaryotic, small non-coding RNAs that control a wide variety of developmental and physiological pathways, and have altered expression in cancer cells. miRs regulate gene expression post-transcriptionally by forming complementary seed-matches with 3'UTRs of specific mRNAs. Some miRs control processes like cell metabolism, differentiation, proliferation, and apoptosis. miRs enable emerging strategies for modulation of biological processes, and improved diagnosis and therapy of diverse disease conditions.

Methods: A software pipeline, miRWoods, was developed to discover, identify and quantify miRs in tissues. miRWoods, a machine learning-based software pipeline, integrates structure predictions, miR biogenesis, and high-throughput sequence data for the discovery of miRs. To test our pipeline and generate reference libraries, we profiled miRs in normal feline tissues using size-selected RNA sequencing. Small RNA was isolated from flash-frozen tissue samples and sequenced on an Illumina HiSeq 2000 and yielded up to 10 million high quality reads per sample.

Results: Of the close to 1000 identified putative miRs, 391 were determined to be orthologs of known mammalian miRs using the miR databases miRBase v21. A comparison of the miR profiles of normal skin versus normal muscle identified close to 100 miRs that were significantly differentially expressed at an FDR of 0.05. Comparison with mRNA expression reveals changes consistent with miR regulation in these tissues.

Conclusions: Our software pipeline, miRWoods, successfully discovered, identified and quantified miRs in feline tissues using an unbiased approach and will be used to examine molecular mechanisms underlying physiological and pathological processes.

December 4, 2016
4:00 PM – 4:15 PM

PATHOLOGIC LESIONS AND PREVALENCE OF TRYPANOSOMA CRUZI IN WILD COYOTES (CANIS LATRANS) OF TEXAS

Carolyn L Hodo, Edward J Wozniak, Erin E Edwards, Rosa M Bañuelos, Elise C Birkner, Sarah A Hamer

Background: *Trypanosoma cruzi* is a vector-borne protozoal parasite of mammals. Infected humans, dogs, and non-human primates may remain asymptomatic or may develop Chagas disease, characterized by lymphoplasmacytic myocarditis with myocardial degeneration and fibrosis, ultimately resulting in heart failure. Although wildlife play important roles as sylvatic reservoirs, investigations into the pathology of *T. cruzi* in wildlife are limited to a few studies documenting histologic lesions in opossums and raccoons. Pathology in coyotes has not been described, despite their recognition as a reservoir and close genetic relationship to domestic dogs.

Objective: We aimed to assess the prevalence and pathology of *T. cruzi* in coyotes and expected to observe similar pathologic lesions to those seen in dogs.

Methods: We collected hearts and blood from 97 hunter-harvested coyotes in central Texas and 25 coyotes harvested for management purposes in south Texas. We examined the hearts grossly and processed tissues for histopathology and PCR.

Results: No significant gross lesions were observed. 5% of 97 coyotes from central Texas and 4% of 25 from south Texas were PCR positive for *T. cruzi*, and a majority of the hearts from positive coyotes had histologic lesions. Lesions ranged from mild to marked, including characteristic lymphoplasmacytic inflammatory infiltrates with varying degrees of myocardial necrosis and fibrosis.

Conclusion: *T. cruzi* infection can result in pathologic lesions in coyotes, in some cases severe enough to suggest cardiac dysfunction. Further work is necessary to assess the population-level impacts of infection on wildlife, as well as the potential risk of exposure to hunters.

December 4, 2016
4:15 PM – 4:30 PM

PNEUMONIA DUE TO CHLAMYDIA PECORUM IN A KOALA (PHASCOLARCTOS CINEREUS)

John T. Mackie, Amber Gillett, Chiara Palmieri, Tianyi Feng, Damien P. Higgins

Chlamydiosis is a common infectious disease of koalas, contributing to population decline in this threatened species. It has been associated with four major syndromes in koalas: keratoconjunctivitis; urinary tract inflammation and incontinence; reproductive tract inflammation and infertility; and rhinitis/pneumonia complex. A causative relationship has been demonstrated for all of these conditions except respiratory disease.

Chlamydial organisms isolated from koalas were initially classified as *Chlamydia psittaci* but chlamydial taxonomy has subsequently undergone revision. *C. pecorum* is the most common chlamydial species infecting koalas and appears to be responsible for most of the clinical disease. Other koala isolates include *C. pneumoniae* and a number of Chlamydia-like organisms of unknown significance.

A juvenile male koala died following an episode of respiratory disease. At necropsy, the lung tissue was consolidated. Histologic changes included pyogranulomatous bronchopneumonia, proliferation of bronchiolar and alveolar epithelium and interstitial fibrosis. Numerous hyperplastic bronchiolar epithelial cells contained large cytoplasmic inclusions, which consisted of aggregates of > 50 small basophilic punctate organisms 0.5-1.0 um diameter.

Immunohistochemical evaluation of formalin fixed, paraffin embedded lung using a rabbit anti-chlamydial LPS antibody demonstrated strong labeling of the cytoplasmic inclusions. Transmission electron microscopic evaluation revealed the inclusions contained chlamydial elementary bodies and reticulate bodies. Real-time PCR of formalin fixed, paraffin embedded lung identified these organisms as *Chlamydia pecorum*.

This report provides the best evidence to date of chlamydial infection causing pneumonia in a koala, and the first evidence that *C. pecorum* is capable of infecting the bronchiolar epithelium of the koala.

December 4, 2016
4:30 PM – 4:45 PM

PATHOLOGIC INVESTIGATION OF FELINE JUVENILE HYPERTROPHIC CARDIOMYOPATHY: FURTHER INSIGHTS.

Sara F Santagostino, Chiara Palmieri, Elisabetta Fabbri, Chiara Locatelli, Ilaria Spalla, Annalisa Forlani, Paola Roccabianca

Background: Feline hypertrophic cardiomyopathy (HCM) can be associated with progressive heart failure in middle-aged to old cats, whereas data concerning this entity in young cats seem fragmentary. We identified a group of young cats with undiagnosed HCM experiencing sudden death.

Objective: To semi-quantitatively characterize the cardiac pathology of suddenly dead young cats with HCM and investigate cardiomyocyte hypertrophy, myocardial disarray, interstitial fibrosis, and expression of Ki-67, STAT3 and PTEN among cats with different disease severity.

Methods: Criteria included cats less than 11-months-old, with no detectable ante-mortem clinical signs. HCM was confirmed post-mortem due to increased heart weight and thickening of the left ventricular free wall (LVFW). A total of 23 cats were enrolled; 3 age-matched healthy cats were used as controls. Hearts were sliced transversely in 2-mm-thick serial sections. The following parameters were histologically evaluated and quantified with Image J software: cardiomyocytes hypertrophy, disarray, vacuolar degeneration, necrosis; vascular wall thickening; interstitial fibrosis. Sections were also stained with primary antibodies against Ki-67, PTEN, STAT3.

Results: LVFW, area and diameter of cardiomyocytes were significantly different in diseased cats compared to controls. The expression of Ki67 and STAT3 did not differ among groups while PTEN increased significantly in HCM.

Conclusions: This study provides a complete evaluation of feline juvenile HCM, and confirms cardiomyocyte hypertrophy and myocardial disarray as typical morphologic features. Both gross and histopathologic data may help to accurately diagnose the disease in young cats in the absence of clinical data. The meaning of PTEN needs further elucidation.

December 4, 2016
4:45 PM – 5:00 PM

STREPTOCOCCOSIS IN LABORATORY-PROPAGATED ADULT GULF KILLIFISH (FUNDULUS GRANDIS)

Elizabeth V. Bamberger

One-year-old adult Gulf Killifish or “cocahoe minnows” (*Fundulus grandis*) from a cohort of wild caught marine baitfish transferred to and propagated in a 300 gallon, closed recirculating quarantine system in a research laboratory at Louisiana State University were presented to the Louisiana Animal Disease Diagnostic Laboratory, Aquatic Section, for acute death and erratic swimming behavior. Approximately 30% mortality had occurred over a one week period at the time of submission. Gross necropsy findings included patchy dermal hemorrhage and ulceration; scale loss; periocular hemorrhage; and exophthalmia. Microscopic examination of Hematoxylin and Eosin-stained sections revealed acute histiocytic meningoencephalitis, pericocular cellulitis, choroiditis, and multiorgan vasculitis with myriad intra- and extracellular coccoid, Gram-positive bacteria. Clusters of similar free and phagositized coccoid bacteria were found sporadically within the liver, pancreas, mesenteric adipose tissue, choroid rete, and spleen. Incidental parasites, including a piscine coccidium consistent with *Calyptospora funduli*, digenetic trematodes (*Ascocotyle* sp.), and myxozoans (*Myxobolus* sp.), were also identified in tissues examined. Select tissues plated on TSAB agar predominantly produced colonies of *Streptococcus agalactiae*. Lancefield group B *Streptococcus* spp. (GBS) have been reported to cause septicemia, including meningoencephalitis, in a variety of fish species and is recognized as an emerging pathogen associated with significant morbidity and mortality in aquaculture operations worldwide. In a recent study, GBS type Ib isolates were identified from streptococcal outbreaks in wild-caught Gulf Killifish. This case demonstrates the importance of characterizing bacterial infections in research laboratory-reared populations of *Fundulus grandis* where high-stocking densities and environmental stressors may induce increased susceptibility to disease.

Natural Disease Posters

N-01: HEPATIC TREMATODE INDUCES STRIKING DUCTULAR REACTION IN A HUBB'S BEAKED WHALE (MESOPLODON CARLHUBBSI)

Shotaro Nakagun, Akira Shiozaki, Mari Ochiai, Ayaka Matsuda, Takashi Matsuishi, Noriyuki Horiuchi, Yoshiyasu Kobayashi

Background: One of the most prevalent parasitic infestations known to occur in cetaceans is caused by trematodes of the genus *Campula* spp., which inhabit the hepatobiliary system and the pancreatic duct. However, reports of hepatic trematodes from Mesoplodont whales are extremely scarce, where only two species, *Mesoplodon stejnegeri* and *M. europaeus*, are known to have been associated so far.

Objective: This study describes a case of hepatic trematode parasitism in an adult male Hubb's beaked whale (*M. carlhubbsi*) that stranded on the Pacific coast of Hokkaido, Japan.

Results: Macroscopic observations revealed a 20 x 10 cm localized area of discoloration in the left hepatic lobe, where spindle shaped, flat helminths were noted within dilated and hypertrophic bile ducts. Histologically, this lesion was characterised by adenomatous hyperplasia of the biliary epithelium, where periductal fibrosis was also found around ducts containing trematodes with tegmental spines. Some areas of the biliary epithelium showed evidence of Goblet cell metaplasia. Findings in the adjacent hepatic parenchyma presented striking ductular reactions, where only a limited number of hepatocytes remained in small foci. Inflammatory cell infiltration was insignificant in these areas. Detailed examinations of the trematodes revealed morphological features consistent to that of the genus *Oschmarinella* spp. PCR amplification and sequencing analyses revealed unique sets of gene sequences, and found no matches with known trematode sequences submitted to the GenBank repository.

Conclusion: Further research in the pathogenesis of hepatic trematode infestations of Mesoplodont whales is anticipated in order to appropriately characterize its health significance to the host.

N-02: HEPATIC ALVEOLAR ECHINOCOCCOSIS IN A BOXER DOG IN SOUTHERN ONTARIO

Britta J Knight, Emily Brouwer, Christopher Pinard, Benoit Cuq, Tom Gibson, Brigitte Brisson, Bruno Gottstein, Andrew Peregrine, Dorothee Bienzle, Brandon Lillie, Brandon Plattner

A 4-year-old male neutered Boxer presented to the Ontario Veterinary College for lethargy, vomiting, and abdominal distention. Peritoneal fluid was highly cellular with 90% degenerate neutrophils. Exploratory laparotomy revealed three 10-25 cm diameter hepatic masses, which were interpreted as abscesses. The two smaller masses were removed but due to extensive hepatic involvement, the largest mass was drained and omentalized.

On cut section, multiple fluid-filled cavities lined by friable tan tissue were observed. Histology revealed multifocal encapsulated expansile cystic structures lined by a thick PAS-positive laminated hyaline membrane with an inner germinal epithelium. Cysts contained abundant fibrin and necrotic debris, as well as many calcareous corpuscles. A presumptive diagnosis of hepatic alveolar echinococcosis caused by the metacestode stage of *Echinococcus multilocularis* was made; this was confirmed by PCR. Daily oral albendazole treatment was recommended; however because this was not tolerated, a modified intermittent therapeutic regimen was initiated, but seven months later the dog re-presented for discomfort and lethargy. One liter of fluid containing large numbers of folded parasitic membranes and numerous calcareous corpuscles was drained from a large cystic cavity in the abdomen that extended from the liver to the urinary bladder. The dog deteriorated clinically, and was subsequently euthanized. Post mortem examination confirmed progression of hepatic alveolar echinococcosis with metastasis to the diaphragm.

This case represents the third confirmed canine case of hepatic alveolar echinococcosis caused by *Echinococcus multilocularis* in southern Ontario in the past 4 years; this represents an emerging disease in southern Ontario with public health significance.

N-03: ANAPLASTIC OLIGODENDROGLIOMA WITH PROMINENT NEUROCYTIC DIFFERENTIATION IN TWO BOXER DOGS

Ingrid Cornax, Gloria E Pluhar, H B Clark, Michael G O'Sullivan

Background: Neurocytic differentiation characterized by mild synaptophysin-immunoreactivity has been described in canine oligodendroglioma; however, the presence of large numbers of neoplastic neural progenitor-like cells within an oligodendroglial tumor has not been previously reported in dogs.

Objective: Our objective is to describe the neural imaging, gross, and microscopic characteristics of two cases of canine anaplastic oligodendroglioma with prominent neurocytic differentiation.

Methods: Cases were identified during routine diagnostic evaluation of dogs enrolled in translational intra-cranial treatment protocols.

Results: Case 1 was a minimally contrast-enhancing right temporal lobe tumor in a 12-year old female-spayed boxer dog. Case 2 was a ring-enhancing left olfactory/frontal lobe tumor in a 5-year old male castrated boxer dog. Grossly, the tumors were poorly demarcated gelatinous masses. The tumors comprised two cell populations. The predominant population displayed the classic oligodendroglial "fried egg" appearance, and the second population comprised clusters of pleomorphic cells with abundant eosinophilic cytoplasm that were concentrated in discrete regions of each tumor. Olig2-immunoreactivity was present in both cell populations, however, synaptophysin-, NSE-, β3-tubulin-, and MAP2-immunoreactivity was limited to the second neoplastic population.

Conclusion: This is the first report of prominent neurocytic differentiation in canine oligodendroglioma. These cases highlight the importance of performing olig2-immunohistochemistry to rule-out neurocytoma and ganglioglioma on tumors displaying neuronal features and support the idea that a common oligodendroglial-neural progenitor cell could be responsible for oligodendroglial tumor histogenesis.

N-04: A CASE OF SYSTEMIC ASPERGILLUS TERREUS IN A DOG

Cheryl A Lawson, Jennifer R Springfield, Kathryn E Rhue, Kelsey A Cline, Elizabeth A Spangler, Pete W Christopherson, Richard C Weiss, Gregory T Almond, Terri L Hathcock

A five-year-old, spayed female, Siberian Husky dog was presented to Auburn University's Bailey Small Animal Teaching Hospital for dialysis after being diagnosed with acute kidney injury (AKI) by the referring veterinarian. Lab work at Auburn was consistent with inflammation, liver injury, and marked renal azotemia.

Abdominal radiographical and ultrasonographical findings were consistent with AKI and included prominent kidneys with normal shape and smooth margins. On abdominal ultrasound, there was moderate bilateral pelvic dilation filled with highly echogenic fluid. The abdominal lymph nodes were enlarged and had irregular margination and decreased echogenicity consistent with lymphadenopathy.

Cytological evaluation of fine-needle aspirates from several lymph nodes revealed purulonecrotic inflammation and branching, septated, hyphal organisms. These organisms stained positively with Gomori methenamine silver stain (GMS).

The patient was treated aggressively, but was ultimately euthanized due to worsening clinical signs. On necropsy, multiple tan masses were identified within the temporalis muscle, spleen, left eye, heart, and kidneys. Histologically there were variably sized, well demarcated foci of macrophages and multinucleated giant cells admixed with neutrophils and mild, multifocal lymphoplasmacytic inflammation surrounding a central core of fibrinonecrotic debris (pyogranuloma) of which many contained clear, refractile, fungal hyphae. GMS highlighted these aggregates of regularly septate hyphae with progressive, dichotomous branching, aleuriospores, and bulbous termini consistent with the culture result of *Aspergillus terreus*.

Systemic aspergillosis most commonly affects young to middle-aged, female, German Shepherd dogs. This case illustrates findings appreciated with the use of several diagnostic modalities in a rare, serious, systemic disease in an uncommonly represented breed.

N-05: RNA-SEQ OF SERIAL KIDNEY BIOPSIES OBTAINED DURING PROGRESSION OF CHRONIC KIDNEY DISEASE FROM DOGS WITH X-LINKED HEREDITARY NEPHROPATHY

Candice P Chu, Jessica A Hokamp, Mia M Aguilar, Mary B Nabity

Background: Dogs with X-Linked Hereditary Nephropathy (XLHN) have a glomerular basement membrane disease leading to juvenile-onset renal failure. These dogs provide a good model of rapidly progressive chronic kidney disease (CKD).

Objective: To identify novel differentially expressed genes (DEGs) and over-represented pathways that contribute to the progression of CKD using high-throughput RNA sequencing (RNA-seq).

Methods: Total RNA from kidney biopsies was isolated at three time points (T1, T2, and T3) from rapidly-progressing affected males (n=3), slowly-progressing affected males (n=3), and age-matched unaffected male littermates (controls; n=2). RNA-seq reads were mapped to the canine genome, and raw read counts were normalized. Differential analysis was used to identify DEGs, and functional and pathway analyses were performed to determine the most prevalent pathways represented by the DEGs.

Results: Sixty-five DEGs (q-value < 0.05) were identified comparing rapidly- and slowly-progressing groups at specific time points. Some of these DEGs have been described in previous CKD studies (e.g., BGN, CD248, COL1A1, COL1A2, COL4A1, COL4A2, COL3A1, FN1, MMP2 (upregulated in rapidly-progressing dogs); NAT8, OAT3/SLC22A8, PRLR (downregulated in rapidly-progressing dogs)), but many have not been reported in CKD. Based on time course analysis comparing all samples with each other, 1158 DEGs were identified over the three time points, revealing upregulation of several inflammatory pathways: integrin signaling, LPS/IL-1-mediated, hepatic fibrosis, and metalloproteases. TGF-beta was identified as the primary upstream regulator, regulating 167 of the 1158 DEGs.

Conclusion: These results provide new insights into the underlying molecular mechanism of disease progression in a canine model of CKD.

N-06: DEREGULATION OF WNT/BETA-CATENIN AND HIPPO PATHWAYS IN HUMAN, FELINE, AND CANINE MAMMARY TUMORS

Alessandro Sammarco, Giorgia Beffagna, Laura Cavicchioli, Silvia Ferro, Maria Elena Gelain, Roberta Sacchetto, Enrico Orvieto, Valentina Zappulli

Background: Mammary cancer is a common neoplasm in women and pets that still represents a therapeutic challenge. Many studies address those molecular pathways that drive tumor progression, scarce cell differentiation, and metastatic potential.

Objective: The aim of this study was to investigate two major pathways involved in cancer aggressiveness (Wnt/beta-catenin and Hippo) in relation to cell differentiation in tumors of the mammary gland.

Methods: Immunohistochemistry (estrogen and progesterone receptors, HER2, cytokeratin - CK - 8/18, CK 5, CK 14, beta-catenin, E-cadherin, TAZ/YAP, vimentin, CD44, and Ki-67) and Western Blotting (beta-catenin and TAZ/YAP) for protein expression/localization and semi-quantitative PCR for gene expression (beta-catenin, CTGF, cyclinD1, TAZ, YAP, and ANKRD1) were performed in a subset of human, feline, and canine mammary carcinomas.

Results: Evidence of scarce differentiation and epithelial-to-mesenchymal transition was particularly observed in feline and human triple-negative samples. Canine carcinomas were more similar to estrogen receptor-positive human samples. In the investigated tumors the Hippo pathway seems to be activated with evidence of nuclear translocation of YAP and TAZ and downstream gene activation. Beta-catenin showed cytoplasmic translocation with mild active nuclear localization. Apparently, Wnt/beta-catenin downstream genes were not overexpressed.

Conclusions: Feline and triple-negative human samples showed some interesting similarities and the results indicated the Hippo pathway as potentially involved in mammary carcinogenesis. Further investigation should better elucidate if some molecules could represent therapeutic targets with beneficial effect both in veterinary and in comparative medicine.

N-07: EVOLUTION OF FELINE IMMUNODEFICIENCY VIRUS IN A NOVEL HOST: IMPLICATIONS FOR CONSERVATION OF THE FLORIDA PANTHER.

Jennifer L Malmberg, Justin Lee, Sahaja Hladky, Ryan Troyer, Melody Roelke, Mark Cunningham, Winston Vickers, Walter Boyce, Erin Boydston, Laurel Serieys, Seth Riley, Kevin Crooks, Sue VandeWoude

Owing to a complex history of host-parasite coevolution, naturally occurring lentiviruses typically exhibit a high degree of species specificity. We previously discovered an exception to this pattern, documenting naturally occurring cross-species transmission of Feline Immunodeficiency Virus, (FIV subtype PLVA), between bobcats (*Lynx rufus*) and mountain lions (*Puma concolor*) in California and Florida. In this study, we investigate host selection pressures, estimate within-host viral fitness, and examine phylogenetic relationships to show that PLVA is under strong pressure to evolve in the novel recipient host – the mountain lion. PLVA fitness approximated by proviral load was severely restricted in mountain lions compared to bobcats, and 3' regions of three of six viral genomes from mountain lions were under diversifying selection. In contrast, diversifying selection was not detected across 20 PLVA genomes from bobcats. Furthermore, transmission from bobcats (reservoir host) to mountain lions (spillover host) is reflected in the phylogenetic relationships among California isolates, where mountain lions apparently fail to produce subsequent infections (dead end hosts). In contrast, our findings strongly suggest that a historic chain of transmission occurred among endangered Florida panthers. Collectively, our data demonstrate that mountain lions are at risk for exposure to viruses of sympatric mesocarnivores, a finding that may be of particular importance to small, isolated populations in which novel viruses could have an apparent propensity to adapt and emerge as virulent pathogens with important conservation implications.

N-08: IMMUNOHISTOCHEMICAL CHARACTERIZATION OF PROCASPASE-3 OVEREXPRESSION AS A DRUGGABLE TARGET WITH PAC-1, A PROCASPASE-3 ACTIVATOR, IN CANINE BRAIN CANCER

Lisa J Schlein, Holly C Pondenis, Edward J Roy, Stephane Lezmi, Peter J Dickinson, Paul J Hergenrother, Timothy M Fan

Introduction: Gliomas are one of the most common intra-axial spontaneous brain neoplasms affecting pet dogs. The management of non-resectable or recurrent brain tumors remains problematic, and warrants the discovery of novel therapies. PAC-1 is a blood-brain barrier penetrant, pro-apoptotic small molecule activator of procaspase-3 (PC-3) that irreversibly commits cells to apoptosis. PC-3 is frequently over-expressed in malignantly transformed tissues, providing an opportunity to selectively induce apoptosis in cancer cells. This study evaluates PC-3 staining in spontaneous canine gliomas and the *in vitro* activity of PAC-1 against immortalized glioma cell lines.

Methods: Tissues from 21 normal control dogs were collected and evaluated with PC-3 immunohistochemical staining, as well as fluorescent double-staining to identify cells that normally express PC-3. Over 150 gliomas were evaluated with PC-3 immunohistochemical staining. Correlation between manual and automated imaging cytometer (iCyte) grading for PC-3 expressions were evaluated. Murine, canine, and human glioma cell lines were evaluated for PC-3 expression and correlated with their *in vitro* sensitivity to PAC-1.

Results: In normal tissue, there is low level PC-3 expression within some neurons and synaptic fibers. PC-3 is generally overexpressed in canine intracranial neoplasms relative to control tissues. There is good agreement between manual and automated grading schema. Immortalized canine, human, and murine glioma cell lines show *in vitro* sensitivity to PAC-1.

Conclusions: PAC-1 shows therapeutic promise in the treatment of canine glioma. Investigation of therapeutic approaches that combine PAC-1 with radiation and/or chemotherapeutics will further elucidate its therapeutic potential in murine models and canine patients.

N-09: CONGENITAL SHORT BOWEL SYNDROME IN AN ADULT DOG

Chad S Clancy, Khrista Jensen, Arnaud J Van Wettere

A three-year-old, neutered male, pit bull was euthanized and necropsied following an approximately one-year history of intractable diarrhea and weight-loss of undetermined cause. At necropsy, the dog was emaciated and the combined small and large intestine length was approximately 40% of the expected intestinal length, consistent with a diagnosis of short bowel syndrome (SBS). Reports from human and veterinary literature have shown that a reduction of intestinal length by 60-70% is sufficient to induce acquired SBS. The dog had no clinical history of intestinal resection since adoption as a puppy and no gross intestinal lesions to suggest surgical reduction had been performed. To the author's knowledge, this is the first case of congenital SBS reported in veterinary literature.

N-10: A NOVEL AMDOPARVOVIRUS IN THE RED PANDA

Charles E. Alex, Steven V. Kubiski, Linlin Li, Raymund F. Wack, Megan A. McCarthy, Eric Delwart, Patricia A. Pesavento

We have identified and characterized a novel parvovirus in the endangered red panda (*Ailurus fulgens*). The virus, tentatively named Red Panda amdoparvovirus (RpAPV), was discovered by metagenomic analyses of tissue from a 20-year-old, captive red panda that died with chronic, effusive peritonitis. By full-length sequence analysis the virus is a member of the genus *Amdoparvovirus* (family *Parvoviridae*, subfamily *Parvovirinae*), along with several other carnivore amdoparvoviruses including the eponymous Carnivore amdoparvovirus 1 (Aleutian mink disease virus). RpAPV is distinct from Carnivore amdoparvovirus 1 (85% sequence identity), and amdoparvoviral species of skunks (72-81% sequence identity) based on partial VP1 alignments. In situ hybridization (ISH) demonstrated RpAPV nucleic acid in the cytoplasm of inflammatory cells in peritoneal and vascular lesions and within germinal centers and sinuses of associated lymph nodes. RpAPV DNA was amplified from multiple fecal samples of in-contact red pandas up to 7 months after the death of the sentinel animal. ISH also uncovered infection by RpAPV in an archived red panda, whose death was otherwise unexplained. Our results indicate that RpAPV can be associated with significant lesions and that asymptomatic shedding can also occur. Additional studies are needed to determine the basis for the range in clinical outcomes.

N-11: GROSS AND HISTOLOGIC LESIONS IN 93 BALD EAGLES WITH SPONTANEOUS LEAD INTOXICATION

Leah K Manning, Arno Wuenschmann, Anibal G Armien, Michelle Willette, Kathleen MacAulay, Jeff Bender, John P Buchweitz, Mike Murphy, Patrick T Redig

Background: This retrospective review describes and documents the frequency of lesions associated with lead toxicity in bald eagles.

Method: The review included bald eagles with blood lead concentration greater than 1.0 ppm that subsequently underwent necropsy between 2004 and 2015.

Results: Of the 93 identified cases 79 were euthanized. Gross cardiac changes were documented in 51 (54.8%) cases and included multifocal, pale tan areas within the myocardium, cardiomegaly with rounding of the cardiac apex and/or hydropericardium. Histologically, cardiac abnormalities consisted of cardiomyocyte degeneration and/or necrosis (66 cases; 80.0%), fibrinoid vascular necrosis (40 cases; 43.0%) and/or myocardial fibrosis (38 cases; 40.9%). Gross changes were infrequent within the brain (12 cases; 12.9%) and included hemorrhages in brainstem and cerebellum or less commonly yellowish discoloration in the brainstem. Histologic changes observed in the brain (55 cases; 59.1%) included hemorrhage, fibrinoid vascular necrosis, edema, pan-necrosis and/or infarction, predominately within the brainstem, mesencephalon or cerebellum. Fibrinoid vascular necrosis occurred within the choroid of the eyes (24 out of 84 cases; 28.6%), and was occasionally present in other tissues.

Conclusion: Heart lesions but particularly brain and eye lesions attributable to lead toxicity are more common in bald eagles than previously reported and include a spectrum of acute to chronic changes in the heart and acute lesions in the brain and eye. Fibrinoid vascular necrosis with parenchymal degeneration, necrosis and/or hemorrhage and subsequent fibrosis may represent the principal underlying lesion and may be a key to understanding the pathogenesis of lead toxicity in bald eagles.

N-12: VERMINOUS ARTERITIS DUE TO CRASSICAUDA SP. IN CUVIER'S BEAKED WHALES (ZIPHIUS CAVIROSTRIS)

Josué Díaz-Delgado, Yara Bernaldo de Quirós, Manuel Arbelo, Aina Xuriach, Eva Sierra, Blanca Mompeo, Lilian Pérez, Marisa Andrada, Juliana Marigo, Jose Luis Catão-Dias, Kátia Regina Groch, John F. Edwards, Antonio Fernández

Background: The vascular system of Cuvier's beaked whales (CBW) (*Ziphius cavirostris*; family Ziphiidae), an extremely deep, prolonged-diving cetacean, is increasingly receiving anatomic and physiologic study due to possible anthropogenic interactions; however, vascular pathology rarely has been reported in this species.

Methods: Thirteen CBW stranded in the Canary Islands from June 2008 to June 2014 were autopsied. A careful dissection of the thoracic and abdominal vasculature was performed on these animals.

Results: All had moderate to severe and extensive chronic fibrosing arteritis with aneurysms, hemorrhages, and thrombosis primarily involving the mesenteric and gastroepiploic arteries and the thoracic and abdominal aorta. Microscopically, the lesions varied from subacute subintimal hemorrhages and severe neutrophilic, eosinophilic, and histiocytic dissecting arteritis with intralesional nematode larvae to marked, chronic, fibrosing arteritis with thickening and distortion of the vascular wall with calcification and occasional cartilage metaplasia. In addition, adult nematodes in renal arteries and veins, renal parenchyma and/or ureter were identified morphologically as *Crassicauda* sp. Nucleic acid sequenced from renal nematodes from 2 animals yielded closest nucleotide identity to *C. magna*. The pathogenesis is proposed to involve a host response to larval migration from the intestine to the kidney through the mesenteric arteries, abdominal aorta, and renal arteries.

Conclusion: Severe consequences for such lesions are possible and could vary from reduced vascular compliance to chronic renal disease and predisposition to the development of disseminated intravascular coagulation and multiorgan failure. Severe chronic arteritis in CBW is associated with renal parasitism by *Crassicauda* spp.

N-13: UTILIZATION OF HIGH-THROUGHPUT TECHNOLOGIES TO INVESTIGATE THE PATHOGENOMICS OF INFECTIOUS DISEASE SUSCEPTIBILITY IN LIVESTOCK

Russell S Fraser, Heindrich N. Snyman, Jutta D. Hammermueller, Ann Meyer, John S. Lumsden, Luis G. Arroyo, Tony A. Hayes, Brandon N. Lillie

Background: Mutations in genes of the innate immune system are associated with an increased susceptibility to infectious disease. Modern technologies, such as next-generation sequencing (NGS), offer powerful, large-scale approaches to investigating the genetic basis of infectious disease susceptibility in livestock.

Objective: Using NGS, we aimed to identify functionally significant mutations in innate immune genes, such as collagenous lectins, associated with infectious disease susceptibility in horses, cattle, and pigs.

Methods: To identify genetic mutations, genomic DNA was obtained from liver collected at postmortem (horses, cattle) or slaughter (pigs). For horses and cattle, animals were classified based on the presence of infectious disease as “normal” or “diseased”. Members of each species were grouped together (3-5/group) based on dominant disease process and equimolar amounts of DNA were pooled. By contrast, all pigs were considered “normal” and sequenced individually. Using probe-based enrichment, candidate innate immune genes and surrounding regulatory DNA were targeted for sequencing.

To investigate the relationship between mutations and gene expression in pigs, hepatic RNA was extracted and gene expression analyzed using a microarray.

Results: In horses, 5145 single nucleotide variants (SNVs) were found (35 missense mutations); 47 SNVs were differentially distributed between "normal" and "diseased" populations. In cattle, 6525 SNVs were identified (54 missense mutations); a separate 54 of which were unevenly distributed between the two populations. In pigs, 41,894 SNVs were identified, 369 of which significantly affected gene expression.

Conclusions: High-throughput sequencing offers an effective and economical approach towards understanding the pathogenomics of infectious disease susceptibility in livestock.

N-14: IMMUNOHISTOCHEMISTRY OF LYMPHOID TISSUE AND PERIPHERAL BLOOD MONONUCLEAR CELLS FROM UNINFECTED AND ENDOTHELIOTROPIC HERPESVIRUS-INFECTED ELEPHANTS

Erin E Edwards, Anna K Blick, Angela Fuery, Rita McManamon, Paul D Ling, Raquel R Rech

Background: Elephant endotheliotropic herpesvirus (EEHV) can cause systemic vascular disease and sudden death in juvenile Asian elephants. Despite its notoriety, little is known about the viral pathogenesis. Elephants infected with EEHV exhibit viremia confirmed by qPCR of whole blood samples. However, it is not known which peripheral blood cell types are infected during these periods of subclinical infection.

Objective: The goal of this study is to optimize immunohistochemistry (IHC) for lymphocytic and monocytic markers from Asian elephants naturally infected with EEHV to help determine which cell type is infected.

Methods: A panel of lymphocytic and monocytic IHC markers for CD3, CD20, and Iba-1 were optimized on formalin-fixed paraffin-embedded (FFPE) sections of lymph node and spleen from uninfected African and Asian elephants. The panel was then applied to both FFPE and cytospin samples of peripheral blood mononuclear cells (PBMCs) from EEHV-infected elephants with and without viremia.

Results: The lymphocytic and monocytic IHC stains were optimized and highlighted the structure of elephant spleen and lymph nodes. The PBMC samples are composed largely of monocytes, including those with bi-lobed and polymorphic nuclei, and far fewer B and T cells.

Conclusions: The optimization of these IHC techniques will allow for co-localization studies with an EEHV antibody or EEHV-specific RNA or DNA. This will help determine which blood cell types are infected during periods of viremia and will also allow for viral tracking in lymphoid tissues of elephants fatally infected with EEHV. These new immunomarkers will help elucidate key steps in the pathogenesis of EEHV.

N-15: GENE EXPRESSION SIGNATURE OF CANINE T ZONE LYMPHOMA

Kelly L Hughes, Janna A Yoshimoto, Jeremy Dossey, Julia D Labadie, Anne C Avery

Background: T zone lymphoma (TZL) is an indolent disease accounting for approximately 10% of all canine lymphomas, diagnosed by flow cytometry due to loss of CD45. Over 10% of dogs with TZL have evidence of immunosuppression. Not only is CD45 central to T cell activation through the T cell receptor (TCR) but it also is a ligand for galectin binding and death in Th1 and Th17 subsets, sparing Th2 cells. We hypothesized that the cell of origin of TZL is a Th2 cell.

Objective: Determine the cell of origin of TZL and investigate mechanisms of immunosuppression and proliferation in this disease.

Methods: TZL cell RNA was isolated from 34 canine samples received through the CSU-Clinical Immunology laboratory and analyzed by NanoString technology. Genes with differential expression in T cell subsets were investigated. TZL blood was cultured to investigate mechanisms of proliferation in this disease.

Results: TZL exhibited a unique gene expression profile. Genes specific to Th2 differentiation including the transcription factor, GATA3, and Th2 specific receptors, CrTh2 and CCR3 were significantly upregulated. Galectins 1 and 3 and the inhibitory cytokine, TGFb were also highly expressed. In vitro, TZL cells showed poor proliferation when stimulated through the TCR.

Conclusions: Gene expression profiling for TZL is suggestive of a Th2 subset origin. Stimulation through the TCR does not appear to be a major mechanism of proliferation in oncogenesis. Furthermore, loss of CD45, along with increased expression of galectins and TGFb may promote an immunosuppressive environment in dogs with TZL.

N-16: 'IRON ACCUMULATION' GENE EXPRESSION PROFILE IN OBESE HARTLEY GUINEA PIG KNEE JOINTS IS ASSOCIATED WITH MORE SEVERE OSTEOARTHRITIS

Lauren B Radakovich, Kelly S Santangelo

Background: Knee joint osteoarthritis (OA) is a leading cause of pain and disability worldwide. Obesity is the largest modifiable risk factor associated with OA. Hartley guinea pigs are prone to obesity and are the only laboratory model of naturally-occurring knee joint OA that mimics human disease. Free radical damage contributes to OA, but the specific role that iron dysregulation plays is an unexplored niche.

Objective: Our aims were: 1) to compare gene expression of iron molecules between obese and calorie-restricted guinea pigs, and 2) to associate those findings with severity of OA.

Methods: Twelve male Hartley guinea pigs were obtained at 2 months of age. The obese group was fed ad libitum. The calorie-restricted group received 25 grams of chow daily. Animals were harvested at 5 months. Left legs were submitted for microcomputed tomography (microCT) and subsequent histopathology. MicroCT images were evaluated using a novel grading scheme. Cartilage and menisci from the right legs were collected for quantitative real time polymerase chain reaction. Non-parametric Mann Whitney analyses were performed.

Results: Compared to calorie-restricted guinea pigs, obese animals had increased gene expression of iron molecules (transferrin receptor, ferritin, interleukin-6). They also exhibited increased expression of cartilage-degrading enzymes and antioxidants induced by free radicals. These findings were associated with worsened OA based on microCT and histologic analyses.

Conclusion: Excess iron accumulation is associated with the development of knee joint OA in obese guinea pigs. Similar finds could be expected in people, and local iron chelation may be a viable treatment option.

N-17: PIGEON CIRCOVIRUS IN A FLOCK OF RACING PIGEONS WITH CONCURRENT VIRAL, BACTERIAL, AND MYCOTIC INFECTIONS

Tiffany A Peterson, Atsushi Kawabata, Ingrid Cornax, Christine Higbie, Mustajab Mirza, Rudy Bauer, Nobuko Wakamatsu

Background: Immunodeficiency within a flock of birds becomes apparent through multifactorial infections. A flock of racing pigeons infected with pigeon circovirus had concurrent viral, bacterial and mycotic infections. Pigeon circovirus replicates in lymphoid tissues, causing destruction of lymphocytes and macrophages, leading to immunodeficiency and subsequent secondary infections.

Methods: Six juvenile racing pigeons from the same flock presented to LADDL for postmortem examination. Pigeons were unvaccinated; however, parents were vaccinated for poxvirus, paramyxovirus-1, and *Salmonella*. Deworming was performed using moxidectin and praziquental. Birds were housed in two large holding pens. Clinical signs included cutaneous nodules of the beak, cere, and eyelids, anorexia, weight loss, and lethargy. Due to disease progression one pigeon was euthanized and subsequent pigeons succumbed to disease naturally. Postmortem examinations were performed with ancillary diagnostic testing including virology, bacteriology and mycology.

Results: Cutaneous nodules were confirmed histopathologically as avian pox with secondary bacterial and/or mycotic infections. In addition, multiple histopathological changes indicated other infectious diseases, including basophilic intracytoplasmic botryoid inclusions in the bursa of Fabricius (PCR confirmed pigeon circovirus), eosinophilic intranuclear inclusions in multiple organs (PCR confirmed columbid herpesvirus-1), bacterial blepharitis (mixed bacteria) and mycotic pneumonia and airsacculitis (aspergillosis) and ingluvitis (candidiasis).

Conclusions: Immunosuppression due to primary infection with pigeon circovirus is believed to have led to secondary viral, bacterial, and mycotic infections, ultimately resulting in death. Infection of circovirus in the rearing loft has been suggested as the major route of transmission in young pigeons; therefore, improved hygiene in rearing lofts would aid in preventing outbreaks.

N-18: CARDIAC MESOTHELIAL PAPILLARY HYPERPLASIA IN FOUR DOGS
Shannon G.M. Kirejczyk, Annabelle Burnum, Corrie C. Brown, Kaori Sakamoto, Daniel R. Rissi

Background: Mesothelial papillary hyperplasia (MPH) has been considered an incidental lesion on the epicardial surface of laboratory Beagle dogs. A proposed pathogenesis for MPH is that it occurs in response to friction between the epicardial surface and pericardium. In humans, MPH may also result from contact of the pericardium with the sternum or with the vertebral bodies in cases of cardiomegaly and kyphoscoliosis, respectively.

Objective: The aim of this study was to characterize MPH in four dogs submitted to our diagnostic service. Pathology reports and histology slides were reviewed and sections were submitted for Masson's trichrome histochemical stain.

Results: Histologically, the epicardial surface exhibited varying degrees of mesothelial-lined papillary proliferation supported by fibrous connective tissue, and occasionally expanded by edema and mononuclear inflammatory cells. Masson's trichrome stains highlighted the MPH and subepicardial fibrosis, which occasionally radiated downward into the superficial myocardium. Two dogs had mitral endocardiosis and cardiac insufficiency, which resulted in surgical intervention, death, or euthanasia. The affected right auricle in one dog was enlarged and covered with a blood clot and fibrin, suggesting a chronic-active process leading to ongoing hemorrhage, and ultimately, death. The clinical significance of MPH in one case remains undetermined (cause of death was hemorrhagic diathesis with hemopericardium).

Conclusion: We describe MPH in four dogs in a diagnostic setting and suggest that these changes may occur concomitantly with other cardiac changes. Additional insight into the pathogenesis of this entity is needed to elucidate the significance of canine MPH in the face of concurrent cardiac disease.

N-19: ASSESSMENT OF THE VALUE OF GANGLION CYTOLOGY FOR THE RAPID POST MORTEM DIAGNOSIS OF EQUINE GRASS SICKNESS
Chiara Piccinelli, Rachel C. Jago, Elspeth M. Milne

Background: Equine grass sickness (EGS or equine dysautonomia) is a frequently fatal disease associated with neuronal degeneration in the autonomic nervous system. It affects grazing horses; the etiology is unknown, but a soil-borne ingested agent is suspected. Gold standard for diagnosis is histopathological examination of hematoxylin-eosin (H&E) sections of autonomic ganglia but processing takes 2-3 days. As space-time clustering of cases occurs, rapid diagnosis is essential, even in fatal cases, to allow early implementation of management changes for other at-risk horses.

Objective: The purpose of the study was to evaluate the accuracy of rapid cytological examination of cranial cervical ganglion scrapings obtained post-mortem for diagnosis of EGS.

Methods: Three cranial cervical ganglion smears were obtained during post-mortem examination from each case (20 controls and 16 EGS cases). For each, the slides were stained with May-Grünwald Giemsa (MGG), H&E and Cresyl Fast Violet (CFV). The slides were examined blind and assigned to EGS or control category; H&E-stained sections were the gold standard.

Results: Considering all 3 stains together, a correct diagnosis was reached for every case (sensitivity and specificity: 100%). However, a small number of individual smears (4/107, 3.7%) were non-diagnostic due to low cellularity. Also, for a small number of individual smears the final diagnosis was correct but not certain (5/36 CVF, 2/35 H&E and 4/36 MGG) due to low cellularity or suboptimal morphology, the latter especially for CFV stain.

Conclusion: Cranial cervical ganglion cytology is reliable for rapid post mortem diagnosis of EGS, particularly using MGG and H&E stains.

N-20: STREPTOCOCCOSIS IN LABORATORY-PROPAGATED ADULT GULF KILLIFISH (FUNDULUS GRANDIS)

Elizabeth V. Bamberger

One-year-old adult Gulf Killifish or "cocahoe minnows" (*Fundulus grandis*) from a cohort of wild caught marine baitfish transferred to and propagated in a 300 gallon, closed recirculating quarantine system in a research laboratory at Louisiana State University were presented to the Louisiana Animal Disease Diagnostic Laboratory, Aquatic Section, for acute death and erratic swimming behavior. Approximately 30% mortality had occurred over a one week period at the time of submission. Gross necropsy findings included patchy dermal hemorrhage and ulceration; scale loss; periocular hemorrhage; and exophthalmia. Microscopic examination of Hematoxylin and Eosin-stained sections revealed acute histiocytic meningoencephalitis, pericocular cellulitis, choroiditis, and multiorgan vasculitis with myriad intra- and extracellular coccoid, Gram-positive bacteria. Clusters of similar free and phagositized coccoid bacteria were found sporadically within the liver, pancreas, mesenteric adipose tissue, choroid rete, and spleen. Incidental parasites, including a piscine coccidium consistent with *Calyptospora funduli*, digenetic trematodes (*Ascocotyle* sp.), and myxozoans (*Myxobolus* sp.), were also identified in tissues examined. Select tissues plated on TSAB agar predominantly produced colonies of *Streptococcus agalactiae*. Lancefield group B *Streptococcus* spp. (GBS) have been reported to cause septicemia, including meningoencephalitis, in a variety of fish species and is recognized as an emerging pathogen associated with significant morbidity and mortality in aquaculture operations worldwide. In a recent study, GBS type Ib isolates were identified from streptococcal outbreaks in wild-caught

Gulf Killifish. This case demonstrates the importance of characterizing bacterial infections in research laboratory-reared populations of *Fundulus grandis* where high-stocking densities and environmental stressors may induce increased susceptibility to disease.

N-21: INCREASED DENSITY OF EPIDERMAL NERVE FIBERS IN DOGS WITH ATOPIC DERMATITIS

Ileana Miranda, Sheila Torres, Marna Ericson

Background: Pruritus is the hallmark of atopic dermatitis (AD). Cutaneous nerve fibers involved in itch consist mostly of free nerve endings intimately associated with keratinocytes and immune cells, allowing recognition and response to various pruritogens. Neuronal hypersensitivity is suggested as a leading cause of chronic pruritus in humans and animal models of AD.

Objective: This study aims to quantify the nerve fibers and evaluate the immunoreactivity of itch-related molecules in AD dogs. We hypothesize that canine AD skin is hyperinnervated and itch mediators are overexpressed compared to healthy dogs.

Methods: Skin biopsies were collected from the interdigital area, groin, axilla and lateral thorax of three AD dogs and three healthy controls. Samples were processed and immunostained with antibodies to PGP9.5, a pan-neuronal marker, and to itch mediators (GRP, SP, CGRP, TSLPR and IL31RA). Images were captured by laser scanning confocal microscopy, density of nerve fibers was quantified using histomorphometric analysis, and an unpaired t-test was performed for statistical evaluation.

Results: AD dogs had significantly increased density of epidermal PGP9.5-immunoreactive nerve fibers at all biopsied sites compared to controls. No difference in immunoreactivity of the itch mediators was detected.

Conclusion: Our results suggest that higher density of epidermal innervation is at least partly responsible for itch sensitization in canine AD, either by hyperinnervation, decreased threshold of activation, or a combination of both. In future studies, a larger cohort of dogs will be examined to focus on mechanisms regulating epidermal innervation; with the ultimate goal of controlling pruritus in dogs with AD.

N-22: ABERRANT EXPRESSION OF 14-3-3-SIGMA PROTEIN IN CANINE RENAL CELL CARCINOMAS

Alejandro Suárez-Bonnet, Ana Lara-García, Alexander L Stoll, Sofia Carvalho, Simon L Priestnall

Background: Renal cell carcinomas (RCCs) are the most common primary canine renal tumours, however little is known about the mechanisms and molecules that contribute to their development and progression. 14-3-3-sigma (FTTs) protein is a cell cycle regulator, induced by the *p53* gene that has been associated with tumour development in human and canine malignancies.

Objective: The aims of this study were to establish if FTTs is expressed in normal canine kidney versus RCCs and to correlate the expression of FTTs with differentiation markers (keratin and vimentin) and proliferation measures (mitotic activity and Ki67 indices).

Methods: Forty-one cases of canine RCC were examined. Tumours were categorized into three main histologic types (tubular, papillary and solid). Mitotic activity index (MAI) was determined by counting the number of mitoses in 10 HPF (2.37 mm^2). Immunohistochemical labelling using antibodies against FTTs, CK AE1/AE3, vimentin, Ki67 and COX-2 was performed and semi-quantitatively assessed.

Results: Normal canine kidney did not express FTTs. 18 (44%) RCCs (representing 12 tubular, 4 papillary and 2 solid types) demonstrated positive expression of FTTs. Mean FTTs values for each type were 2.7, 2 and 0.6 respectively. FTTs expression was correlated with a high Ki67 index (20.96), high MAI (21.72) and higher mean COX-2 score (5.93). These values were approximately twice those of FTTs-negative cases.

Conclusion: FTTs protein is neo-expressed in a significant number of RCCs and its presence is positively associated with cell proliferation. Further studies are required to address the value of FTTs as a prognostic factor in canine RCCs.

N-23: NANNIZZIOPSIACEAE SPP. INFECTION IN AQUATIC CHELONIANS
Daniel B Woodburn, Matthew C Allender, Carol W Maddox, Andrew N Miller, Katherine Haman, Jennifer N Langan, Caryn P Poll, Stefanie Bergh, Drury R Reavill, Karen A Terio

Background: The family Nannizziopsiaceae comprises significant reptile pathogens including *Ophidiomyces ophiodiicola* and *Nannizziopsis vriesii*. While these fungi are well-documented pathogens of lizards, snakes, and crocodilians, there are no known reports of disease in turtles. However, a novel presentation of shell disease has recently been recognized in free-ranging and captive aquatic chelonians associated with fungi that are morphologically and genetically consistent with Nannizziopsiaceae spp.

Objective: The aim of this study was to describe the histologic lesions of Nannizziopsiaceae spp. shell infections and characterize the fungal organism.

Methods: A retrospective review of chelonian cases submitted to the Zoological Pathology Program identified 38 individuals with shell lesions and histologic sections available for review. Of these, 18 individuals had similar lesions and intralesional fungi compatible with Nannizziopsiaceae spp. Infection was confirmed via PCR and culture in a subset of these cases. Cultured isolates were characterized by morphology and sequence analysis.

Results: Histologic lesions associated with Nannizziopsiaceae spp. infection were unique and were similar across the 18 affected individuals. The lesions consisted of necrotizing dermatitis and osteomyelitis with dermal inclusion cysts lined by keratinized squamous epithelium and containing morphologically compatible fungal hyphae. Sequence data obtained from frozen tissue samples and phylogenetic analysis of the fungal ITS region indicated that fungi infecting chelonians comprise a monophyletic clade within the Nannizziopsiaceae distinct from species infecting other reptiles.

Conclusion: Nannizziopsiaceae spp. infection results in characteristic shell lesions in aquatic chelonians, and the causative agent is distinct from other related reptile pathogens. Additional morphologic and molecular characterization is ongoing.

N-25: FELINE LEUKEMIA VIRUS ASSOCIATED GIANT CELL DERMATOSIS AND ATTEMPTED TREATMENT IN A CAT – A CASE REPORT

Darren J Berger, Amanda J Fales-Williams, Jodi D Smith, James O Noxon

Background: Giant cell dermatosis is a rare manifestation of Feline Leukemia virus (FeLV) infection. Prognosis in affected individuals is poor. This case report describes the clinical, histologic, and treatment aspects of a case from the Iowa State University Lloyd Veterinary Medical Center dermatology service.

Case Report: A 15-month-old castrated male domestic long hair cat was presented for evaluation of non-resolving periocular crusting and paronychia. Prior to presentation the patient was diagnosed as FeLV positive via ELISA and IFA serology. Symmetric periocular, perioral, periauricular and perianal alopecia, erythema, and crusting were noted, along with some degree of marked crusting paronychia, onychomadesis, pachyonychia, and onychodystrophy affecting all digits. Thickened irregular hyperkeratotic metacarpal and metatarsal pads were also present.

Results: Skin biopsies from multiple sites revealed marked orthokeratotic hyperkeratosis, altered layering of squamous epithelium, multifocal, large, rounded, multinucleated cells in all layers of surface and follicular epithelium. There was splitting of the altered keratin layer and follicular distension. The cat was treated with amoxicillin-clavulanic acid and zidovudine (AZT). Temporary significant clinical improvement was noted until AZT was discontinued due to medication-associated adverse events. The cat deteriorated due to FeLV associated complications and was euthanized. Bone marrow was hyperplastic and positive by PCR and IFA for FeLV on post-mortem evaluation.

Conclusion: FeLV-associated giant cell dermatosis is a rare condition in cats with few described cases in the literature. This report expands on the limited number of presented cases.

N-26: RNA-SEQ TRANSCRIPTOME ANALYSIS OF FORMALIN-FIXED PARAFFIN EMBEDDED CANINE MENINGIOMA
Polly A Foureman, Erica A Sloma, Jennifer Grenier, Andrew D Miller

Meningiomas are the most commonly reported primary intracranial tumor in dogs and humans. Canine meningiomas closely resemble human meningiomas in histological appearance as well as biological behavior. Because of the similarities between canine and human meningiomas, dogs have been proposed as models and surrogates for studying and treating human cancers including meningioma. However, little is known about specific pathways and individual genes that are involved in the development and progression of canine meningioma. In addition, there is a paucity of suitable fresh material available for studies of differential expression (DE) of genes in cases of veterinary neoplasia. We report here the use of formalin-fixed paraffin embedded (FFPE) specimens as much as a decade old, to provide RNA suitable for transcriptome analysis using next-generation sequencing (NGS). RNA was extracted from 13 canine meningiomas – 12 from (FFPE) and one flash-frozen. These represented 6 grade I and 7 grade II/III. RNA was also collected from flash-frozen meningeal tissue (arachnoid) from 3 control dogs. RNA from FFPE was of sufficient quality to successfully identify 70 significantly DE genes, the majority of which were related to oncogenic processes. 12 genes (AQP1, BMPER, FBLN2, FRZB, MEDAG, MYC, PAMR1, PDPN, PECAM1, PERP, ZC2HC1c) were validated using qPCR. Among the DE genes were oncogenes, tumor suppressors, transcription factors, VEGF-related genes, and members of the WNT pathway. Our work demonstrates that RNA of sufficient quality can be extracted from FFPE samples to provide biologically relevant transcriptome analyses using a NGS technique, such as RNA-seq.

N-27: NATURAL MORTALITY IN THE CAPTIVE CHIMPANZEE (PAN SPP.): A 32 YEAR REVIEW
Hannah M Laurence, Michael A Owston, Robert E Lanford, Shyamesh Kumar, Edward J Dick

Background: We report the spontaneous causes of mortality for 137 chimpanzees that died due to natural causes or were euthanized for humane reasons at the Southwest National Primate Research Center at Texas Biomedical Research Institute, San Antonio, Texas, over a 32 year period.

Methods: For each chimpanzee, the individual records were reviewed and a primary cause of death or indication for euthanasia was determined. The associated morphologic diagnosis, system, organ, and etiology, the animal's age at diagnosis, and sex were recorded and analyzed.

Results: Lesions involving the cardiovascular system accounted for half of all mortality (51.1%), followed by lesions of the gastrointestinal (13.9%), respiratory (9.5%), multi-systemic disease (8.8%), urogenital (7.3%), and nervous (5.1%) systems. Cardiovascular-related deaths were typically seen in middle aged to geriatric males and cardiomyopathy was the most frequent diagnoses. An etiology was identified for 85 of the 137 chimpanzees. The most common etiologies related to mortality were degenerative disease (49.4%), bacterial (14.1%), traumatic (9.4%), physiologic (8.2%), neoplastic (7.1%), and mycotic (4.7%). Bacterial and traumatic causes of mortality were most often seen in perinatal, infant, and juvenile chimpanzees. Degenerative causes of mortality were most often observed in adult, middle aged, and geriatric chimpanzees.

Conclusion: Cardiovascular disease was the most frequent cause of death, and degenerative disease was the most frequently identified etiology associated with mortality in captive chimpanzees. Knowledge of the natural pathology of chimpanzees is required so that veterinary and husbandry staff can continue to provide high quality care to these animals.

N-28: IDENTIFICATION OF FAS-LIGAND DEFECTS IN CATS WITH EARLY-ONSET SEVERE FATAL LYMPHOPROLIFERATION. CONFIRMATION OF A FELINE AUTOIMMUNE LYMPHOPROLIFERATIVE SYNDROME.

John S Munday, Danielle Aberdein, Barbara Gandolfi, Keren Dittmer, Leslie Lyons, Richard Malik

Background: A severe fatal lymphoproliferative syndrome was recently identified in British shorthair (BSH) kittens in multiple litters. The disease had phenotypic similarities to human autoimmune lymphoproliferative syndrome (ALPS) which is caused by inherited defects in FAS-mediated lymphocyte apoptosis.

Objective: To determine if affected BSH kittens had defects in genes coding for proteins of the FAS-mediated apoptosis pathway.

Methods: The whole genome of two affected kittens was sequenced and compared to 82 cat genomes.

Results: Both BSH kittens had homozygous insertions of adenine on chromosome F1 within exon 3 of the *FAS-ligand* gene. The resultant frame-shift and premature stop codon was predicted to result in a severely truncated protein that is unlikely to be able to activate FAS. Specific primers were designed and used to amplifying this region of the *FAS-ligand* gene. The presence of homozygous variants was identified in an additional three affected BSH kittens. Additionally, heterozygous variants were detected in 10 of 15 unaffected, but closely related BSH cats that were interpreted to be carriers. In contrast, none of 515 non-BSH cats carried the defective gene. All BSH cats in the study were from a population with significant inbreeding.

Conclusion: Identification of a genetic defect in the FAS-mediated apoptosis pathway confirms that the lymphoproliferative disease in BSH cats fulfills the diagnostic criteria for ALPS in humans. These cats are a potential animal model for human ALPS due to homozygous defects in the FAS apoptosis pathway. Additionally, the results allow genetic testing to detect carrier animals.

N-29: REPORT OF CYSTICERCOSIS IN SHEEP OF BALLESTA, BOLIVAR, COLOMBIA.

Jahnier Caicedo Caicedo, María Angélica A Ávila Rubiano, Byron Abdel A Hernandez Ortiz

Focus: Natural Disease Sheep are the intermediate hosts of Cysticercus tenuicollis, the metacestode of Taenia hydatigena. Cysticercosis produces economic losses associated with loss of condition and offal condemnation at slaughter. The aim of this study was to determine the prevalence of C. tenuicollis cysts in sheep, the most common sites for these cysts, and the histology of livers affected by this parasite in Bolivar, Colombia. This study included 23 sheep (4 females and 19 males) and it was carried out between August and November 2015. The prevalence of C. tenuicollis cysts was 39.13% (9/23). Fifty six percent (5/9) of the animals had body condition (BC: scale of 1 to 5) of 2; 22% (2/9) 3, and 22% (2/9) 4. A single cyst was found in 78% (7/9) of the animals, and 22% (2/9) had two cysts. Cysts were located in the rumen serosa: 44% (4/9); abomasal serosa: 22% (2/9); liver: 22% (2/9); bladder serosa 22% (2/9); and diaphragm: 11% (1/9). The C. tenuicollis metacestodes found belong to the species Taenia hydatigena. Lesions in the liver of all affected animals consisted of severe infiltration of eosinophils, lymphocytes and plasma cells. Also present were areas of periportal fibrosis. In conclusion, cysticercosis damages the liver of sheep and potentially have a deleterious effect on the productivity of the flock. This is the first report describing post-mortem diagnosis of Cysticercus tenuicollis infection in sheep in Colombia. Speakers: Jahnier Caicedo Caicedo;María Angélica A Ávila Rubiano;Byron Abdel A Hernandez Ortiz

N-30: STUDY OF LUNG LESIONS IN SHEEP SLAUGHTERED FROM BOLIVAR, COLOMBIA

Jahnier Caicedo Caicedo, María Angélica A Avíla, Danny W Sanjuanelo

Introduction: ovine respiratory complex is caused by the interaction of an infectious agent, host immunity, and the environment. This syndrome results in reduction in growth rate and in reduction in growth rate and lung condemnations. **Objective**: The main objective of this study was to determine at slaughter the occurrence of disease conditions affecting the lungs. **Methods:** A cross-sectional descriptive study was done on clinically healthy 26 sheeps slaughtered in Bolivar, Colombia. The frequency of gross and microscopic lung lesions were recorded, the association between severity of gross lesions and body condition was measured through chi-square test. **Results.** 92% of sheep revealed lung lesions. 32 Gross lesions were diagnosed: pleuritis (34%), interstitial pneumonia (32%), embolic pneumonia (13%), granulomatous pneumonia (6%), emphysematous bulla (6%), verminous pneumonia (3%), suppurative bronchopneumonia (3%) and atelectasis (3%). Pleuritis showed a highly significant relationship with body condition (p value **Conclusion**: This is the first study in Colombia describing lung lesions in slaughtered sheeps and they were highly frequent since most respiratory complex presentation is subclinical; the diagnosis is postmortem. Almost certainly, they are involved multiple etiological agents responsible for the dynamic of the disease leading to substantial economic losses due to reduced sheep growth rates, lung condemnations and decreased carcass value.

N-31: SNAKE FUNGAL DISEASE: UPDATE AND FIRST DESCRIBED CASES IN ONTARIO, CANADA

Nicole M. Nemeth, Lenny Shirose, Douglas D. Campbell, Hugh Cai, Claire M. Jardine

Background: Snake fungal disease (SFD) is a recently described condition of wild snakes in North America, and is caused by infection with *Ophidiomyces ophiodiicola*. The disease was first documented in wild snakes in New Hampshire in 2006, and has since been reported in 16 U.S. states in a variety of snake species. Grossly, scabs and crusts are on scales or skin, sometimes with underlying pustules or ulcers, subcutaneous nodules (often facial), ocular cloudiness, and dysecdysis. Histopathology includes hyperkeratosis, granulomatous dermatitis, and ulceration. Lesions may extend to underlying muscle and bone.

Objectives: This study seeks to detect *O. ophiodiicola* in snakes in Ontario and describe pathology in infected snakes. Opportunistically collected and diseased snakes from as broad a geographic range as possible are examined.

Methods: Affected tissues are tested by PCR and examined microscopically.

Results: The first documentation of SFD in Canada occurred in a free-ranging eastern fox snake (*Pantherophis vulpinus*) from southwestern Ontario in 2015; two additional fox snakes have since been diagnosed. Diagnostic evaluation revealed characteristic lesions that tested PCR-positive for *O. ophiodiicola*. Based on these findings, additional studies to assess the prevalence and distribution of this fungus in Ontario have been initiated. These include testing of opportunistically-collected skin swabs from live snakes, carcasses and the environment, along with continued diagnostic evaluations.

Conclusions: Ontario represents the northern extent of the geographic ranges of numerous endangered snake species that are likely susceptible to SFD. Therefore, a better understanding of the prevalence and potential effects of SFD in Ontario is needed.

N-32: GENOME-WIDE COPY NUMBER PROFILE ANALYSIS OF CANINE CUTANEOUS MAST CELL TUMORS AND ITS APPLICATION TO PREDICT HIGH-RISK TUMOR PHENOTYPE

Hiroyuki Mochizuki, Rachael Thomas, Scott Moroff, Matthew Breen

Background:

Canine mast cell tumor (MCT) is a common skin malignancy characterized by heterogeneous biological behavior. Knowledge regarding the underlying molecular aberrations in the development and progression of MCTs are largely unknown. Characterization of genomic alterations in the tumors may identify genome regions and/or genes responsible for the malignant alteration of canine MCTs, facilitating the development of new therapeutic strategies and improved clinical management of this cancer.

Methods:

Genome-wide DNA copy number profile of 109 canine primary MCTs was obtained using oligo array comparative genomic hybridization (oaCGH). Tumor *KIT* mutation status was also evaluated and assessed for correlation with specific copy number aberrations (CNAs). Digital PCR was employed to detect four CNAs associated with high-risk tumor phenotypes and evaluated in 147 MCTs as a means to predict aggressive tumor phenotype.

Results:

A stepwise accumulation of numerical chromosomal alterations was demonstrated as tumor grade increases. Tumors with *KIT* mutations showed genome-wide aberrant copy number profiles, with frequent alterations of genes in the p53 and RB pathways, whereas CNAs were less common in tumors with wild-type *KIT*. Presence of four CNAs was able to predict high-risk phenotypes with a sensitivity of 75–91% and specificity of 86–93%, when using oaCGH and digital PCR platforms.

Conclusion:

CNAs associated with high histological grade and the presence of *KIT* mutations were identified in canine MCTs. Further investigation of these genomic lesions may lead to the development of a molecular tool for classification and prognosis, as well as identification of therapeutic target molecules.

N-33: CHOLESTEROL DEFICIENCY IN HOLSTEIN CATTLE - A NATURAL LARGE ANIMAL MODEL OF FAMILIAL HYPOBETALIPOPROTEINEMIA?

Thomas Mock, Kemal Mehinagic, Fiona Menzi, Eveline Studer, Anna Oevermann, Michael H. Stoffel, Cord Drögemüller, Mireille Meylan, Nadine Regenscheit

Background: Cholesterol deficiency (CD), a newly identified autosomal recessive genetic defect in the apolipoprotein B gene (*APOB*) in Holstein cattle is associated with clinical signs of diarrhea, failure to thrive and hypocholesterolemia. In human patients, truncating mutations in *APOB* give rise to familial hypobetalipoproteinemia (FHBL), a genetic disease impairing the fat metabolism.

Methods: Six Holstein cattle, five calves with a clinical history of chronic diarrhea and one heifer with erosions in the buccal cavity and neurological symptoms were clinicially and pathologically investigated to describe the so far unknown clincopathological phenotype. Clinical examination included a complete blood count, blood chemistry and measurements of cholesterol and triglycerides. The animals were euthanized and necropsied. A PCR-based direct gene test was applied to determine the *APOB*genotype.

Results: All six animals were inbred, confirmed homozygous for the *APOB* mutation. The clinical phenotype included poor development, underweight and intermittent diarrhea in the calves, and neurological signs in the heifer included hypermetria and pacing. Hypocholesterolemia and low triglycerides concentrations were present in all animals. The pathological phenotype of all animals was steatorrhea with enterocytes of the small intestine containing intracytoplasmic lipid vacuoles. The peripheral nervous system of the heifer displayed degenerative changes.

Conclusion: Suspicion of CD in Holstein cattle is based on the presence of chronic diarrhea with no evidence of primary infections. Confirmation of the associated *APOB* gene mutation is needed. The described phenotype support a chylomicron formation defect due to the lack of the APOB-48 protein, as described in FHBL.

N-34: ROLES FOR NLRX1, NF-KB, AND AKT SIGNALING IN CANINE HISTIOCYTIC SARCOMA

Sheryl L Coutermarsh-Ott, Nikolaos G Dervisis, Irving C Allen

Background: Histiocytic sarcoma is a highly malignant neoplasm of macrophage or dendritic cell origin. It is not uncommon in dogs and can originate in a variety of different organs including spleen, lung, skin, and others. Treatment options are limited and often unsuccessful. Despite this, very little is known about the pathogenesis of this disease. Previous studies in our lab have suggested a role for NLRX1, a member of a recently characterized sub-group of NLRs, in histiocytic sarcoma development in mice. Moreover, these studies suggested that dysregulation of NF-kB and AKT signaling pathways was also involved. The current study aimed to test the hypothesis that those pathways found to be dysregulated in murine histiocytic sarcoma would also contribute to the disease in canines.

Methods: To test this, we used archived, formalin-fixed, paraffin embedded samples from canine patients diagnosed with histiocytic sarcoma at the VMCVM VTH. This was converted to cDNA and used in a qualitative RT-PCR reaction to evaluate expression levels of *NLRX1*, *AKT1*, *IL1B*, *PIK3R*, *IL6*, *TNF*, and *MTOR*. Additionally, samples from a different cohort of canine histiocytic sarcoma patients were used in a microarray analysis.

Results: Results confirm a significant downregulation of *NLRX1* in canine histiocytic sarcoma samples as well an upregulation of genes associated with both NF-kB and AKT signaling.

Conclusions: These data suggest that similar pathways are dysregulated in canine histiocytic sarcoma as have been previously identified in mice. A better understanding of the mechanisms associated with this disease is necessary for development of better treatment options.

N-35: FATAL PNEUMOTHORAX IN A DOG CAUSED BY A LESION RESEMBLING CONGENITAL PULMONARY AIRWAY MALFORMATION (CPAM) OF CHILDREN

Cameron G. Knight, Lorenza Malaguti, Sophie Rajotte, Nicolas Rousset, Brielle V. Rosa, James R. Wright

An 8 month old Boxer dog was referred for management of acute spontaneous pneumothorax. Computed tomography (CT) showed multiple variably-sized bullae in the right middle lung lobe, and partial or complete atelectasis of all remaining lung lobes. This was assumed to be a case of congenital lobar emphysema (CLE). The dog was anesthetized for surgery but died from cardiopulmonary arrest during pre-surgical thoracocentesis. Necropsy examination confirmed that ruptured bullae in the right middle lung lobe were responsible for the dog's pneumothorax and death. In addition, a malformation of the right subclavian artery was present. Histologically, the right middle lung lobe and primary bronchus were malformed, with microscopic changes resembling those of human congenital pulmonary airway malformation (CPAM), rather than CLE. CPAM has not been reported in the veterinary literature and should be considered in the differential diagnosis for spontaneous pneumothorax in dogs.

N-37: PATHOLOGY AND CAUSES OF DEATH OF STRANDED HUMPBACK WHALES (MEGAPTERA NOVAEANGLIAE) IN BRAZIL

Kátia R Groch, Josué Díaz-Delgado, Milton C. C. Marcondes, Elitieri B Santos-Neto, Adriana C Colosio, Vítor L Carvalho, Gisele S Boos, Ana Carolina O Meirelles, Antonio Fernández, José L Catão-Dias

Background: Humpback whales (*Megaptera novaeangliae*) migrate annually from feeding areas in the Scotia Sea to the Brazilian coast to breed and nurse their calves in the first months of their lives. Strandings have been recorded and efforts to gather information on causes of death through autopsy examination have increased in the last decade.

Objectives: This study investigates causes of stranding and death of 21 humpback whales found along the Brazilian coast from 2004 to 2014.

Methods: Data collected included date and location of stranding, sex, morphometrics and stranding condition (alive or dead). Autopsy was performed; tissues were collected and processed for histopathology. Immunohistochemical labeling for *Toxoplasma gondii* and morbillivirus antigen was performed on tissue sections of brain, lung, liver, lymph nodes and intestine.

Results: Fifteen (71.4%) whales were found stranded alive whereas six died at sea and were washed ashore. Of the 21, 16 (76.2%) were calves, four (19.0%) juveniles, and one (4.8%) adult. The most probable cause of stranding and/or death was determined in 19/21 (90.5%) individuals, and were as follows: neonatal/perinatal pathology (13/19; 68.4%); emaciation (2/19; 10.5%); discospondylitis (1/19; 5.3%); septicemia (1/19; 5.3%); vessel strike (1/19; 5.3%); and trauma of unknown origin (1/19; 5.3%).

Neonatal/perinatal pathology coexisted with trauma in 3 cases. Morbilliviral and *Toxoplasma gondii* antigen testing was negative in all animals assessed (n=19).

Conclusion: This study initiates the delineation of the baseline pathology and causes of death of a South Atlantic population of humpback whales and aids in the assessment of the population health.

N-38: DEVELOPMENT OF REAL-TIME PCR ASSAY FOR DETECTION OF VERONAEA BOTRYOSA IN ENVIRONMENTAL SAMPLES AND STURGEON TISSUES

Matthew F Sheley, Brian Murphy

Background: It has recently been determined that infection of cultured subadult (5-6 year old) white sturgeon with the fungus *Veronaea botryosa* results in severe morbidity and mortality within captive affected populations, which further results in a significant economic loss for the producers.

Objective: The aim of this study was to develop and validate a real-time PCR for the detection of the fungal organism in aquaculture-associated environmental samples to determine the source of the infection.

Methods: *Veronaea botryosa* DNA was extracted from subadult sturgeon tissues confirmed to be infected by the fungus based on histology, culture, and DNA sequencing. Two previously reported primer sets were acquired and used to amplify the extracted fungal DNA, which was then confirmed by sequencing and GenBank analysis. The fungal DNA was replicated in a plasmid, serially diluted, and evaluated by real-time PCR.

Results: A standard curve was created based on the various cycle threshold (Ct) values attained from serial dilution of the *V. botryosa* plasmid, which can be used to quantify the fungal DNA from various samples.

Conclusion: Real-time PCR is an efficient and sensitive way of detecting and quantifying *V. botryosa* DNA, and will be used to evaluate environmental and tissue samples from aquaculture raised sturgeon in order to detect the source of the fungus and to better understand the age at which the sturgeon are infected.

N-39: COMPARISION ANALYSIS OF IRON METABOLISM GENES OF THREE DIFFERENT LEMUR SPECIES, WITH INITIAL FOCUS ON HEPCIDIN IN BLOOD

Maja Ruetten, Hanspeter W Steinmetz, Vidhya Jagannathan, Sandro Altamura, Lloyd Vaughan, Max Gassmann

Background: We aim to measure hepcidin levels, the major iron regulating hormone, in *Lemur catta, Hapalemur griseus* and *Varecia rubra* with different sensitivities to iron storage disease (ISD) and identify all genes involved in iron uptake and metabolism to provide a comparative molecular basis for delineating the pathogenesis of ISD.

Methods: Hepcidin of *L. catta* and *V. rubra* was measured in 20 µl heparin-serum by Hepcidin-25 (bioactive) ELISA, DRG GmbH, Germany. DNA was isolated from blood or liver tissues using Qiagen kitÒ and the genomes sequenced by IluminaÒ HiSeq . Genomes were mapped to mouse lemur genome (Burrows wheeler aligner, SAM Tool, Velvet Columbus). Sequences of hepcidin were compared (CLC main workbench, swiss-model.expansy.org).

Results: The mean value of hepcidin of the *L. catta* was 11.7 (3.5-29.3) compared to 13.1 (10.0-18.0) ng/ml in *V. rubra*. The sequences of all lemurs differed from their nearest primate relatives by ten amino acids (aa) over the full length 84 aa proteins. The mouse lemur sequence differed at eight positions from the other lemurs, which also differed. Even within the 25 aa active peptide, *H. griseus*, *L. catta* and *V. rubra* differ from *M. murinus* at 2 positions and from the primates at three positions. In regions where mutations have been described in humans causing haemochromatosis, there are no changes in the lemur sequences.

Conclusions: The higher levels of hepcidin in *V. rubra* may induce higher sequestering of iron. The sequences of the precursor and cleaved proteins is unlikely to cause functional disruption of hepcidin.

N-40: OSTEOSARCOMA AND MAMMARY CARCINOMA CELLS ACTIVATE PLATELETS IN DOGS

Sandra C. Bulla, Peres R. Badial, Kari Lunsford, Stephen Pruett, Camilo Bulla

Background: Platelet-cancer cell interactions have long been studied in human and mice. It is well known that tumor cells can interact with and activate platelets, and that this interaction promote cancer malignancy and progression in those species. In dogs, there is no information regarding how cancer cells affect platelets.

Objective: The objective of the present study was to evaluate whether canine osteosarcoma and mammary carcinoma cell lines can activate platelets.

Methods: Platelet-rich plasma was obtained from heparinized blood of healthy dogs. Platelet-rich plasma samples were adjusted to a platelet concentration of approximately 200,000 platelets/uL. Aggregation was evaluated in response to three osteosarcoma cell lines (OSCA-8, OSCA-40, and OSCA-78) and one mammary carcinoma (CMT-28) cell line, in a double-channel optical light transmittance aggregometer. Pre incubations with an antagonist of ADP receptor P2Y12 were also done. Aggregation with collagen and ADP were included as positive controls. Leukocytes were isolated from the same dogs by density centrifugation and used as negative controls of aggregation. The resulting averages of percent maximum aggregation, slope, and area under the curve were evaluated in all samples. Pair-wise comparisons were performed using ANOVA followed by Tukey's or t-test, at the significance level of 0.05.

Results: All cell lines resulted in significantly increased percent aggregation and slope when added to the platelets. Pre incubation with P2Y12 antagonist significantly inhibited the aggregation by cancer cells.

Conclusion: Similarly to humans and murine cell lines, canine cancer cells activate platelets. This activation appears to involve the P2Y12 ADP receptor.

N-41: DETECTION AND SEQUENCING OF PPMV-1 IN PARAFFIN-EMBEDDED TISSUES FROM WILD PIGEONS BY NEXT-GENERATION SEQUENCING

Ying He, Marcos Isidoro-Ayza, Salman L Butt, Poonam Sharma, Kiril M Dimitrov, Claudio L Afonso, Hon S Ip, James B Stanton

Pigeonparamyxovirus 1 (PPMV-1) is a genetically distinct member of avian paramyxovirus serotype 1 (APMV-1). APMV-1, is synonymous with Newcastle disease virus, with PPMV-1 being part of class II genotype VI. PPMV-1 is highly contagious and can result in enteric, visceral, and neurologic disease in infected birds, including poultry. As with other NDVs, PPMV-1 is genetically diverse, with the diversity represented by its division into nine sub-genotypes. The detection of different PPMV-1 strains and the relationship between mutation and virulence is still not fully defined, which hampers effective surveillance of Newcastle disease. Next-generation sequencing (NGS) is a powerful tool in genomic research. While other studies have used formalin-fixed, paraffin-embedded (FFPE) tissues in NGS, the use of NGS on FFPE tissues is relatively limited in veterinary medicine. The use of FFPE in NGS allows for retrospective studies, and for molecular investigations of highly regulated infectious organisms (e.g., foreign animal diseases such as NDV) outside of containment. In this study, we used Illumina MiSeq-based NGS to detect and sequence PPMV-1 from FFPE samples of wild pigeons. RNA extraction and library preparation protocols were optimized. KAPA Stranded RNA-Seq kit was used to prepare the sequenced libraries, which resulted in detection of PPMV-1, with up to 99% coverage of the entire viral genome. This study indicates the feasibility of using NGS to detect and sequence pathogens in FFPE tissues from wild birds.

N-42: INFESTATION OF HOPILAS MALABARICAS WITH THE NEMATODE PARASITEEUSTRONGYLOIDES SPP. IN THE ARENA DAM OF TRINIDAD, WEST INDIES.

ROD B SUEPAUL, RYAN S MOHAMMED, ASOKE BASU

Background: *Hopilas malabaricas,* is found in Trinidad, locally called 'guabine,' but referred to as 'wolf fish' in the ornamental trade. After discovering "worms" in the flesh, a local fisherman brought seven *H. malabaricus* for necropsy.

Methodology: Necropsy was performed and the helminths were preserved in 70% alcohol for identification. Tissue samples were fixed in 10% buffered formalin for 48 hours, embedded in paraffin and stained with haematoxylin and eosin for histological examination. Sections of the gastrointestinal tract were also examined for parasite.

Results: In 4 out of 5 fish, 14 to 18 worms approximately 30.0 mm long with a 2.0 mm diameter were isolated. Amongst the muscle fibres, there were multiple sections of large helminth parasites within a dense fibrous capsule which was infiltrated by granulocytes and macrophages. The nematodes were identified as *Contracaecum* sp. and *Eustrongyloides* sp.

Conclusions: This is the first report of *Eustrogyloides* sp. in Trinidad. The parasites were found in a similar anatomical location as other fish and, typical histological lesion were present. The guabine were caught during a metrological drought. *Eustrongyloides* sp. are capable zoonotic agents.

N-43: DETERMINATION OF IMPORTANT CAUSES OF MORBIDITY AND MORTALITY IN CAPTIVE PSITTACINES SUBMITTED TO THE ONTARIO VETERINARY COLLEGE TEACHING HOSPITAL

Thisuri Eagalle, Nicole M. Nemeth, Hugues Beaufrere, Leonardo Susta

Background: Psittacines are increasingly common in households, aviaries and zoological collections. With the advancement of avian medicine, diagnostic evaluations (i.e., necropsy and biopsy) are increasingly important for early and accurate diagnoses and to predict disease outcomes. However, comprehensive information about disease conditions in birds, including pathological descriptions, is often limited.

Objective: We retrospectively reviewed diagnostic data from psittacines submitted to the Ontario Veterinary College (OVC) to assess baseline prevalence and pathological features of the most commonly diagnosed diseases and conditions.

Methods: Diagnostic data from psittacine carcasses (necropsy) and samples (biopsy) submitted to the OVC and Animal Health Laboratory from 1995-2015 were reviewed. Data included signalment and morphologic diagnoses, which were categorized and ranked to identify primary cause(s) of death.

Results: Eighty-three species (n=1,149) were represented, with the African grey, cockatiel, and green-winged macaw as the most common. Infectious agents were most commonly identified as the primary cause of death (n=409; 35.6%), including 216 viral (52.8%; some suspect), 132 bacterial (32.3%), and 61 fungal (14.9%) infections. The most common respective etiologies were bornavirus (n=154), *Mycobacterium* spp. (n=18), and *Aspergillus* spp. (n=18). Neoplasia was diagnosed in 100 cases (8.7%), which most commonly involved the alimentary tract (n=18); squamous cell carcinoma was the most frequently diagnosed neoplasia (n=9).

Conclusions: Better recognition and understanding of diseases that impact psittacines will aid clinicians and pathologists to formulate more effective differential diagnoses and targeted diagnostic procedures, ultimately leading to better treatment, management and prophylactic protocols to improve the health of captive psittacines in Ontario and elsewhere.

N-44: Immunohistochemical characterization of feline lymphoplasmacytic anterior uveitis

Lawrence W Crossfield, Emma Scurrell, Màrian Matas Riera, Charlotte Dawson, Rachel Hampel, Caroline Thaung, Simon Priestnall, Oliver Garden

Background: Lymphoplasmacytic anterior uveitis is a common, often idiopathic ocular disease of cats.

Hypothesis: Feline idiopathic anterior uveitis is characterized by an induction *in situ* of pro-inflammatory T cells expressing IL-17, associated with a perturbation of the balance of regulatory T cells in the anterior uveal tract.

Methods: Immunohistochemical labeling for CD3, FoxP3 and IL-17 was performed on feline globes with idiopathic anterior uveitis. Cases of feline infectious peritonitis (FIP) and cats euthanized for non-ocular disease were used as controls. Fifty percent of the sectional area of the ciliary body or iris was randomly selected to quantify the cellular infiltrate by means of a macro, expressing the data as number of cells/100,000μm^2.

Results: In both uveitis and FIP samples, membranous CD3 labelling of mononuclear cells was detected. Median areal density of $CD3^+$ cells was greater in the ciliary body than iris (FIP: ciliary body 230, iris 157; uveitis: ciliary body 252, iris 193; p^+ cells was significantly higher in the FIP group (FIP: ciliary body 90, iris 104; uveitis: ciliary body 7, iris 7; $p<0.05$). There was no significant difference in expression of FoxP3 between FIP and anterior uveitis. No CD3, IL-17 or FoxP3 immunoreactivity was detected in the healthy eyes.

Conclusions: IL-17 is unlikely to play a primary role in feline lymphoplasmacytic anterior uveitis, in contrast to FIP, in which this cytokine shows greater expression. $FoxP3^+$ cells were present in both diseases.

N-45: IDENTIFICATION OF A NOVEL PATHOGENIC MUTATION IN GLOBOID CELL LEUKODYSTROPHY (KRABBE DISEASE) IN A FAMILY OF MIXED-BREED DOGS

Moeko Kohyama, Hinako Yamazoe, Kazuyuki Uchida, Akira Yabuki, Mariko Shima-Sawa, Osamu Yamato

Background: Globoid cell leukodystrophy (Krabbe disease) is a neurodegenerative leukodystrophy that affects the central and peripheral nervous systems, which is caused by the deficiency of lysosomal galactocerebrosidase encoded by the *GALC* gene. Canine Krabbe disease has been reported in several breeds, and two pathogenic mutations have been identified. Recently, two mixed-breed littermate dogs were histopathologically diagnosed as having Krabbe disease caused by the massive infiltration of globoid cells in the white matter throughout the forebrain. However, these dogs did not have gene mutations reported previously.

Objective: To identify a novel pathogenic mutation of the canine *GALC* gene.

Methods: Direct sequence analysis was performed by using genomic DNA of the two affected dogs. Breed ancestry was analyzed in one of the dogs by using Wisdom Panel®. A genotyping survey was conducted in the randomly-collected dog population of 13 Golden Retrievers, 582 Shiba Inus, and 54 Chihuahuas in Japan by using the real-time PCR method.

Results: A candidate pathogenic mutation was identified as a homozygous missense substitution in the *GALC* gene in the two affected dogs. The ancestry analysis revealed that the breed ancestry of the dogs included Golden Retriever, Shiba Inu, and Chihuahua in the latest three ancestral generations. However, the mutant allele in these breeds has not been identified yet in the survey.

Conclusions: A novel pathogenic mutation was identified in canine Krabbe disease. This new finding will be important not only in canine veterinary medicine but also as an animal model for the human disease.

N-46: NATURAL PATHOLOGY OF THE CAPTIVE CHIMPANZEE (PAN SPP.): A 32 YEAR REVIEW

Hannah M Laurence, Michael A Owston, Robert E Lanford, Shyamesh Kumar, Edward J Dick

Background: We report the spontaneous lesions observed in 246 chimpanzees at the Southwest National Primate Research Center at Texas Biomedical Research Institute, San Antonio, Texas, over a 32 year period.

Methods: A pathology database search for all the morphologic diagnoses for chimpanzees was performed that included all biopsies and all animals that died due to natural causes or were euthanized for humane reasons. For each morphologic diagnosis, the associated system, organ, and etiology, the animal's age at diagnosis, and sex were recorded and analyzed.

Results: A total of 1361 macroscopic or microscopic morphologic diagnoses were identified in 246 chimpanzees. The most frequent diagnoses in descending order of occurrence were cardiomyopathy, hemosiderosis, hyperplasia, nematodiasis, edema, hemorrhage, fibrosis, pneumonia, congestion, nephritis, necrosis, stillborn, colitis, cyst, amyloidosis, arteriosclerosis/atherosclerosis, leiomyoma, hepatitis, ascites, atrophy, and myocarditis. These 21 diagnoses accounted for 57.4% of the lesions observed. Lesions were most frequent in the gastrointestinal system (25.7%), followed by the cardiovascular (21.8%), urogenital (15.7%), respiratory (11.6%), and lymphatic/hematopoietic (9.4%) systems. An etiology was identified for 770 of the 1361 diagnoses. The most common etiologies were degenerative disease (35.1%), physiologic (21.4%), neoplastic (11.0%), parasitic (10.6%), bacterial (7.7%), traumatic (4.7%), and mycotic (2.9%).

Conclusion: Cardiomyopathy was the most common morphologic diagnosis, the gastrointestinal system was the most often affected system, and degenerative diseases accounted for the most frequently identified etiology of all submissions. Knowledge of the natural pathology of chimpanzees is required for better husbandry practices and understanding of the ageing process in captive chimpanzees.

N-47: THE ROLE OF NEOSPORA CANINUM IN WILDLIFE DISEASE

Shannon L. Donahoe, David N. Phalen, Scott A. Lindsay, Mark Krockenberger, Jan Šlapeta

Neospora caninum is an apicomplexan parasite that is the etiologic agent of neosporosis, a devastating infectious disease regarded as a major cause of reproductive loss in cattle and neuromuscular disease in dogs worldwide.To date, an extensive number of wildlife species have been investigated for their possible role in the *N. caninum* life cycle and many have been implicated as intermediate hosts on the basis of serologic and/or molecular evidence. However, the occurrence and importance of disease due to infection in these nondomestic animals remains poorly understood. Most reports of positive *N. caninum* exposure in wildlife are in asymptomatic animals and, in many instances, investigations of possible associated morbidity, mortality, and pathology have been neglected. In order to improve our understanding of the significance of *N. caninum* infection in nondomestic species, we review the clinicopathologic features described for the surprisingly low number of cases of neosporosis reported in wildlife and present best practice guidelines to follow for reporting wildlife cases of neosporosis. While current data would suggest *N. caninum* infection does not adversely impact wildlife populations, there is a need for greater international awareness and uniformity in the diagnosis of *N. caninum* infection and neosporosis in nondomestic species in order to assess the true consequences of parasite infection.

N-48: EXPRESSION OF MAST CELL GENES IN CANINE TRANSMISSIBLE VENEREAL TUMOR

Hannah S Bender, Elizabeth P Murchison, Andrea Strakova

Canine transmissible venereal tumor (CTVT) is one of only three naturally occurring transmissible cancers, and is the oldest known clonal cell lineage in the world having arisen from a single founder animal approximately 11,000 years ago. CTVT is transmitted between animals during sexual contact, overcoming immune barriers to 'non-self' in order to propagate large, ulcerative masses on the external genitalia. Spontaneous regression occurs in some animals and metastasis is occasionally reported. The factors that facilitate CTVT transmission and regression are poorly understood and may relate to the tumor cell of origin. Although it is widely accepted that CTVT is derived from a histiocytic cell, data supporting this assumption is ambiguous. Here we demonstrate CTVT expression of a suite of mast cell genes including mastin, chymase, carboxypeptidase A3, histidine decarboxylase and FcεR1β. In validating this data, PCR primers were designed around tumor-specific variants identified in the CTVT genome in order to exclude the possibility of host mast cell contamination. The relevance of this finding to CTVT histogenesis, and interactions between the tumor and host will be discussed.

N-49: DISEASES IN FREE ROAMING CHICKENS IN THE CARIBBEAN

Pompei Bolfa, Trista Cheng, Hieuhanh Huynh, Jennifer Ketzis, Silvia Marchi, Aspinas Chapwanya, Taurai Tasara, John Joseph Callanan

Background: Chickens free-roam on Caribbean islands and represent opportunities for pathogen exchange between people and other animals, in addition to being sentinels for island-based diseases and those introduced by migratory birds that travel between North America and the Caribbean. Our study aimed to fill in the gap of information on diseases of free roaming chickens in the Caribbean.

Methods: 81 clinically healthy free-roaming chickens were obtained from 9 areas throughout St. Kitts. Serology for NDV, IBDV, IBD, WNV, Avian influenza, salmonellosis and mycoplasmosis was undertaken in addition to a parasitological examination and a histopathological evaluation of 19 different organs. Furthermore the microbiota of the small and large intestines was evaluated.

Results: Histopathologically the most affected organ was the liver (86% of the chickens) followed by duodenum (45%) skin (41%) kidney and spleen. Correlations between serology and histopathology revealed only NDV and IBDV associated bursal pathology. There was no evidence of WDV exposure or IBDV associated pathology. However, pulmonary pneumoconiosis (10% of the chickens) suggestive of environmental exposures was observed. Serologically IBDV (86%) and IBD (84%) were the most prevalent followed by mycoplasmosis (37%) and NDV (30%). Seropositivity for 2 or more infectious agents was seen in 83% of the chickens. Some human pathogens detected in the intestinal microbiome were several *Escherichia coli* strains, *Clostridium perfringens* or *Campylobacter jejuni*.

Conclusion: Free roaming chickens in the Caribbean while a potential reservoir for various human pathogens harbor predominantly food borne pathogens within their intestinal tract.

N-50: FIRST IDENTIFICATION OF SETARIA SPP. MICROFILARIAE IN FREE-RANGE MAMMALIAN WILDLIFE FROM BRAZIL.

Marina Mitie Monobe, Rodrigo Costa da Silva, Felipe Fornazari, Helio Langoni, Regina Kiomi Takahira, Joao Pessoa Araujo-Junior

Background: *Setaria* spp. is a worldwide vector-borne thread-like parasitic worm transmitted by haematophagous arthropods from bovines and buffalos to other unusual mammalian hosts, i.e., mammalian wildlife and humans. In some countries such as Finland, Sweden and Norway setariosis causes substantial economic losses and increase workload associated with meat processing.They are known for their harmless effects on their definitive hosts, but are still poorly studied and reported in wildlife, especially in South America. Any report of epidemiological characterization of this parasite has been previously registered in wildlife animals from Brazil.

Objective: The aim of the present study was to identify and determine the prevalence of *Setaria* spp. in some wildlife species living in Brazil.

Methods: This study included hematological evaluation of peripheral blood from 160 mammalian wildlife individuals captured from urban, farm and forest areas in Brazil. The microfilaria detection included a modified Knott's technique and direct analysis of blood smears. A multiplex PCR protocol was performed to detect and differentiate the most common filaroids infecting dogs. Sequencing of PCR products confirmed the parasite's specie.

Results: Microfilaremia was observed in 17.31% of captured white-eared opossum (*Didelphis albiventris*), 83.33% of coati (*Nasua nasua*), 20% of lesser grison (*Galictis cuja*) and 50% hoary fox (*Lycalopex vetulus*). The DNA matched to *Setaria labiatopappilosa* [GenBank (AJ544872)].

Conclusion: The high prevalence of this parasite and the closed contact to humans emphasize the importance of this vector-borne neglected zoonosis and its impact on Brazilian ecosystems, since the studied area is endemic for dengue disease, with same vectors.

N-51: CLINICAL FINDINGS ENTEROLOBIUM CYCLOCARPUM POISONING IN CATTLE IN THE EASTERN PLAINS OF COLOMBIA

LEONARDO ROA, HECTOR RINCÓN, BENJAMIN DONCEL

Background: In Colombia the *Enterolobium cyclocarpum* was identified as the plant species with greater potential toxic livestock in the eastern plains; however, clinical and pathological findings of their exposure are unknown in cattle.

Objective: The aim of this study was to evaluate the toxic potential of *E. cyclocarpum* experimentally in cattle in the eastern plains of Colombia.

Methods: Twelve animals were distributed as in the following groups: Group 1, control animals consumed fresh natural grazing and other exposed groups consumed the fruit in doses previously established, Group 2, 10 g/kg BW / *E. cyclocarpum* fruit in a single dose, Group 3, 10 g/kg BW daily for 15 days and Group 4, 20 g/kg BW daily for 15 days. Clinical examination, clinical pathology (hematology and serum chemistry), necropsy and histopathology were conducted at the end of the feeding trial.

Results: The animals developed one of two clinical presentations: 1) Acute toxicity with digestive, skin, kidney and nervous disorders. 2) Subchronic toxicity with alterations in the same systems but less severe and some transient. There was increased AST, BUN, creatinine and fibrinogen. Clinical signs, pathological lesions and clinical pathology alterations were observed in animals exposed to *E. cyclocarpum.* These included inflammation, hepatic lesions and azotemia. The clinical and pathological lesions were more severe in animals exposed to high doses and frequency of *E. cyclocarpum* fruit consumption.

Conclusion: *E. cyclocarpum* is toxic to cattle under the ecological conditions of the Eastern Plains of Colombia.

Veterinary Student Posters

VSP-01: METASTATIC SEMINOMA IN A DOG

Megan Zalek, Kurt Zimmerman, Sabrina Barry, Thomas Cecere

A four-year-old, intact male, mixed breed dog was presented to the Virginia-Maryland Veterinary Teaching Hospital for an abdominal mass. Upon physical examination, a large firm abdominal mass and an 11x6x5 cm left testicular mass were palpated. Clinicopathological findings revealed a mild regenerative, microcytic hypochromic anemia, lymphopenia, basophilia, and mild hypoalbuminemia. Radiographic and ultrasound imaging revealed a left abdominal mass that arose from the retroperitoneal space and a suspected neoplastic left testicular mass. Computed topography (CT) scans showed a cranial mediastinal mass in the thorax. Needle aspirates from all masses were cytologically similar and consistent with anaplastic round cell sarcoma. Two preliminary differentials included seminoma with metastatic disease and histiocytic sarcoma. A castration and scrotal ablation were performed and both testes were submitted for histopathologic evaluation. Histopathology revealed diffusely effaced testicular architecture in the left testis that was replaced with dense sheets of round to polygonal neoplastic cells separated by collagenous septa. Neoplastic cells were characterized by moderate amounts of eosinophilic cytoplasm, large round to ovoid nuclei with stippled chromatin, and 1-4 nucleoli. Mitoses were frequent, often bizarre, and multinucleated neoplastic cells were present. Multifocal areas of necrosis occurred throughout the neoplasm and intravascular emboli of neoplastic cells were within the pampiniform plexus. Within a focally extensive area of the right testis, seminiferous tubules were distended by the same neoplastic cells described previously. These findings supported the diagnosis of malignant seminoma. Although unusual, several cases of malignant seminomas with extensive metastases, including to the lung, have been reported in dogs.

VSP-02: HISTOLOGICAL EFFECTS OF ELECTROCHEMICAL DISSOLUTION IN ANIMAL STUDIES OF CARDIOVASCULAR DEVICES

Jiajie Xu J Xu, Elizabeth Crowling B Crowling, Andrew Nguyen V Nguyen, Brad Weeks Weeks

The in situ examination of metallic stents is common in cardiovascular pathology. Current methods of evaluating stented vessels include gross dissection (resulting in artifactual destruction) and plastic embedded micro-ground histology (time consuming and expensive). A recent cardiovascular pathology manuscript proposed electrochemical stent dissolution using lemon juice or 5% citric acid in order to process the sample through classical histology. Dental literature proposed similar methods of dissolving endodontic files using 0.5 % NaF and 0.1 % NaCl. This study compares the histological effects of previously validated solutions on the effects of tissue quality. We hypothesized that there would be no significant differences between the tissue samples.

Solutions composed of 100 % lemon juice, varying concentrations of citric acid (2.5, 5, and 10 %), and the combination of 0.5 % NaF + 1 % NaCl. In another set of solutions, NaCl concentrations in all solutions were brought up to 5g/100ml solution to protect the tissues from osmotic damage. Each solution had canine cardiac tissue blocks immersed for 1 and 24 hours. Samples were processed using paraffin fixed histology and stained with H&E. The results showed solutions with the 5g/100 ml of NaCl preserved tissue quality better than those without, samples soaked for 24 hours preserved tissue quality better than samples soaked for 1 hour, and lemon juice, followed by 5% citric acid showed the best preservation of tissue. These results will be applied to optimize electrochemical dissolution of cardiovascular stents and other implanted metallic devices while preserving histological quality.

VSP-03: PROLIFERATIVE GRANULOMATOUS CYSTITIS IN AN ENGLISH BULLDOG: A NOVEL PRESENTATION OF ESCHERICHIA COLI INFECTION

Timothy K. Wu, Andrew D. Miller, Kenneth Simpson, Jonathan Schnier

A 4-month- old, female intact English bulldog presented with hematuria, stranguria,and polyuria. Urinalysis revealed a urinary tract infection and ultrasound revealed a mass arising from the cranioventral aspect of the urinary bladder mucosa and protruding into the lumen. The bladder mass was removed via partial cystectomy. Histopathology revealed that the lamina propria was markedly expanded by sheets of macrophages with abundant eosinophilic to finely granular cytoplasm. The cytoplasm of the macrophages stained strongly positive with periodic-acid Schiff (PAS) and FISH analysis revealed large numbers of Escherichia coli within cells in the bladder wall. The lesions in the urinary bladder appeared very similar to E. coli associated granulomatous colitis of boxers and other breeds in the mastiff cluster including English bulldogs. Extraintestinal manifestations of aberrant E. coli infection are uncommon but have been reported in cats (bladder and vagina) and a boxer dog (kidney and colon). Recent studies have related a region on chromosome 38 encoding SALm molecules to susceptibility in boxers and French bulldogs. Deletion of SALm in mice is associated with a failure to kill E.coli.

VSP-04: HYALURONAN PROCESSING AND FUNCTION IN THE PROGRESSION OF BREAST CANCER

Patrice M Witschen, Polly Chuntova, Nicholas J Brady, Jaime F Modiano, Kathryn L Schwertfeger

Hyaluronan (HA) is a large, soluble, glycosaminoglycan of the extracellular matrix that has anti-inflammatory effects under physiologic conditions. However, HA is cleaved into low molecular weight (LMW) fragments under conditions of cellular or organismal stress, acting as a molecular “switch” promoting inflammation. In breast cancer, a decrease in HA synthesis has been correlated with decreased tumor cell proliferation and migration. However, the roles of HA fragmentation in the progression of breast cancer are unknown. We predict that HA fragmentation increases during this transition, promoting inflammation through LMW HA-CD44 interactions. To test our hypothesis, the presence/absence of HA fragmentation was determined using gel electrophoresis in breast cancer cell lines. Additionally, qRT-PCR was performed to examine gene expression of the three major hyaluronan synthases (HAS) 1-3 and the two major hyaluronidases (HYAL) 1 and 2. Our data suggest as a cancerous lesion progresses, HMW HA production increases (primarily through HAS2), but HA fragmentation does not occur until the tumor acquires a more aggressive phenotype (primarily through HYAL1). We also verified the presence of CD44 protein in normal and cancerous cell lines via flow cytometry and found an increase in CD44 cell surface expression in aggressive tumor cells when compared to normal cells. Following characterization of HA fragmentation and machinery within our system, we found changes in inflammatory cytokines (such as IL8) as downstream effects of CD44 signaling in response to HA synthesis inhibition. By targeting CD44 signaling associated with inflammation, new therapeutic approaches can be developed for the treatment of breast cancer.

VSP-05: COMBINATION TREATMENT WITH CHEMOTHERAPY AND AN AUTOPHAGY INHIBITOR IN CANINE OSTEOSARCOMA CELLS

Kadi T.N. White, Courtney R. Schott, Geoffrey A. Wood

Background: Autophagy is the degradation and recycling of cytoplasmic components. This process, which occurs at basal levels in normal cells, improves cell survival during adverse conditions such as starvation and hypoxia. Osteosarcoma is the most common primary malignancy of bone in dogs. Osteosarcoma cells may be using autophagy to improve survival during exposure to chemotherapy, contributing to its chemoresistance. Objective: To assess the effects of combination treatment of a chemotherapeutic drug and an autophagy inhibitor (spautin-1) on osteosarcoma cell viability and expression of autophagic proteins. Methods: Metastatic osteosarcoma cells were cultured in a combination of four different conditions for 6 hours: starvation, 40.83 μM doxorubicin (IC50), 15 μM or 120 μM spautin-1 (IC15 or IC25), and 75 μM hydroxychloroquine. An LC3 western blot was performed to assess autophagic flux. Differences in cell viability between conditions were assessed by one-way ANOVA followed by Tukey post-tests. Results: Treatment with doxorubicin or 120 μM spautin-1 alone decreased cell viability by 51% and 37%, respectively ($p < 0.05$). Combination treatment with 120 μM spautin-1 and doxorubicin decreased cell viability by and additional 9% ($p < 0.05$). On western blot, spautin-1 lowered LC3 levels, indicating a reduction in autophagic flux, under all conditions except when combined with doxorubicin. Conclusion: Spautin-1 inhibits autophagy in canine osteosarcoma cells in vitro. At the current dosages, combination treatment with doxorubicin and spautin-1 significantly increases osteosarcoma cell chemosensitivity but the mechanism does not appear to be the result of autophagy inhibition.

VSP-06: AGE-RELATED DIFFERENCES IN COLONIC STRUCTURE AND FUNCTION IN VERVET MONKEYS

Magan N. Wells, Chrissy Sherrill, Ashley Davis, Matthew Brown, Anthony Fodor, Kylie Kavanagh

Microbial translocation (MT) is the movement of bacteria from the gut lumen to extra-intestinal sites. A dysfunctional intestinal mucosal barrier has been recently recognized as a feature of aging, which can cause increased MT and incite systemic inflammation. The purpose of this study was to search for reasons underlying decreased intestinal mucosal barrier function that occurs in aging by comparing the colonic histological, immunological, and microbiome characteristics of healthy young and old vervet monkeys (n=19; Chlorocebus aethiops). It was hypothesized that there would be decreased structural protection as fewer mucin-producing goblet cells and tight junction protein abundance, and shallower crypts in older monkeys. Innate immunodeficiency was expected in older monkeys as measured by fecal and plasma IgA, and Paneth cell defensin concentration, leading to a different microbiome. Blinded observers quantified histological sections, conducted ELISA assays and quantified the microbiomic profile. MT burden was significantly elevated in old monkeys despite no structural changes in the colon. Circulating defensin levels were significantly higher and comparable IgA levels suggest competent innate immunity with healthy aging. Furthermore, associative relationships suggest the colon structure does alter positively to counteract factors related to MT. There was greater colonic bacterial diversity and abundance of selected bacterial species in the older monkeys, suggesting that increased MT could result loss of adaptive immunity and the ability to modulate microbial type and amount in the colon niche. Future work will investigate gut-associated lymphoid tissue and innate lymphoid cells as targets for age-related MT increase.

VSP-07: The effect of FOXA2 inactivation on mucus overproduction in canine respiratory infections

Michelle A Waltenburg, Woosuk Choi, Carol Maddox, Brendan McKiernan, Richard Fredrickson, Gee Lau

Dogs naturally develop several pulmonary diseases similar to humans. Older dogs, especially of small breeds, develop infectious canine tracheobronchitis (kennel cough), chronic bronchitis (CB), and chronic obstructive pulmonary disease (COPD) with excessive mucus due to chronic exposure to environmental pollutants or infectious agents. Previous clinical studies involving 766 canine lower respiratory tract infection cases showed that major pathogens included those on the ESKAPE list: Pseudomonas aeruginosa, Staphylococcus, E. coli + other Enterobacteriaceae and Enterococcus. FOXA2 is a key transcriptional regulator that maintains airway mucus at healthy levels. FOXA2 is inactivated in human airways with respiratory infection, resulting in excessive mucus. Previously, our laboratory demonstrated that pyocyanin (PCN), a virulence factor of P. aeruginosa, causes goblet cell hyperplasia and metaplasia and mucus hypersecretion by repressing the expression of FOXA2. However, the role of FOXA2 regulation in canine respiratory diseases remains unexplored. In this study, we retrospectively examined cases of canine respiratory infection received by the University of Illinois Veterinary Diagnostic Laboratory. We used immortalized canine pulmonary bronchial epithelial carcinoma (BACA) cells to examine the mechanism of FOXA2 inactivation by PCN. Finally, we examined bronchial sections from dogs with CB and COPD by immunohistochemistry to determine whether microbial pathogens found in these airways inactivate FOXA2. Our results indicate that inactivation of FOXA2 by Bordetella spp. and P. aeruginosa bacteria leads to goblet cell hyperplasia and metaplasia and mucus hypersecretion in the airways of dogs. Moreover, in vitro studies with BACA cells show that PCN inactivates FOXA2 in a time- and concentration-dependent manner.

VSP-08: DIETARY IRON AND THE RISK OF NONALCOHOLIC STEATOHEPATITIS
Katherine N Turnbull, Lipika Salaye, Sandy Sink, Donald McClain

Background: Nonalcoholic steatohepatitis (NASH) is a common but serious condition that often accompanies diabetes. Among the many factors that affect the development of both diabetes and NASH is high tissue iron. Objective: With large ranges of iron intake in humans, and the association of nonalcoholic fatty liver disease at both high and low iron levels, we sought to examine the dose-dependent interaction between dietary iron levels and hepatic lipid metabolism. Methods: We used a mouse model of NASH, namely males fed a Western "fast food" (FF) diet. Groups (N=5-6 each) were fed either regular chow or FF with one of four iron levels, 4mg, 35mg, 500mg, or 2g per kilogram of chow. Results: The two diets did not affect iron intake, and iron did not affect the initial development of NAFLD. Lower levels of iron protected mice on FF from NASH. When the rate-limiting enzyme for heme synthesis was bypassed by adding aminolevulinic acid to the water, mice on the protective low iron FF diet developed levels of liver damage comparable to higher iron diets, suggesting that some of the effects of iron may be mediated by heme. Female mice appear to be protected from the effects of the fast-food diet, with liver damage enzymes trending up with iron but at lower levels than males. Conclusion: Iron and gender are factors in the development of NASH, consistent with human epidemiologic observations. The model can be used to dissect mechanisms of protection from NASH in females and with low tissue iron.

VSP-09: JQ-1 DECREASES VIABILITY IN CANINE HEMANGIOSARCOMA CELL LINES INDEPENDENT OF APOPTOSIS
Erin E Trageser, Kristen R Richter, Cory Nielsen, Kimbal E Cooper, Brian P Wellensiek, Mark Olsen, Valerie M Wong

Hemangiosarcoma is a common, aggressive, and usually fatal neoplasm in dogs. Standard doxorubicin-based chemotherapy regimens offer minimal survival benefits. To evaluate novel treatment options, this study aimed to (1) determine whether or not canine hemangiosarcoma expressed BRD4, a bromodomain-containing protein, and (2) determine the effects of JQ-1, a bromodomain inhibitor, on two canine hemangiosarcoma cell lines (FROG and EFS). The BRD expression sequence was amplified from mRNA expressed in the FROG cell line. Additionally, the nuclear expression of BRD4 was detected in patient-derived tissues using an anti-human BRD4 antibody by immunohistochemistry. FROG and EFS cells were incubated with a concentration gradient of JQ-1 ranging from 62.5 nM to 2000 nM, with or without doxorubicin (100 nM). Cell viability and caspase-3/7 activities were determined at 48 h and 72 h. Compared to cells treated with the vehicle only, JQ-1 decreased cell viability significantly at both time points (P

VSP-10: Comparison of Region of Interest (ROI) Scanning vs Whole Slide Imaging (WSI) Using the PanoptiqTM System for the Preclinical Assessment of Medical Devices.

Teo Rousselle, Liam Rousselle, James Stanley, Krista Dillon, Dane A Brady, Serge D Rousselle, Armando Tellez

Introduction:Slide scanning systems require a large capital investment,space,maintenance cost,specialized staff and high capacity data storage.Once the scan is acquired,a pathologist may have to select specific regions of interest (ROI). We compare a novel dynamic microscope imaging acquisition system(Panoptiq™ViewsIQ,Richmond,Canada)to a conventional high-throughput scanning platform. Methods:Ten slides were scanned in a specific ROI with the Pantoptiq™ equipped with a microscope camera(Point Grey) on a BX51 Olympus microscope. The whole slides were also scanned with a VS120(Olympus Life Science) high-throughput scanner. An experienced technician performed the full-section image acquisition on the VS120 system while the Pantoptiq™ acquisition was performed by an individual not routinely familiar with the slide scanning process. The parameters evaluated were: set up time required, scanning time and resulting file sizes. Results:The time to set up the VS120 system was of 57.6" on average. This setup is comprised of three parts: slide metadata (8.4±2"),complete scan overview (28.9±3.5"),and setting of the scan area(20.3±4.5").Comparatively, the Pantoptiq™ is very direct,taking 17.6" from slide placement on the stage,selection of the ROI and being ready to scan.The Pantoptiq™ also required 30% less time to perform the image acquisition, including archiving the image, when compared to the VS120 system (44.2±8.3"vs.72±9.44",respectively).Furthermore,the size of the files that resulted from the Pantoptiq™ was only ~19% of the size that resulted from VS120 (6.5±1.3vs.34.5±8.4 MB). Conclusion:The Panoptiq™ ViewsIQ software provides an intuitive, versatile, economical and user-friendly platform which allows a pathologist to quickly acquire easy to manage images directly during slide evaluation without sacrificing the quality of the scanned images.

VSP-11: HYPERCALCEMIA OF MALIGNANCY IN A MINIATURE HORSE WITH SQUAMOUS CELL CARCINOMA

Jere K Stern, Amy S Stewart, Sarah Beatty

A 16-year-old Miniature Stallion was presented to a veterinary referral center with a draining wound over the right facial crest and concurrent swelling of several months duration. Previously reported treatment included lancing and flushing the area; detailed response to therapy was unknown. Physical examination revealed moderate tachycardia, decreased borborygmi and enlargement of the right mandibular lymph node. Complete blood count showed a stress leukogram with mild neutrophilia and lymphopenia. Serum biochemistry revealed severe hypercalcemia and mildly increased liver enzymes. Mild hyperbilirubinemia and hypertriglyceridemia were also present consistent with recent anorexia. Skull radiographs revealed severe bone lysis of the maxilla deep to the swelling. Cytologic interpretation of the mass and regional lymph node was squamous cell carcinoma with metastasis. A calcium panel at Michigan State University confirmed hypercalcemia (including ionized) along with severely increased

PTHrP, minimally increased 25-Hydroxyvitamin-D, and low normal PTH. Hypercalcemia of malignancy is an uncommon finding in equids with squamous cell carcinoma.

VSP-12: IMMUNOPHENOTYPIC CHARACTERIZATION OF CANINE SPLENIC FOLLICULAR DERIVED B-CELL LYMPHOMAS

Leah R Stein, Matti Kiupel, Cynthia Bacmeister

Background: Splenic follicular derived B-cell lymphomas belong to a subgroup of indolent lymphomas that arise on a background of lymphoid follicular hyperplasia. While both, marginal zone lymphomas (MZLs) and mantle cell lymphomas (MCLs) have been reported in dogs, there is little information on their immunphenotypical characteristics. Objective: The goal of this study was to characterize the immunophenotype of canine splenic MZLs and MCLs. Methods: Thirteen MCLs and 35 MZLs were selected based on their morphologic features and tissue micro arrays were generated to evaluate expression of CD3, CD45, CD20, CD79a, Pax-5, Bcl-2, Bcl-6, CD10, MUM-1 and cyclinD1. Results: MZLs were characterized by proliferating intermediate sized B-cells with abundant, lightly stained cytoplasm and mildly vesiculated nuclei with marginated chromatin and a large single, central nucleolus that formed homogenous cuffs around fading germinal centers. While MCLs also surrounded fading germinal centers, neoplastic B-cells were intermediate sized cells with scant cytoplasm and chromatin dense nuclei with variable degrees of indentation and angulation and inconspicuous nucleoli. Neoplastic cells in all cases were positive for CD45, CD20 and CD79a and negative for CD3, CD10, Bcl-6 and cyclin D1. All cases except one MCL and 3 MZLs were positive for Pax-5 and all except one MZL were positive for Bcl-2. Three MCLs and 27 MZLs were positive for MUM-1. Conclusion: The observed immunophenotype for canine MZLs and MCLs is similar to human counterparts with the exceptions that human MCLs overexpress cyclin D1 due to a translocation mutation. A similar mutation has not been reported in dogs.

VSP-13: A NEW INCAPACITANCE METER FOR THE MEASUREMENT OF STATIS WEIGHT-BEARING FORCES IN THE HIND LIMBS OF MICE.

Samantha Sommer, Michael Nolan

BACKGROUND: Radiation dermatitis decreases the likelihood of obtaining a positive outcome for both veterinary and human cancer patients. We have developed a device can measure pain caused by radiation dermatitis via hindlimb lameness. Our new incapacitance meter (IM) measures hind limb (HL) static weight-bearing forces in mice. AIMS: The aims of this investigation were three-fold: 1) calibrate the IM; 2) describe HL weight distribution for normal mice; and, 3) assess the ability of the IM to quantify experimentally-induced HL lameness in mice. METHODS: Calibration was performed using standardized weights. HL weight distribution data were collected using the IM in normal mice. Acute HL lameness was induced in a second cohort of mice via a subcutaneous injection of capsaicin (0.25 µL 0.1% w/v) into the plantar surface of the left hind paw. Baseline IM measurements were recorded and these measurements were repeated at various time points post-injection. These measurements were compared to visual lameness scores. RESULTS: Calibration remained stable for 12 days and from 17 to 35 degrees Celsius. In normal mice 50.2% +/- 1.2% of HL force was borne on the left HL as measured by the IM. This distribution did not vary significantly over time. In mice with capsaicin-induced lameness, IM weight-bearing measurements correlate well with visual lameness scores ($r2 = 0.93$). CONCLUSION: IM weight-bearing measurements accurately reflected the observed capsaicin-induced lameness. These results provide proof of concept for the use of this IM to objectively measure static weight-bearing forces in the hind limbs of laboratory mice.

VSP-14: Developing a preclinical mouse model of acute myeloid leukemia
Alexandria M. Schauer, Anh Diep, Yu Miao, Jonathan Nagel, Amato J. Giaccia

Acute myeloid leukemia (AML) presents as a large healthcare burden in the U.S. with 20,000 newly diagnosed cases and 10,000 deaths annually. Adults older than 60 are most commonly affected by AML and have low tolerance for the cytotoxic effects of chemotherapy, the current standard of care. Furthermore, 30% of patients with AML have a mutation in the receptor tyrosine kinase FLT3, a marker associated with poor prognosis and chemoresistance. The current cure rate is 5-15% in the elderly. To improve patient outcomes, we seek better treatments for AML. Axl is a receptor tyrosine kinase that is overexpressed in AML and plays a role in its proliferation and pathogenesis. We have previously demonstrated that an Axl decoy receptor significantly reduces tumor burden of AML in vivo in subcutaneous xenografts of wild type strains of AML in mice. A leukemic bone engraftment pilot study was conducted to create a better clinical representation of disease for future drug comparison studies. Four mice were inoculated with varying numbers of FLT3-ITD mutant MV4:11 LucNeo AML cells. Bioluminescence and post mortem examination results demonstrate that AML cells successfully engraft in bone marrow of long bones and circulate hematogenously. A cell inoculation dose of 0.5 million was determined to best represent clinical disease while maintaining overall health of mice. Our pilot study creates the foundation for testing the efficacy of the Axl decoy receptor in vivo in preclinical AML mouse models using a FLT3-ITD mutant AML strain.

VSP-15: CLINICAL, PATHOLOGICAL, IMMUNOHISTOCHEMICAL AND MOLECULAR FINDINGS IN WEST NILE VACCINATED ALLIGATOR MISSISSIPPIENSIS HATCHLINGS
Dana M Romano, Javier G Nevarez, Kanako Sakaguchi, Ingeborg M Langohr, Del L Phillips, Jacqueline D Ferracone, Fabio Del Piero

West Nile Virus (WNV) is able to induce a mortality rate of up to 60% in captive reared alligators. Alligators can serve as a reservoir and amplifier of this zoonotic flavivirus. Clinical signs include “star gazing”, head tilt, ataxia, swimming in circles, and muscle tremors. Lesions may include encephalitis, hepatitis, colitis, splenitis, interstitial pneumonia and dermatitis. Encephalitis is a very frequent finding in deceased animals. We examined 16 hatchling alligators vaccinated with WNV modified-live vaccine between 0-7 days of age. No clinical signs were observed. Necropsy was performed at two weeks of age and revealed no significant gross findings. RT-PCR performed on brain and liver pool revealed 14/16 viral RNA expression in hatchlings. Histologic lesions were moderate to severe histiocytic myositis and focal lymphocytic dermatitis in the tail region vaccination sites. We also observed a few very small non-significant lymphocytic foci within liver and lung. IHC revealed one area of WNV antigen in the tail region at the level of the vaccination site is association with histiocytic myositis. The absence of significant lesions and antigen localization in the organs examined suggests that the viral RNA identified via RT-PCR in these hatchlings belong to the vaccine WNV. The vaccine caused a moderate inflammation within the inoculation only.

VSP-16: MOLECULAR EPIDEMIOLOGY OF BORRELIA BURGDORFERI CARRIED BY TICKS ON HEDGEHOGS IN SCOTLAND

Amy Robinson L Robinson, Melissa Yates Yates, Caroline Millins Millins, David Walker Walker, Romain Pizzi Pizzi, Adrian Philbey W Philbey

Lyme disease, caused by spirochaetal bacteria in the Borrelia burgdorferi sensu lato complex, is the most common tick-borne disease in the Northern Hemisphere. Throughout the United Kingdom and Northern Europe, B. burgdorferi is principally transmitted by generalist ticks in the Ixodes species. The organism persists in a wide range of wildlife reservoirs. The European hedgehog (Erinaceus europaeus) has been shown to be a competent reservoir by larval xenodiagnosis and is thought to be significant in the distribution and abundance of B. burgdorferi in Europe; however little data is available on the prevalence of B. burgdorferi in hedgehogs in the United Kingdom. The role of hedgehogs as reservoir hosts for Lyme disease in Scotland is unknown, but their presence in suburban gardens makes them potentially important as a source of human infection. In this study, tissue samples from hedgehogs (n = 105) submitted to two Scottish Wildlife rescue centres in 2014 and 2015 were tested by the polymerase chain reaction (PCR) for B. burgdorferi. The organism was detected in 9 of 105 (8.6%) ear skin samples. Sequencing confirmed the presence of B. burgdorferi genospecies B. afzelii (n = 7) and B. bavariensis/B. garinii (n = 2). These genospecies can cause zoonotic infection in humans and are commonly associated with small mammal hosts.

VSP-17: CHEMODECTOMA IN A MEERKAT (SURICATA SURICATTA)

Christina L. Peck, Lani R. Bower, Mee-Ja M. Sula

A 12-year-old intact female meerkat (Suricata suricatta) was evaluated at the Chattanooga Zoological gardens for lethargy and respiratory distress. The meerkat was eating and acting normally the day prior to presentation. On lateral and ventrodorsal radiographs all of the lung fields, except the right caudal lung field, were obscured by a fluid opacity. Complete blood count and chemistry were performed and the pertinent abnormalities were severe thrombocytopenia, hyperbilirubinemia, and hypoglycemia. On thoracocentesis, non-clotting blood was found bilaterally, leading to a presumptive diagnosis of rodenticide intoxication . The meerkat died, and the body was submitted to the University of Tennessee for necropsy. Gross necropsy confirmed severe hemothorax bilaterally with an irregular, 9x9x2mm gray tan, soft firm nodular mass admixed with blood clots surrounding the heart and filling the mediastinum. Histologically, the mass was composed of round to polygonal cells arranged in loose packets markedly separated by blood. Mediastinal or heart base masses have not been described in meerkats. The two most common differentials for a mass in this location are an ectopic thyroid carcinoma or a chemodectoma. A sample of the mass was sent to the Indiana Animal Disease Diagnostic Laboratory and stained with thyroid transcription factor-1 (TTF-1) and chromograninA. The neoplastic cells were immunopositive for chromograninA and immunonegative for TTF-1 leading to the diagnosis of a chemodectoma. Although this is an unusual presentation, chemodectomas should be considered as a differential for heart based or mediastinal masses in meerkats.

VSP-18: CLINICAL AND PATHOLOGICAL CHARACTERIZATION OF OPHTHALMIC DISEASE IN A CANINE MODEL OF MUCOPOLYSACCHARIDOSIS TYPE I

A.S. Nenninger, G. Ben-Shlomo, J.K. Jens, N.M. Ellinwood, P.I. Dickson, J.D. Smith

Mucopolysaccharidosis (MPS) type I is a rare, pediatric lysosomal storage disease caused by a defect in a-L-iduronidase. This enzyme deficiency results in glycosaminoglycan (GAG) accumulation in various cell types, including ocular tissues. Currently available treatments for MPS I are suboptimal and do not address ophthalmic disease. A canine model of MPS I exists, but with limited characterization of ocular pathology. The purpose of this study was to further characterize ocular changes in MPS I-affected dogs. Complete ophthalmic examinations, including slit-lamp biomicroscopy, indirect ophthalmoscopy, rebound tonometry and ultrasonic pachymetry, were performed on twelve hound mixes, 6 affected and 6 unaffected (mean age 19 months, range 13-25 months), by a single board certified veterinary ophthalmologist. Corneal edema, neovascularization, and fibrosis, as well as corneal thickening (773± 214 vs. 584± 13 μm) and a slight increase in intraocular pressure (18 ± 4 vs. 16 ± 2 mmHg) were detected in MPS I-affected dogs. Eyes from 2 MPS I-affected dogs (aged 13 and 20 months) and 2 littermate controls were evaluated microscopically. Vacuolated cells were identified in the corneal stroma, iris, sclera, and optic nerve meninges of affected dogs. Positive staining with alcian blue (pH 2.5) was consistent with primary intracytoplasmic lysosomal GAG accumulation. Intracytoplasmic glycosphingolipid accumulation was demonstrated with luxol fast blue stain in rare retinal ganglion cells, consistent with lysosomal ganglioside accumulation, a known secondary accumulation product. Results of this study further characterize ocular pathology in the canine model of MPS I, and provide foundational data for future therapeutic efficacy studies.

VSP-19: INHALATION CHAMBER IN A PIG MODEL: DEVELOPING A MODEL OF REINKE'S EDEMA

Allison Mustonen, Preeti Sivasankar, Abigail Durkes

Approximately 7.5 million Americans suffer from difficulty using their voice. Phonatory difficulties are often multifactorial with Reinke's edema being a common underlying pathology. Common causes of Reinke's edema include smoking, voice overuse, and reflux of stomach acids into the larynx. There is currently no reproducible animal model of this common human laryngeal disease. The purpose of this study was to test the feasibility of an in vivo model of Reinke's edema by challenging pigs with cigarette smoke in a customized inhalation chamber. The pig was chosen as the animal model because the pig larynx offers the greatest structural, cellular, immunologic, and neuroanatomical similarity to the human larynx. Six Sinclair miniature pigs were exposed to 15 cigarettes per day for 20 days. A week prior to treatment, each group of pigs was habituated to the chamber and human contact to reduce stress and ease treatment administration. Behavior modification was accomplished using positive reinforcement. The week of habituation for each group of pigs successfully increased the efficiency, reliability and reproducibility of treatments. The pigs willingly entered in the chamber and tolerated enclosure for the allotted time. The pigs demonstrated no observable signs of stress which suggests that challenge time could be increased greatly in future chronicity studies. This study offers a potential novel experimental methodology to simulate the development of Reinke's edema in healthy pigs and will impact future research on prevention, improving early diagnosis, and therapeutic options for voice disorders.Student support: Merial Veterinary Scholars ProgramResearch support: NIH R01DC011759

VSP-20: TRANSMURAL GALLBLADDER HEMORRHAGE AS AN INDICATOR OF ANAPHYLAXIS IN DOGS

Kathleen Mulka, Colleen Monahan, Dalen Agnew

Anaphylaxis is an immediate systemic type-I hypersensitivity reaction that can be difficult to confirm at post-mortem examination. In anaphylaxis, antigenic stimulation results in release of chemical mediators, which lead to cardiovascular collapse, respiratory distress, and circulatory shock which can be life-threatening. In dogs, the liver is the shock organ. In cases of shock, hepatic arteries vasodilate and venous outflow is obstructed, resulting in marked hepatic congestion. Gallbladder vascular outflow is also impaired due to profound portal hypertension, leading to thickening of the gallbladder wall with congestion, hemorrhage, and edema. From 2005 – 2016, three cases of canine anaphylaxis were identified at the Diagnostic Center for Population and Animal Health (DCPAH) that had gallbladder lesions compatible with anaphylaxis. All cases were in small breed dogs of varying ages that died suddenly shortly after vaccine administration. Each case had gallbladder mural hemorrhage or edema, severe hepatic congestion, and pulmonary congestion and edema. In dogs, transmural hemorrhage and edema in the gallbladder should suggest anaphylaxis to the diagnostician.

VSP-21: SPONTANEOUS MEDIASTINAL MYELOID SARCOMA IN A COMMON MARMOSET (CALLITHRIX JACCUS)

Danielle T. Morosco, Curtis R. Cline, Michael A. Owston, Shyamesh Kumar, Edward J. Dick

Background: Myeloproliferative disease is defined by the proliferation of one or more bone marrow cell lineages excluding lymphomas and lymphoid leukemias. Myeloid sarcoma is an exceedingly rare manifestation of a myeloproliferative disease defined in both human and veterinary literature as an extramedullary focal mass composed of myeloid precursor cells. Case Report: A mediastinal mass was discovered during necropsy of a 9-month old common marmoset with a history of increased respiratory effort. Grossly, the mass was grey to tan with a slight greenish appearance and expanded and effaced the thymus. Microscopically, the thymus and heart were expanded and replaced by a pleocellular infiltrate composed of granulocytes, lymphoid cells, nucleated erythrocytes, megakaryocytes, and hematopoietic precursors of indeterminate cell lineage. Immunohistochemical staining with lysozyme, CD117, CD61,

CD3, CD20, and hemoglobin demonstrated results suggestive of myeloid sarcoma. Conclusion: While various myeloid leukemias have been reported in other non-human primates, myeloid sarcomas have rarely been described. Idiopathic myeloid fibrosis is the only other myeloproliferative disease that has been reported in a common marmoset (Calithrix jaccus). We report the first case of myeloid sarcoma in a marmoset, which shares similarities with mediastinal myeloid sarcoma in humans.

VSP-22: ADULT POLYCYSTIC LIVER DISEASE IN A DOG
Caroline S Moon, Noelle Muro, Sabrina Barry, Thomas Cecere

A 10-year-old spayed female dog was referred to the Virginia-Maryland College of Veterinary Medicine Veterinary Teaching Hospital for evaluation of a liver mass and elevated liver enzymes. Physical examination revealed numerous subcutaneous lipomas but was otherwise unremarkable. A right-sided cavitated hepatic mass was found on abdominal ultrasound, and an abdominal exploratory revealed multiple 2-3 mm diameter cystic nodules in all liver lobes, as well as a 7x7x5cm mass in the quadrate lobe. The mass from the quadrate liver lobe was fluctuant, multilocular, cystic and exuded serosanguinous fluid from the cut surface. Histologically this mass consisted of multiple, variably sized, cystic spaces lined by a single layer of cuboidal to attenuated epithelium and occasionally filled with lightly eosinophilic proteinaceous fluid or blood. Punch biopsy samples from other liver lobes revealed similar unilocular or multilocular biliary cysts and multifocal biliary hyperplasia. These findings support a diagnosis of adult polycystic liver disease, which is a morphologic subtype of congenital cystic disease of the liver. This complex group of diseases are thought to result from anomalous development of the intrahepatic bile ducts, so called ductal plate malformations. Although the kidneys were not examined histologically, there was no gross or ultrasonographic evidence of renal cysts in this patient.

VSP-23: VALIDATING THE CELL OF ORIGIN OF CANINE T-ZONE LYMPHOMA
Zachary G Millman, Kelly L Hughes, Anne C Avery

T-zone lymphoma (TZL) is an indolent, non-effacing subtype of T-cell lymphoma characterized by the loss of expression of CD45, unique cytological and histological features, and in some cases signs of immunosuppression. Due to their rarity, T-cell lymphomas are poorly studied in human patients compared with B-cell lymphomas. TZL is common in dogs and has a strong breed predilection, making it a useful model for studying the origins and function of one type of T-cell lymphoma. Previous gene expression profiling performed in our laboratory using NanoString technology, implicated the Th2 subtype of T helper cells as the cell of origin for TZL. This gene expression data also revealed significant expression of the proteins galectin-1 and galectin-3, which can drive T helper cell differentiation toward a Th2 phenotype. Based on this data, we hypothesized that TZL should express high levels of the Th2 transcription factor GATA-3, and low levels of the Th1 transcription factor T-bet. We also suspect that the loss of expression of the CD45 antigen may cause upregulation of galectins-1 and 3, which normally bind CD45 on the cell surface to induce apoptosis. Using immunohistochemistry (IHC), it was determined that there was abundant positive staining for GATA-3 in canine TZL lymph nodes, and only sparse positive staining in the non-neoplastic control nodes. Likewise, IHC showed expression of both galectins-1 and 3 was greater in TZL nodes compared to the control nodes. Western blot analysis showed no expression of T-bet in TZL cases, but did show expression of T-bet in controls.

VSP-24: DETECTION OF AKI IN CATS: OPTIMIZATION OF KIDNEY INJURY MOLECULE 1 (KIM-1) IMMUNOHISTOCHEMISTRY

Marisa A. Maglaty, Susan K. Bland, Rachel E. Cianciolo

There is no gold standard for the histologic detection of acute kidney injury (AKI) in cats. Standard hematoxylin and eosin (H&E) staining cannot always identify mild to moderate acute tubular epithelial injury, and subtle lesions can be obscured by autolysis. The novel biomarker kidney injury molecule 1 (KIM-1) is a transmembrane protein expressed by renal proximal tubules after ischemia or nephrotoxicity. The current study sought to optimize KIM-1 immunohistochemical staining to determine whether it provided additional evidence to support the diagnosis of acute tubular epithelial injury, with the ultimate goal of incorporating this technique into the workflow of the International Veterinary Renal Pathology Service. Renal tissue was obtained within 24 hours of death from cats presenting to the OSU Autopsy Service. Clinical histories were reviewed, and cats were placed into one of three groups: highly suspect AKI, possible AKI, and unlikely AKI. Renal tissue was evaluated for KIM-1 expression by immunohistochemistry. Samples that had been allowed to autolyze for 24 hours from the same subjects were stained to determine whether KIM-1 antibody binds non-specifically to autolyzed tissue, producing a false positive result. Expression of KIM-1 in feline renal specimens correlated with clinical history and group placement. Positive samples showed staining confined to proximal tubules as expected. Autolysis did not result in false positive staining – autolyzed samples matched KIM-1 staining patterns from non-autolyzed samples from the same patients. Results suggest that KIM-1 immunohistochemistry is a more sensitive biomarker than standard H&E staining for evaluating AKI on feline renal specimens.

VSP-25: BOVINE MUCOSAL GAMMA-DELTA T CELLS IN HOST IMMUNITY AND DEFENSE

Latasha A. Ludwig, Rebecca Egan, Monica M. Baquero, Kevin J. Stinson, Brandon L. Plattner

Mucosal surfaces are the frontlines of host defense against many pathogens. It is at these surfaces where success or failure of host defense is thought to influence establishment of infectious disease in the host. Our laboratory is working to understand innate lymphocytes, especially gamma-delta (gd) T cells, during early host-pathogen interactions, and how they contribute to host defense. Our hypothesis is that gd T cells are differentially distributed along the gastrointestinal (GI) mucosal surface, where they play an important role during early intestinal infections. We tested this by first characterizing gd T cell distribution at GI mucosal surfaces of healthy calves using spectral microscopy. Approximately 40-60% of T cells in GI mucosa are gd T cells, but the ileum has significantly more gd T cells compared to other GI segments. Intestinal gd T cells are present in the lamina propria and epithelium, but are primarily WC1-, in contrast to peripheral blood. We then used spectral microscopy and flow cytometry to show that gd T cells are significantly recruited into the bovine ileum after intestinal Mycobacterium avium subspecies paratuberculosis (Map) infection. The majority of gd T cells recruited into the ileal mucosa are WC1-, though a significant number of WC1+ gd T cells are also recruited to the epithelium. These data support our hypothesis, that WC1- gd T cells, are uniquely positioned along GI mucosa for host defense, and that WC1- and WC1+ gd T cells play a critical role in early recognition and response to intestinal Map infection in calves.

VSP-26: INDUCTION OF GI TRACT MICROBIAL DYSBIOSIS DOES NOT ACCELERATE DISEASE PROGRESSION IN SIV-INFECTED ASIAN MACAQUES

Mackenzie E Long, Alexandra M Ortiz, Miriam Quinones, Jason M Brenchley

Disruption of the intestinal epithelium during acute HIV infection results in the translocation of microbial products from the intestine to other areas of the body. These microbial products then stimulate the immune system and exacerbate disease progression. Recent data have suggested that microbial translocation is associated with alteration of the composition of the GI tract microbiome in HIV-infected individuals. However, this microbial dysbiosis is not observed in SIV-infected rhesus macaques and the dysbiosis observed among HIV-infected individuals is possibly attributed to risk factors for HIV and not the virus infection itself. To empirically assess the hypothesis that microbial dysbiosis contributes to disease progression via increasing systemic immune activation, a vancomycin treatment protocol was established to induce sustained dysbiosis in a non-human primate model for HIV infection. To ensure the subjects of the study exhibit sustained dysbiosis using the vancomycin protocol, 16S sequencing is used to identify fecal microbial members in samples collected at different time points prior to and post SIV infection for each subject of the study. Despite induction of significant GI tract microbial dysbiosis, we observed that animals treated with vancomycin had similar SIV viral loads, CD4 T cell loss, and inflammation compared to control animals. Moreover, the animals with GI tract microbial dysbiosis progressed to simian AIDS at a similar rate to control animals. Results from this study suggest that microbial dysbiosis does not significantly contribute to SIV disease progression.

VSP-27: EXPERIMENTAL MODELING OF THE NONSPECIFIC PROTECTIVE EFFECTS OF MEASLES VIRUS VACCINATION.

Sarah C Linn, Stefan Niewiesk

The administration of a vaccine can have non-specific protective effects against unrelated pathogens in an infant patient and can, therefore, be protective against pathogens for which currently no vaccines exist. Respiratory syncytial virus (RSV) and Streptococcus pneumoniae are two of the most common causes of acute respiratory tract infections in infants and children. Recent work in Denmark, however, demonstrated that children whose most recent vaccine was the live measles-mumps-rubella vaccine had a lower rate of RSV hospitalization compared to children who had inactivated DTaP-IPV-Hib3 as their most recent vaccine. It has also been shown that there is an epidemiological link between measles vaccination and reduction in S. pneumoniae load.

The aim of this study is to provide an experimental model for the nonspecific protective effects of measles virus immunization against infection with RSV or S. pneumoniae. Three groups of cotton rats immunized with measles virus intranasally or subcutaneously and unvaccinated controls were established. After challenge with RSV, we measured viral titers in lung and nasal turbinate homogenates. After challenge with S. pneumoniae, we measured bacterial titers from nasopharyngeal washes. Measles vaccination did not influence RSV titers 1, 3, and 5 weeks or S. pneumoniae titers 1, 2 and 3 weeks post vaccination. Measles virus vaccination did, however, lower bacterial load 3-fold in animals colonized with S. pneumoniae indicating that cotton rats may be a model to investigate the unspecific effect of measles vaccination on bacterial colonization.

VSP-28: IBA1 IMMUNOHISTOCHEMISTRY AIDS IN DIFFERENTIATING CANINE SUBCUTANEOUS GRANULOMATOUS STEATITIS FROM LIPOSARCOMA

Kristen M Leipzig, Elizabeth Driskell

Three different immunohistochemical markers were evaluated for differentiating canine subcutaneous liposarcoma from subcutaneous granulomatous steatitis. Twenty-two cases of canine subcutaneous masses diagnosed as granulomatous steatitis or liposarcoma were selected. Hematoxylin and eosin slides were reviewed to confirm the diagnosis and cases were examined using both CD18 and ionized calcium binding adapter molecule (Iba1) immunohistochemistry. Additionally, four of the liposarcoma cases were stained with desmin immunohistochemistry. Cases diagnosed as granulomatous steatitis typically exhibited a pattern of sheets of large, foamy macrophages with the macrophages lining the circumference of large clear vacuoles. In contrast, cases of liposarcoma exhibited primarily sheets of vacuolated polygonal cells and didn't circumscribe clear vacuolar areas. Granulomatous steatitis cases had nearly the entire cell population composed of cells strongly expressing intramembranous and intracytoplasmic CD18 or Iba1. Liposarcoma cases had numerous scattered leukocytes throughout the neoplasm that strongly expressed intramembranous and intracytoplasmic CD18 but Iba1 expressing cells were more limited. Importantly, patterns observed on hematoxylin and eosin staining were enhanced with Iba1 immunohistochemistry. Three cases originally diagnosed as liposarcoma were reclassified as granulomatous steatitis based on Iba1 immunohistochemistry. None of the four liposarcomas stained with desmin immunohistochemistry expressed desmin. We conclude that canine subcutaneous masses that are granulomatous steatitis may be misdiagnosed as liposarcoma and Iba1 immunohistochemistry can be helpful in classifying the lesion. Desmin immunohistochemistry was not useful for diagnosis of liposarcoma in the limited number of cases we examined.

VSP-29: Canine Leptospirosis caused by Leptospira interrogans serovar Copenhageni

Chris R Larson, Michelle Dennis, Sree Rajeev

Background: This report describes a case investigation of illness in a 14 week old male, orphan puppy. The puppy presented to RUVC with icterus, anemia, weakness and anorexia, a heavy infestation of Rhiphicephalous sanguineus and multifocal alopecia were also noted. Due to the grave prognosis, humane euthanasia was elected. Objectives: 1. To evaluate clinical abnormalities manifested by this patient and correlate it with gross and histopathological findings. 2. To identify and characterize etiologic agents related to the illness in this puppy. Methods: Postmortem examination, histopathology, and laboratory diagnostics were conducted and compared to ante mortem clinical findings in order to evaluate the likely pathophysiology. Leptospirosis was considered high in the differential diagnosis list; PCR, DFA, and culture were performed to confirm this etiology. Results: Clinical pathology data confirmed azotemia, hyperbilirubinemia, thrombocytopenia, and hypoalbuminemia. Necropsy findings included severe generalized jaundice, accentuated hepatic pattern, yellow discoloration of the kidneys and acute ileocecocolic intussusception. Histopathology findings included marked diffuse hepatocellular dissociation, acute renal tubular necrosis, lymphoplasmacytic interstitial nephritis, pancreatitis, and necrotizing enteritis. PCR and DFA was positive for Leptospira. Leptospira was isolated from blood collected ante-mortem. Whole genome sequencing identified this isolate, relating to (99%) Leptospira interrogans serovar Copenhageni str. Fiocruz. Additional findings included an antibody test positive for Ehrlichia exposure and histopathological lesions suggestive of canine parvovirus infection. Conclusion: Leptospira infection was confirmed. This is the first isolation of Leptospira from the island of Saint Kitts. Further studies are underway to estimate seroprevalence, reservoir status, and potential sources of infection.

VSP-30: RENAL LESIONS IN THE SMALL INDIAN MONGOOSE IN THE CARIBBEAN ISLAND OF SAINT KITTS

Kristen L LaCroix, Luis Cruz-Martinez, Sean Callanan, Sree Rajeev

Background: The Small Indian Mongoose (Herpestes auropunctatus) is an introduced species in Saint Kitts which poses threat to local wildlife such as sea turtles and avian species and their population is flourishing having no natural predators. Leptospirosis is a widespread zoonotic disease affecting humans and animals and mongoose may serve as a reservoir host. As Leptospira infection can result in tubulointerstitial nephritis in animals and humans, we examined mongoose kidney samples for the presence of any histopathologic lesions that may have been associated with Leptospira infection. Objectives: To conduct a histopathological examination of mongoose kidney samples. Methods: Kidneys were collected from trapped and euthanized mongoose following an RUSVM IACUC approved protocol. The kidneys were fixed in 10% neutral buffered formalin immediately after collection. Kidney sections were trimmed and routinely processed for histopathology. The hemotoxylin and eosin (H&E) stained sections were examined by light microscopy and lesions were recorded. Results: Mild to severe histopathological changes were observed in 25 of 75 (33.3%) mongoose kidney sections examined. The lesions included changes in tubules, glomeruli and interstitium of both the cortex and medulla. Tubular changes included loss of tubules, dilation of tubular lumen, degeneration or attenuation of the tubular epithelium with occasional regeneration, presence mineral deposits in medullary tubules, mononuclear interstitial nephritis and interstitial fibrosis. Glomerulosclerosis was observed in severely affected kidneys. Conclusions: Microscopical evidence of renal pathology was evident in several mongoose examined. However, at this time, we do not have any convincing evidence that these lesions are caused by Leptospira infection.

VSP-31: PANCREATIC INSUFFICIENCY AND DIABETES MELLITUS ATTRIBUTED TO ZINC TOXICOSIS WITH SELECTIVE CEREBELLAR DEGENERATION AND NECROSIS IN AN ADULT TOULOUSE GOOSE (ANSER ANSER DOMESTICUS)

Angela Jugan, Angela Jugan, Nicholas Crossland, Peter DiGeronimo, Catherine Barr, Javier Nevarez, Thomas Tully, Jr., Dawn Evans

Background: A five year old male Toulouse goose (Anser anser domesticus) presented for an acute history of lethargy, weight loss, and trouble lifting his head. On physical exam, he exhibited ataxia, head tremors, difficulty balancing while preening, prominent keel bone, and was 5-10% dehydrated. Treatment with Glipizide (10 mg/mL) and subcutaneous fluids was initiated. The goose was found dead in the cage three days later. Methods: Antemortem CBC, serum chemistry, and glucose levels were evaluated. A postmortem examination with routine histopathologic evaluation was performed. Ancillary tests included fecal flotation, hepatic vitamin E and zinc levels, and PCR for West Nile Virus, exotic avian Newcastle, and avian influenza viruses. Results: Significant CBC and chemistry findings included anemia, heteropenia with a left shift, and persistent hyperglycemia. Histologically, the pancreatic acini were replaced by mature fibrous tissue, with concurrent exocrine atrophy and degeneration. Concurrent vacuolar degeneration of islet endocrine cells was observed. Multifocal cerebellar Purkinje cell loss, degeneration, and necrosis, with Bergmann's astrocytosis and hypocellularity of the molecular layer were observed. Regional malacia of a cerebellar folium with gitter cell infiltration was evident. Liver zinc value (wet weight) was twice outside normal reference range. All other ancillary tests within normal limits. Conclusion: Clinical and postmortem findings are consistent with zinc toxicosis manifesting as pancreatic insufficiency/diabetes mellitus and anemia. The selective cerebellar degeneration/necrosis are attributed to the anemia resulting in cerebellar hypoxia. Furthermore, diabetes mellitus may have played a contributory role in cerebellar changes through mitochondrial dysfunction as has been reported in rats.

VSP-32: Magnetic Resonance Imaging and Pathologic Findings in a Case of Ataxia of Unknown Origin in Two Mixed Breed Canines

Aimee Jones, Chris Levine, Dennis O'Brien, Gayle Johnson

Two Maltese-Poodle mixed breed dogs presented to the Sarasota Companion Animal Neurology Hospital for a primary complaint of ataxia. The other four litter mates demonstrated no neurologic symptoms. The neurologic examination of both dogs revealed cerebellar and vestibular dysfunction. At 15 months of age, one of the canines underwent Magnetic Resonance Imaging (MRI) after the ataxia progressed further. The MRI found T-2 weighted hyperintensity present within the cerebellar nuclei, peduncles, brainstem and midbrain. Analysis of the cerebrospinal fluid indicated vacuolated cells. Humane euthanasia was performed at 15 and 18 months on both canines after losing complete ambulatory function. Further testing was completed postmortem. On histopathology, vacuolated cells were also present in the cytoplasm of brain and spinal cord neurons, as well as in the lung, liver, spleen and pancreas. The hypothesis was that the vacuoles were lipid containing. However, Oil Red O stain was performed and came back negative. In addition to the vacuolated cells, marked astrocytosis was noted.

Electron microscopy results are still pending. In conclusion, the diagnosis for both canines was ataxia of unknown origin.

VSP-33: RADIATION-INDUCED BONE LOSS IN MACACA MULATTA

Brendan Johnson, Jeffery Willey, Greg Dugan, Catherine Okoukoni, Mark Cline

Radiation exposure produces pathologies in a number of organ systems including bone. The exact mechanism by which radiation induces bone loss is unknown. Systemic effects such as growth hormone deficiency impact bone growth. Other studies have shown impaired chondrogenesis is a cause of impaired bone development, suggesting a local effect, but whether local irradiation causes systemic changes in bone is unclear. The focus of this study was to evaluate the effect of radiation on long bone length, vertebral bone mineral density (BMD) and vertebral cortical thickness in rhesus monkeys (Macaca mulatta). The cohort consisted of 18 males, aged 3-10 years, which received a whole-lung single dose radiation exposure of 10 Gy to study mitigation of radiation pneumonitis. CT scans were taken one month prior to irradiation, and 3, 5, 7, and 9 months following irradiation and all animals received a complete necropsy with histology. Long bone growth was not adversely affected by whole lung irradiation. In-field and out-of-field vertebral heights did not significantly change over time in growing animals or in adult animals. Bone mineral density decreased significantly over time in irradiated animals compared to age-matched controls (p

VSP-36: CONSTRUCTION AND EXPRESSION OF FULL-LENGTH CANINE CIRCOVIRUS MOLECULAR CLONE

Steven TC Hsu, Rie Watanabe, Steven Kubiski, Patricia Pesavento

Since the discovery of canine circovirus (CaCirV) in 2012, there have been multiple reports describing its association with enteric and systemic disease in dogs and other canids. CaCirV can also be detected in the feces of normal, healthy dogs. Much remains unknown about the factors that determine the severity of an infection. While continuing research on natural cases is unequivocally valuable to determining these factors, the establishment of a viral culture is also critical in studying the host-pathogen interaction Therefore, we developed an in vitro research model by constructing a full-length molecular clone of CaCirV . The construct was transfected into Madin-Darby canine kidney (MDCK) and human embryonic kidney (HEK) cell lines. Polymerase chain reaction (PCR) was used to confirm successful transfection as well as propagation of the viral DNA from passage-to-passage. Western blot was used to detect the production of viral nucleocapsid. Finally, immunofluorescence assay (IFA) for viral nucleocapsid protein as well as transmission electron microscopy (TEM) were used to localize viral production to specific cellular compartments. This project will be the first of our knowledge to demonstrate production of canine circovirus from an infectious clone, providing an in vitro model for the study of canine circoviral pathogenesis.

VSP-37: DETECTION AND CHARACTERIZATION OF FELIS CATUS PAPILLOMAVIRUS 2 (FCAPV2) E6 AND E7 ONCOGENE TRANSCRIPTION IN FELINE CUTANEOUS SQUAMOUS CELL CARCINOMAS THROUGH IN SITU HYBRIDIZATION

Nathan K Hoggard

Felis catus papillomavirus 2 (FcaPV2) is implicated in development of some feline cutaneous squamous cell carcinomas (SCCs) from non-solar exposed sites. This association is based upon detection of FcaPV2 DNA and RNA within feline SCCs using PCR based methods. However, these methods cannot differentiate among by-standard infections, infections of peri-tumoral skin, and virally-driven cancers. A key event in the pathogenesis of papillomavirus-driven cancer is overexpression of viral E6 and E7 oncogenes. Therefore, evidence of their expression within tumor cells would support a causative role for FcaPV2 in feline SCCs. To this end, RNAscope in situ hybridization was performed on eighteen formalin-fixed paraffin-embedded samples of feline cutaneous SSC to identify the presence and localization of FcaPV2 E6/E7 mRNA. Positive hybridization signals were present within 5/9 (56%) samples from non-solar exposed sites and 0/9 samples from solar-exposed sites. In hyperplasic skin (n=4), the hybridization signal pattern was characterized by intense nuclear staining within the superficial epidermis and punctate staining within the basal epithelial layers. As the lesions progressed to in situ and invasive SCC (n=5), punctate staining was present within all layers of the epidermis, with progressive loss of intense nuclear staining within the superficial epidermal layers. This staining pattern is consistent with human papillomavirus-driven cancers, where loss of intense nuclear staining and increase punctate staining throughout all epithelial layers occurs as lesions progress from hyperplastic (productive infections) to in situ cancer (transformation). These findings support a causative role for FcaPV2 in the pathogenesis of some feline SCCs from non-UV exposed sites.

VSP-38: MODELING THE EFFECTS OF BOLUS FLUID ADMINISTRATION ON CANINE PLATELET FUNCTION: A COMPARISON OF DIFFERENT FLUID FORMULATIONS

Emily L Hipp, Elizabeth Spangler

The critical patient is at significant risk for hemostasis disturbances resulting in both hemorrhage and thromboembolism. These patients are often candidates for large volume intravenous fluid replacement. The goal of this study was to evaluate the effect of blood dilution in vitro on platelet aggregation, using volumes intended to model the administration of an IV fluid bolus. Sixteen healthy dogs between 1-12 years of age participated in the study. Fluid effects on platelet function were assessed using 0.9% isotonic saline, Plasma-lyte, Vetstarch, and 7.5% hypertonic saline. The Multiplate Analyzer® was used to assess the impact each had on platelet aggregation. ADP was used as the agonist to stimulate platelet activation. In vitro blood dilution was performed by combining 200uL of each solution with 1mL of citrated blood to simulate a 15 ml/kg bolus (16.7% volume replacement). Platelet aggregation was measured for six minutes in arbitrary aggregation units (AU) and plotted over time to obtain the area under the curve (AUC). Platelet aggregation in unaltered blood was tested for comparison and each blood solution was run as both an unactivated control, without an agonist, and after addition of ADP to trigger platelet activation. Isotonic saline was intended to measure the impact of blood dilution on platelet function, and resulted in a statistically significant decrease in AUC from the unaltered blood. Trends in measured AUC with each solution demonstrated that some fluids had effects that could not be explained by dilution alone, even though statistical significance was not always achieved.

VSP-39: OPTIMIZING DIAGNOSTICS FOR RABBIT HEMORRHAGIC DISEASE VIRUS-2--AN EMERGING FATAL HEPATITIS THREATENING RABBITS AND HARES IN EUROPE

Dana Hill, Karen Moran, Alexa Bracht, Karyn Havas, Fernando Torres-Velez, Lorenzo Capucci, Fawzi Mohamed

Rabbit hemorrhagic disease (RHD) is an acute fatal hepatitis restricted to the European rabbit (Oryctolagus cuniculus), characterized by depression, fever, disseminated intravascular coagulation, epistaxis, and death within 24-72 hours following exposure. RHD is caused by rabbit hemorrhagic disease virus (RHDV), a non-enveloped single stranded positive-sense RNA virus in the Lagovirus genus of the family Caliciviridae. Also included in the Lagovirus genus are European brown hare syndrome virus (EBHSV), which causes clinical disease indistinguishable from RHD in a number of European hare species, as well as non-pathogenic rabbit calicivirus. The inability to cultivate RHDV in vitro makes it difficult to study and diagnose, as current methods rely upon history, clinical signs, RT-PCR, ELISA, hemagglutination, and electron microscopy. RHDV and RHDVa, a highly-pathogenic strain which appeared in 1996, represent a single serotype and are both protected by current RHDV vaccinations. Rabbits older than 6-8 weeks are considered susceptible to RHDV/RHDVa, and subclinical disease is often observed in kits under 6 weeks. In 2010, RHD outbreaks were observed in RHDV-vaccinated rabbits in France. This new virus, characterized as RHDV-2, has been proposed as a new genetic group within Lagovirus owing to VP60 sequence variation. Additionally, it shows partial protection from current vaccinations as well as a lower mortality rate and distinct host profile. In this study, we investigated the pathogenicity of RHDV-2 in New Zealand white rabbits and optimized current RHDV diagnostic assays for the detection of RHDV-2 at the USDA APHIS Foreign Animal Disease Diagnostic Laboratory.

VSP-40: DETECTION OF PYTHIOSIS BY CYTOLOGIC EXAMINATION OF A RECTAL SCRAPE

Kelly A. Hayes, Sarah Beatty, Jennifer Owen

A 3 year-old neutered, male Bullmastiff presented with a 2-week history of bloody and mucoid diarrhea and hyporexia that remained unresolved after treatment with antimicrobials and antiparasitics. A complete blood count revealed an inflammatory leukogram. Hypoalbuminemia was found on serum biochemistry and may be multifactorial due to gastrointestinal loss and possibly negative acute phase response to inflammation. Rectal examination revealed a thickened, irregular dorsal rectal wall and frank blood within the colon. No parasite ova were seen on fecal flotation. A rectal scrape was performed and consisted mostly of mixed inflammatory cells, with mature neutrophils predominating, and an increased numbers of eosinophils, with fewer epithelial and mesenchymal cells. Rarely, the inflammatory cells were observed in dense aggregates outlining non-to poorly staining, branching hyphal structures, with blunted to occasionally rounded edges. Fungal or pseudofungal sepsis was suspected, specifically oomycete infection, with Pythium insidiosum as the highest differential based on history, geographical location, and morphologic appearance on cytology. Serology was subsequently performed at Louisiana State University School of Veterinary Medicine, which confirmed an active pythiosis infection. To our knowledge, this is the first reported case of canine gastrointestinal pythiosis detection on a rectal scrape. This provides supportive evidence for the usefulness of rectal scraping as a diagnostic tool.

VSP-41: CLINICAL, RADIOGRAPHIC AND HISTOLOGICAL FEATURES OF MAXILLARY FIBROSARCOMA IN 25 CATS

Alexandra Harvey, Cynthia Bell

This retrospective review summarizes the clinical, radiographic, and histopathological features of maxillary fibrosarcoma in 25 cats. Study cases were surgical biopsy specimens submitted to the Center for Comparative Oral and Maxillofacial Pathology at the University of Wisconsin-Madison between February 2015 and July 2016. Case selection was based on the histopathological diagnosis of fibrosarcoma for lesions that involved the gingiva, all of which occurred on the maxilla. There were 7 neutered male cats and 18 spayed female cats. Age ranged from 4 to 16 years (mean, 11.7 years). The lesions were infiltrative and often caused swelling with variable mass effect. The majority of these neoplasms were extensive and involved more than one quadrant of the maxilla. Ten were multifocal. Approximately 50% of cases had involvement of the palate as well as the gingiva. Further infiltration to involve the lip or orbit were documented in one case each. Maxillary fibrosarcoma in cats most often presents as smooth, pink, firm, and lobulated masses. Radiographic bone loss is absent to mild, except when the area is concurrently affected by severe periodontal disease. Histologically, the neoplastic tissue is moderately cellular with short intersecting bundles of spindle cells, moderately abundant fibrous matrix and mild inflammation. Histological grades I, II, and III were assigned in 4, 17, and 4 cases, respectively. While grade is variable, maxillary fibrosarcoma has a consistent presentation in cats and occurs with moderate frequency. A subset of the low-grade neoplasms in cats closely resembles the histologically low-grade, biologically high-grade maxillofacial fibrosarcoma of dogs.

VSP-42: AN UNUSUAL PRESENTATION OF SALMONELLOSIS IN A GROUP OF YOUNG KHAKI CAMPBELL DUCKLINGS (ANAS PLATYRHYNCHOS DOMESTICUS)

Brenna Hanratty, Gavin R Hitchener

Seven, 3 week old Khaki Campbell ducklings (Anas platyrhynchos domesticus) were received for necropsy evaluation at the Cornell University Duck Research Laboratory with a history of increasing mortality in a group of 5000 ducklings recently moved to an old broiler house. These ducklings were being raised for egg production in the fall of 2016. Antemortem clinical disease findings included ataxia, head shaking, leg paddling, tremors, sneezing, and coughing, with sudden death following premonitory signs. Gross findings included massive fibrinous ceolomitis, fibrinous pericarditis, and air sacculitis. Histologically lesions contained massive fibrin deposition along visceral surfaces and air sac epithelium intermixed with degenerate heterophils with fewer macrophages, lymphocytes, and plasma cells. The clinical history, gross, and histologic findings were most compatible with Riemerella anatipestifer or Escherichia coli infection. Preliminary bacteriology results revealed Salmonella group D organism. This preliminary result lead to state mandated confirmatory testing at the National Veterinary Services Laboratories in Ames, Iowa where Salmonella enteritidis was cultured. This is a reportable disease in egg laying poultry species and the flock was depopulated as a result of this. This case represents an unusual presentation of Salmonellosis in young ducklings and should be considered a differential diagnosis in young ducklings presenting with clinical disease characteristic of Riemerella anatipestifer infection.

VSP-43: STRESS LEUKOGRAM INDUCED BY ACUTE AND CHRONIC STRESS IN ZEBRAFISH (DANIO RERIO)

Agata K Grzelak, Daniel J Davis, Susan M Caraker, Marcus J Crim, Jan M Spitsbergen, Charles E Wiedmeyer

The use of the zebrafish, Danio rerio, as an animal model for experimental studies of stress has been increasing rapidly over the years. While many of the physiological and behavioral characteristics associated with stress have been defined in zebrafish, the effects of stress on hematological parameters have not been identified. The purpose of our study was to induce a rise in endogenous cortisol through a variety of acute and chronic stressors and compare the effects of these stressors on peripheral white blood cell (WBC) populations. Acutely stressed fish underwent a series of two stressors over 90 minutes. Chronically stressed fish underwent a series of alternating stressors twice daily over a period of five days. Following the last stressful event, fish were sacrificed and whole blood and plasma were obtained. A drop of whole blood was used to create a blood smear, which was subsequently stained with a modified Wright-Giemsa stain and a 50-WBC differential count determined. Plasma cortisol levels were determined by using a commercially available enzyme-linked immunosorbent assay (ELISA). Endogenous cortisol concentrations were significantly increased in both stress groups compared to control fish. In the acutely stressed fish, a significant lymphopenia, monocytosis, and neutrophilia were noted compared to unstressed, control fish. Chronic stress also induced a lymphopenia and monocytosis, but no significant changes were seen in relative neutrophil populations. These changes are most likely the result of increases in endogenous cortisol concentrations and can be identified as the first stress leukogram to be reported in zebrafish blood.

VSP-44: INTRACEREBRAL BLASTOMYCOSIS IN THE ABSENCE OF CONTEMPORANEOUS VISCERAL LESIONS IN A DOG

Christopher K Gow, Bernard S Jortner

Blastomycosis is a systemic fungal infection of dogs often affecting the lungs, lymph nodes, eye, skin, and bone; neurologic signs are uncommon. The latter were noted in the present case of an eight month old, castrated male, mixed-breed dog that presented for respiratory distress, fever, and anorexia. These signs, the clinical examination and the Mvista Blastomyces quantitative urine antigen test indicated fungal pneumonia with Blastomyces dermatitidis. The dog was treated with itraconazole for three months resulting in resolution of clinical signs. Three months later the dog exhibited ataxia, dysphagia, and dull mentation, shortly followed by death. The necropsy demonstrated pulmonary edema and hemorrhage, pericardial effusion, and multiple intracerebral pyogranulomatous lesions. The latter contained fungal organisms consistent with Blastomyces dermatitidis. In some lesions the purulent component was dominant. This is an unusual case -- at necropsy prominent blastomycotic brain lesions were noted in the absence of pulmonary or other visceral involvement. The pulmonary blastomycosis had been clinically evident three to six months prior to death and was likely cured by the itraconazole therapy. In the presence of and following the therapy, blastomycotic organisms were able to invade the brain, persist, and elicit pyogranulomatous inflammation.

VSP-45: STREPTOCOCCUS EQUI SUBSP. ZOOEPIDEMICUS OUTBREAK IN DOGS IN A NORTH TEXAS ANIMAL SHELTER

Chloe C Goodwin, Barbara Lewis, Andrés de la Concha, Eric Snook

Streptococcus equi subsp. zooepidemicus are Gram positive, facultative anaerobe cocci most often associated with horses. The bacterium is a commensal of the upper respiratory and lower genital tracts in horses that lead to endometritis and severe respiratory disease. In recent literature, it has identified as an emerging canine respiratory pathogen, causing hemorrhagic pneumonia and severe pleural effusion in large confined populations. Texas A&M Veterinary Medical Diagnostic Laboratory received 10 dogs for necropsy from an animal shelter found dead with bloody nasal discharge--9 of the 10 within 34 days. Ages ranged from 1 to 9 years old with 7/10 under the age of 3 years old. Most were pit bull mixes (6/10) and intact males (5/10). Their history prior to intake is unknown. Upon acquisition, the shelter vaccinates for canine distemper, canine adenovirus type 2, parainfluenza, Bordetella bronchiseptica, and rabies as well as administering a flea and tick preventative. Clinical signs included coughing, labored breathing, crackled lung sounds, nasal discharge, lethargy, anorexia, and vomiting. The clinical course from initial observation to death ranged between 2 to 10 days. The most consistent necropsy findings were hemothorax and bronchopneumonia. Histologically, common findings included acute suppurative bronchopneumonia with coccoid bacteria. Culture of lung or thoracic fluid yielded S. zooepidemicus in all cases that were cultured (9/10). Molecular diagnostics were performed in 8/10 cases and showed positives for canine coronavirus (2/8), canine adenovirus type-2 (3/8), and canine herpesvirus (1/8). A source of the outbreak has not been identified to date, but an investigation is ongoing.

VSP-46: CELL-INTRINSIC AND ENVIRONMENTAL FACTORS CONTRIBUTE TO SEX-SPECIFIC DIFFERENCES IN THE CD8+ T CELL RESPONSE TO INFECTION

Elizabeth W. Goldsmith, Jocelyn Wang, Neva B. Watson, Brian D. Rudd

Following infection, naive CD8+ T cells differentiate into more terminally differentiated effector T cells, which kill infected cells by releasing cytotoxic molecules, or less terminally differentiated memory T cells, an important component of immunologic memory. Previous studies have shown that memory cells are increased in males after infection. Therefore, we asked the question, are there sex-linked differences in CD8+ T cell behavior? To address this question, we first examined CD8+ T cells from male and female transgenic mice and found that naïve, unstimulated CD8+ T cells shared similar phenotypic profiles. When stimulated in vitro, female CD8+ T cells produced more interferon gamma in response to pro-inflammatory cytokines, and they proliferated more rapidly in response to T cell receptor stimulation. To investigate intrinsic sex-related differences in vivo, we adoptively co-transferred CD8+ T cells from male and female donors into male recipients and examined the responses to infection. To examine if differences in hormonal environments affect CD8+ T cell responses in vivo, we adoptively transferred female CD8+ T cells into male and female recipients and observed the response post-infection. In each of these in vivo experiments, more female CD8+ T cells became terminally differentiated at the peak of infection. Our results suggest that female CD8+ T cells respond more vigorously to inflammatory signals at the cost of creating memory cells. We conclude that differences in both cell-intrinsic factors and the host hormonal environment contribute to differences in male and female CD8+ T cell behaviors.

VSP-47: miR-146a is an endogenous regulator of both hematopoiesis and bone mass

Jennifer A Geisler, Blake E Hildreth III, James Lee, Albert de la Chapelle, Prosper Boyaka, Michael C Ostrowski, Sudarshana M Sharma

MicroRNAs (miRNAs), non-coding RNAs, regulate cellular activity by binding to protein-coding RNAs, suppressing translation or causing RNA degradation. One microRNA, miR-146a, is a key regulator of inflammation and is a physiologic break on immune activity. In the skeleton, osteoclasts share a common progenitor with macrophages of the myeloid lineage within the hematopoietic hierarchy. miR-146a has been shown to negatively regulate osteoclast differentiation and function by our laboratory and others in vitro. Because of this, we wanted to investigate the role of miR-146a in bone biology and hematopoiesis in vivo. Two transgenic mouse models were used for this purpose: a knock-out (KO) mouse model with global deletion of miR-146a and a knock-in (KI) mouse model overexpressing miR-146a. Male and female mice were aged 6-7 months and, at necropsy, tissue samples were collected for phenotyping. Spleen and liver weights were obtained, femurs isolated for radiographic evaluation, and blood, spleens, and bone marrow collected for flow cytometric evaluation. In both male and female KO mice and male KI mice there was a significant increase in spleen weight compared to wild-type controls. Female KO mice had significantly greater liver weights. These findings suggest altered hematopoietic cell number and/or cellularity. KO mice had decreased bone density and KI mice had increased density, suggesting that miR-146a also negatively regulates osteoclast function in vivo. These findings indicate that miR-146a regulates both hematopoiesis and bone mass. Further phenotyping is ongoing to provide insight into the role of miR-146a in these processes.

VSP-48: Whole genome phylogenetic analyses of bluetongue virus in Colorado ruminants reveals genomic diversity and viral dissemination across the Western U.S.

Teresa M Garcia, Justin Lee, Jennifer Kopanke, Christie Mayo

Bluetongue virus (BTV) is the cause of bluetongue (BT), an economically important disease of ruminants transmitted by Culicoides midges. Recent incursions of the disease in northern Europe have highlighted the risk of a similar BT epidemic in the US, either from the emergence of a virulent virus (by genetic shift or drift) from endemic viruses, or via translocation of a novel virus from an adjacent ecosystem. The specific contributions of viral genetics, host and vector characteristics, or climate change to BTV transmission remain poorly characterized. Whole genome sequencing, using next generation sequencing (NGS) platforms, offers a method to comprehensively analyze the genetic composition of BTV. The goals of this study were to: 1. characterize the whole genome sequences of BTV strains isolated from ruminants in Colorado; and 2. compare with whole genome sequences of historic BTV strains isolated from ruminants in California. A total of 16 isolates from animals positive for BTV via serology and qRT-PCR were obtained from ongoing surveillance in 2015. Samples were processed for NGSon an Illumina NextSeq platform. Phylogenetic trees were generated for each viral segment. The trees confirm diversity among viral strains across Colorado, and clustering patterns suggest that some Colorado isolates are more closely related to California isolates than to other Colorado isolates. Clustering patterns also suggest reassortment has occurred. Coupling the design of active surveillance with the tools of whole genome sequencing has allowed us to completely characterize the genetic diversity of Colorado isolates and compare them to isolates from another endemic region.

VSP-49: SURVEY OF GIARDIA GENETIC ASSEMBLAGES IN DOGS IN THE PHOENIX METROPOLITAN AREA

Miranda Frohlich, Kristen R Richter, Denise Olivas, Sudhindra R Gadagkar, Lauritz A Jensen, Valerie M Wong

Giardia is a protozoan parasite that causes diarrhea and affects animals and humans worldwide. Host specificity of Giardia spp. is related to the genetic assemblages. Assemblages A and B affect humans, dogs, and cats. Assemblages C and D are host-specific to dogs and other canids. Assemblage F is specific to cats. We had previously found that Giardia is common in shelter dogs in the Phoenix Metropolitan Area, but prevalence data on genetic assemblages of the Giardia spp. were unavailable. This study aimed to determine the genetic assemblages of Giardia spp. found in shelter dogs in the Phoenix Metropolitan Area, thereby determining the zoonotic potential of the Giardia organisms found in these animals. Fecal samples (n=82) were obtained from shelter dogs and tested for Giardia by detection of a Giardia-specific protein using an ELISA. DNA was extracted from the 29 ELISA- positive stool samples and used for genetic testing. The sequences for 18S, beta-giardin, and glutamate dehydrogenase were determined and used for genetic assemblage identification. When compared against sequences of known Giardia genetic assemblages, assemblages C and D (of low zoonotic risk) were detected most frequently. Assemblages A and B (of high zoonotic risk) were not detected in any of the samples. Assemblage F was identified in one sample. This is the first report describing the identification of assemblage F from the amplification of 18S rDNA in a canine fecal sample. These results suggest that shelter dogs in the Phoenix Metropolitan Area do not carry Giardia spp. of significant zoonotic potential.

VSP-50: VIRAL-MEDIATED ONCOLYSIS OF CANCER CELLS ISOLATED FROM CANINE TUMORS

Kirsha B Fredrickson, Amy L MacNeill

The three classic arms of cancer therapy are surgical resection, chemotherapeutics, and radiation. As these fields have progressed we have greatly improved cancer survival rates in both companion animals and humans. However, with the complexities of neoplastic disease and their ability to further mutate due to genomic instability, these classic approaches are not always successful and often have undesirable adverse effects. Oncolytic virotherapy has the potential to be highly effective with minimal adverse effects due to targeted lysis of neoplastic cells. Studies were performed to evaluate the oncolytic effects of recombinant myxoma virus (MYXV) on canine osteosarcoma and soft tissue sarcoma primary cell cultures. The tumor samples were obtained following surgical excision. Once cultured, cells were evaluated for vimentin using immunocytochemistry and for alkaline phosphatase activity using nitroblue tetrazolium chloride/5-bromo-4-chloro-3- indolyl phosphate toluidine solution. The cancer cell-lines were then inoculated with recombinant myxoma virus expressing a red fluorescent protein and collected after 48-hours. Cytopathic effects were assessed at 24-hour and 48-hour time points via fluorescent microscopy. Ability of the virus to replicate within the cancer cells was confirmed via Western immunoblot detection of late MYXV protein production. The findings of this study suggest that canine osteosarcoma and soft tissue sarcoma primary cell cultures are lacking antiviral mechanisms and susceptible to MYXV infection leading to oncolysis.

VSP-51: PARVOVIRUS DETECTION BY PCR AND ISH IS ASSOCIATED WITH MYOCARDITIS AND CARDIOMYOPATHY IN YOUNG DOGS

Jordan Ford, Alex Molesan, Laura McEndaffer, Kathleen Kelly

Perinatal parvoviral (PV) infection of pups born to naïve dams causes necrotizing myocarditis resulting in cardiac failure, sudden death, and high mortality at 3-4 weeks of age. Acute infections are characterized by the presence of viral inclusions, necrosis, and colocalization of PV antigen to cardiomyocytes. Viral inclusions and antigen are rapidly lost in pups surviving acute infection. Given the heart's stereotypic response to injury, we hypothesized myocarditis or cardiomyopathy in dogs

VSP-52: NASAL CRYPTOCOCCUS MYCETOMA IN A GOAT

Cosette Faivre, Ana Alcaraz, Spring Halland, Phil Carbert

A 5-year-old female goat presented with signs of severe upper respiratory distress after being treated for pneumonia by another veterinarian for one a week. The goat died despite attempts by the attending veterinarian to investigate the cause of a severe dyspnea. The animal was submitted to the Veterinary Pathology Center at Western University of Health Sciences for complete necropsy. Upon gross examination, there was a firm, pale mass with multifocal gritty areas and a necrotic core which obliterated the lumen of the nasal cavity and replaced all the structures on either side of the nasal septum. The mass was descending through the soft palate and obstructing the pharynx airway. Cytologic examination of the mass revealed multiple round yeast-like fungal structures with a thin wall and a clear halo. There was an occasional narrow based budding. Histologic examination from nasal samples revealed numerous solid areas of fungal organism similar to the ones observed in cytology with superficial ulcerations and minimal inflammatory response. PCR confirmed the organism as Crytococcus spp. This is the first incident of a caprine with a mycetoma in Southern California, as felines are the most common species affected.

VSP-53: Histopathologic Findings in a Colony of Lined Seahorses (Hippocampus erectus)

Jennifer Engelhard, Christiane Löhr, Timothy Miller-Morgan, Jerry Heidel

This study reports findings from the histologic analysis of 28 lined sea horses, Hippocampus erectus, from a single colony. The seahorses were housed in a 100 gallon tank amongst a mixed population of tropical marine species, as part of a display with a rotating theme at the Oregon Coast Aquarium. Individual mortalities began in January of 2012 and continued until late December 2014 at which point the remaining animals were quarantined and subsequently euthanized. Whole fish were fixed in toto in 10% neutral buffered formalin and submitted to the Oregon Veterinary Diagnostic Laboratory for microscopic examination. Mycobacteriosis was diagnosed in 12 of 28 seahorses, followed by fungal infections (presumably Fusarium solani) and epidermal algal proliferation (n=5), scuticociliatosis (n=2) and sepsis (n=2). Poor body condition and hepatic lipidosis were also common. The different diseases left various lesions across predictable areas of each animal forming a complex profile. For example, animals with scuticociliatosis caused by Uronema spp. had lesions confined to the head and branchial arches while animals with algal proliferation had lesions confined to the epidermis. Fungal disease was likely F. solani as it has been reported in sygnathids before and appeared to have similar morphologic characteristics. Animals with mycobacteriosis had granulomas in the kidney, liver, intestinal mucosa, and ceolomic cavity and ranged from severe disseminated disease to mild localized infections. In summary, mycobacteriosis was the most common disease identified in this colony of seahorses. This is of importance, as Mycobacteria have zoonotic potential and infections may go unnoticed.

VSP-54: IMMUNOHISTOCHEMICAL LOCALIZATION OF 3 ZIKA VIRUS STRAINS IN LABORATORY MOUSE SPLEENS

Julia Ellison, William Tang, Kenneth Kim, Sujan Shresta

Zika virus (ZIKV) research is in its infancy, and immunohistochemical detection of replicating ZIKV in formalin-fixed, paraffin-embedded (FFPE) tissues will be an invaluable instrument for research and clinical investigations going forward. The goals of this study were to localize ZIKV antigen in murine spleens and determine if the primary antibody would be reactive to both African and Asian lineages of ZIKV. AG129 mice were infected retro-orbitally with either ZIKV strain MR766 (East African lineage), strain FSS13025 (Asian lineage), or an Asian lineage clinical isolate. Spleens were harvested at 48 hours post infection. An antibody against ZIKV NS2B (GeneTex), a replication protein, was used for primary antibody. The expression of NS2B was observed across all three ZIKV strains. Most immunoreactive cells were in the red pulp and had round to stellate morphology consistent with histiocytes. Few immunoreactive cells with dendritic morphology were also observed in the white pulp. No to little non-specific chromogenic reaction was observed in isotype controls. These results indicate that immunohistochemistry can be used to detect East African and Asian lineages of ZIKV in FFPE mouse tissue.

VSP-55: OLFACTORY ADENOCARCINOMA IN A TASMANIAN DEVIL: A CASE REPORT

Jessica Elbert, Sarah Peck, Dane Hayes, Jim Taylor

A 7-year-old 4.12 kg intact male captive Tasmanian devil presented for necropsy after being euthanized on ethical grounds 3 days prior due to hind limb ataxia and generalized muscle wasting. On necropsy, an irregular, 1 cm nodule was detected on the dorsal surface of the left olfactory bulb. The nodule was composed of an expansive mass of cells with areas of necrosis. Histologically, the cells were round to columnar with abundant lightly eosinophilic cytoplasm and distinct cell borders. Cells were variably arranged in nests, acini or forming a ductular or sinusoidal arrangement. There was also palisading and rosette formation around small vessels. Immunohistochemically, the tumor cells were positive for vimentin and cytokeratin AE1/AE3 and negative for S-100, calretinin, neuron-specific enolase, chromogranin, synaptophysin, and neurofilament protein. Based on the immunohistochemical results and histologic morphology, the neoplasm was classified as an olfactory bulb adenocarcinoma. This is the first documented case of an olfactory bulb adenocarcinoma in a Tasmanian devil.

VSP-56: DETERMINATION OF IMPORTANT CAUSES OF MORBIDITY AND MORTALITY IN CAPTIVE PSITTACINES SUBMITTED TO THE ONTARIO VETERINARY COLLEGE TEACHING HOSPITAL

Thisuri Eagalle, Nicole Nemeth, Hugues Beaufrere, Leonardo Susta

Background: Psittacines are increasingly common in households, aviaries and zoological collections. With the advancement of avian medicine, diagnostic evaluations(i.e., necropsy and biopsy) are increasingly important for early and accurate diagnoses and to predict disease outcomes. However, comprehensive information about disease conditions in birds, including pathological descriptions, is often limited. Objective: We retrospectively reviewed diagnostic data from psittacines submitted tot he Ontario Veterinary College (OVC) to assess baseline prevalence and pathological features of the most commonly diagnosed diseases and conditions. Methods: Diagnostic data from psittacine carcasses (necropsy) and samples (biopsy)submitted to the OVC and Animal Health Laboratory from 1995-2015 were reviewed.Data included signalment and morphologic diagnoses, which were categorized and ranked to identify primary cause(s) of death. Results: Eighty-three species (n=1,149) were represented, with the African grey,cockatiel, and green-winged macaw as the most common. Infectious agents were most commonly identified as the primary cause of death (n=409; 35.6%), including 216 viral(52.8%; some suspect), 132 bacterial (32.3%), and 61 fungal (14.9%) infections. The most common respective etiologies were bornavirus (n=154), Mycobacterium spp.(n=18), and Aspergillus spp. (n=18). Neoplasia was diagnosed in 100 cases (8.7%),which most commonly involved the alimentary tract (n=18); squamous cell carcinoma was the most frequently diagnosed neoplasia (n=9). Conclusions: Better recognition and understanding of diseases that impact psittacines will aid clinicians and pathologists to formulate more effective differential diagnoses and targeted diagnostic procedures, ultimately leading to better treatment, management and prophylactic protocols to improve the health of captive psittacines in Ontario and elsewhere.

VSP-57: Characterization of Gpnmb Expression in Mouse Models of Lysosomal Storage Diseases

Jose J Diaz Torres, Kelly Cavallo, Jennifer Johnson, Errin Roberts, Jennifer Nietupski, Sue Ryan, John Marshall, Dinesh Bangari

Lysosomal storage diseases (LSDs) are inherited disorders of lysosome function resulting in progressive accumulation of metabolic byproducts and subsequent tissue pathology. Mouse models of human LSDs recapitulate the disease pathophysiology and serve as critical drug discovery tools. To further explore the translational utility of LSD mouse models, we studied the expression of glycoprotein nonmetastatic melanoma B(Gpnmb) protein as a disease biomarker in Niemann-Pick type C (NPC), Sandhoff's disease, metachromatic leukodystrophy (MLD), and mucopolysaccharidosis type I (MPS I) mouse models. Plasma and tissues (liver and brain) samples of diseased mice and age-matched wild type (WT) mice were evaluated by quantitative real-time PCR, ELISA and immunohistochemistry for murine Gpnmb, We observed increased Gpnmb mRNAlevels in diseased tissues (brain and liver) of all model mice relative to respective WT controls. Increased plasma Gpnmb levels were observed in NPC and MPS I mice but not in Sandhoff's and MLD models. By immunohistochemistry, Gpnmb expression was localized within substrate-laden foamy macrophages in the liver and brain of NPC and MPS I mice, and brain of MLD and Sandhoff mice. WT mouse brain and liver were devoid of Gpnmb immunopositive cells. In case of NPC mice, there was an age-dependent increase in plasma and tissue Gpnmb expression indicating its role as a marker of disease progression. Overall, our results validate the involvement of Gpnmb-immunopositive macrophages in LSD pathophysiology. As such Gpnmb expression may be used as a tissue biomarker for tracking LSD disease progression or assessing the efficacy of experimental therapeutics.

VSP-58: GRANULAR CELL HEMANGIOSARCOMA IN A DOG WITH FEATURES OF HUMAN GRANULAR CELL ANGIOSARCOMA

Lusan DellaGrotte, Anibal Armien, Teresa Weronko, Pompei Bolfa

Background: A five year old neutered male mixed breed canine from Nevis presented to the Ross University Veterinary Clinic with a bleeding mass. Records showed the mass was originally documented one year ago. The cutaneous mass (6.4x4.5x2.8 cm), had recently erupted and was actively bleeding. Located laterally left to the penis, the mass was surgically excised and submitted for histological examination. Methods: Histopathological examination was performed together with the following histochemical stains: Periodic Acid Schiff, Perls' Prussian Blue, Luxol Fast Blue, and Silver Stain. Immunohistochemistry was performed using endothelial markers CD31 and Factor VII, as well as S100, lysozyme, CKMNF116, and Vimentin. Transmission electron microscopy (TEM) was also performed. Results: Histopathology revealed of contiguous blood filled cavities on a sparse collagenous stroma. Neoplastic cells had large numbers of intracytoplasmic hypereosinophilic granules ranging 1-3 micrometers. The granules were PAS positive, blueish-green on Luxol Fast Blue and negative for Perls' Prussian Blue and Silver Stain. Immunohistochemistry was positive for endothelial marker CD31 and negative for Factor VII. Furthermore, S100, lysosomes, and CKMNF116 were negative while Vimentin was positive. On TEM the neoplastic cells displayed intracytoplasmic granules consisting of accumulated proteins lined by a single membrane, forming an electron dense vesicle. A small proportion of these granules were autophagolysosome and phagolysosomes. Conclusion: We present the first animal case of cutaneous granular cell angiosarcoma, an exceptionally rare sarcoma of endothelial lineage in humans. With only few cases reported, there is very little literature that explicitly identifies this type of malignancy and its pathogenicity.

VSP-59: IDENTIFICATION OF THE HUMAN ENDOGENOUS RETROVIRUS K CELLULAR RECEPTOR

Samantha Darling, Charles Bailey

Background: Human endogenous retroviruses (HERVs) comprise approximately 8% of the human genome and are derived from ancient exogenous retroviral infections. Retroviral elements were subsequently incorporated into the human genome over millions of years. The HERV-K family is the most recently integrated and most biologically active. Accumulating evidence suggests that HERV-K may play a role in various pathological conditions including autoimmune diseases and certain cancers. Objective: The purpose of this study was to identity the HERV-K cellular receptor. Methods: We generated a fusion construct consisting of a CD5 signal peptide, the receptor binding domain (RBD) of HERV-K as described by Robinson and Whelan (2016), and the Fc domain of human IgG1. We expressed the HERV-K RBD-Fc fusion protein in mammalian cells and purified it from culture supernatants via protein G chromatography. Immunoprecipitation of lysates from several human cell lines and one mouse cell line was performed using the HERV-K RBD-Fc construct as a bait protein. Results: The HERK-K RBD monomeric construct ran at the predicated molecular weight of ~80 kDa on gel electrophoresis but proteins were precipitated by the HERV-K RBD construct during immunoprecipitation experiments. Conclusion: One explanation for the HERV-K RBD construct not binding its cellular receptor may be because our fusion construct used the signal peptide from CD5 rather than the natural HERV-K signal peptide. We will therefore generate a new fusion protein consisting of the natural HERV-K signal peptide and receptor binding domain and repeat the immunoprecipitation experiments.

VSP-60: Chronic Brucellosis Induces Persistent Morphological Changes to the Liver and Spleen

Juliane Daggett, Alexandra Brower, Marina Diioia

Hypothesis: There are morphological differences in the microarchitecture of the murine liver and spleen during acute vs. chronic Brucella infection. Introduction: Brucella are gram-negative, facultative, intracellular coccibacilli from the Proteobacteriaca family. Incidental hosts develop an acute flu-like infection, which transforms into a chronic cyclic infection if not cleared. Here, we seek to further validate that BALB/c mice can develop chronic infection with immunohistochemical investigation for signs of infection and changes to splenic and liver microarchitecture. Materials and methods: Using tissues stained with hematoxylin and eosin to show the tissue morphology, we identified morphological differences in liver and spleen between the acutely infected and chronically infected mice. Additionally, splenocytes from acutely and chronically infected mice were transferred to naive mice prior to challenge to help us understand the differences in immune cell responses upon challenge. Results: Both the acutely and the chronically infected livers show areas of inflammation and immune cell infiltration, yet they differ in presentation. The spleen from acute mice show cellular evidence of active infection, while the chronic mice display signs of increased hematopoiesis. The recipients of adoptive transferred cells follow a slightly different granuloma pattern. The granuloma size increases faster while the number of granulomas stays more consistent throughout the infection. Conclusion: Morphological differences exist between acutely and chronically infected livers and spleen. Also, cells from acutely vs. chronically infected animals respond differently upon challenge. Further research is needed to identify differences in the regulation of proteins involved in inflammation to better evaluate acute and chronic brucellosis.

VSP-61: CHARACTERIZATION OF NEURONAL NECROSIS FOLLOWING SEIZURES SECONDARY TO PORTOSYSTEMIC SHUNT (PSS) LIGATION

Catharine R Cowan, Kelli M Kopf, Miranda Vieson, Otto I Lanz, Bernard Jortner

Neurological complications following PSS surgical occlusion are reported to occur in 12 percent of surgical patients, more commonly in dogs >19 months of age. Most serious of these are generalized motor seizures, which have a poor prognosis. However detailed descriptions of the resulting brain lesions are rare, a deficiency we seek to address. Cellophane band ligation of an extrahepatic PSS was performed on a two-year-old male castrated Shih Tzu with signs of hepatic encephalopathy. The day after surgery, the patient developed generalized seizures. Subsequently there were signs of aspiration pneumonia, and the patient struggled to maintain normal oxygenation. After the seizures ceased (following medical treatment), the patient suffered from depressed mentation, apparent blindness and other neurological deficiencies. After two weeks without significant improvement, euthanasia was performed. Brain lesions at necropsy were extensive bilateral, often cavitary, cerebral cortical necrosis. Other affected regions were the thalamus, nuclei of the midbrain and cerebellar hemispheres and vermis. The neuronal/gray matter necrosis was likely related to the effects of profound seizures, possibly complicated by pneumonia-induced hypoxemia. Severe seizures can elicit hypoglycemia, hypoxemia and excitotoxicity to injure neurons. Hepatic encephalopathy also may have been a factor. We present a case of profound bilateral cerebral cortical, brainstem and cerebellar cortical necrosis, representing a neuropathologic complication of surgical correction of a congenital PSS.

VSP-62: AUTOIMMUNE PANCREATITIS IN A BENGAL CAT
Sarah Coe, Wallaya Manatchaiworakul, Matti Kiupel

A four year old male castrated Bengal cat presented for necropsy with a history of anorexia, weight loss, and diarrhea. Clinically, the pancreas was edematous on ultrasound, and the cat had decreased cobalamin and folate levels. Upon gross examination, pancreatic tissue was markedly diminished. Only the pancreatic tail was grossly recognizable and measured approximately 0.5x5 cm. Histologically, there was marked loss of exocrine pancreatic tissue. Few remaining exocrine acini were surrounded by large numbers of adipocytes that also surrounded often isolated islets of Langerhans. The remaining exocrine tissue had moderate interstitial fibrosis and marked infiltrates of lymphocytes and plasma cells that surrounded and infiltrated exocrine acini and ducts. Pancreatic ducts had undergone necrosis and were replaced by severe fibrosis. Fibrotic ductual remnants were surrounded by sclerosis and extensive neocapillarization. Chronic pancreatitis, a common condition in older cats, was considered an unlikely diagnosis because of the selective nature of the tissue destruction that targeted exocrine cells and pancreatic ducts, spared endocrine cells and did not induce ductual proliferation or ectasia. There was no evidence of cholangiohepatitis or inflammatory bowel disease commonly associated with chronic pancreatitis. The lymphoplasmacytic nature of the pancreatitis and duct destruction as well as the young age supports the diagnosis of autoimmune pancreatitis, a very rare condition in cats.

VSP-63: Effects of environmental toxicants on inducible innate immunity using bovine tracheal epithelial cells as an in vitro model
Carmon Co, Laura Bourque, Stephen Raverty, Brandon Lillie, Pierre-Yves Daoust, Jeff Caswell

Pneumonia is a common cause for morbidity and mortality in stranded and free ranging cetaceans. The airway innate defense system can be hindered by multiple factors such as stress, concurrent infection, and environmental factors. The reduction in innate immunity allows opportunistic pathogens to cause disease. Cetaceans exposed to oil spills have higher prevalence of bacterial pneumonia, but the mechanisms are not known. This study explored the relationship between selected environmental contaminants and their effects on an innate immune defense - tracheal antimicrobial peptide (TAP). TAP is a β defensin whose expression is induced by immunostimulants and serves as a first line of defense against bacterial infection. We hypothesized that environmental contaminants would suppress the expression of TAP in vitro. These effects were modelled in bovine tracheal epithelial cells, which were exposed to increasing doses of crude oil extracts, or to benzo-a-pyrene, naphthalene or phenanthrene. A dose response and time course was carried out to characterize the effect of these toxicants. Crude oil and benzo-a-pyrene suppressed LPS-induced expression of TAP, but naphthalene and phenanthrene did not. This is the first characterization of the effect of oil contaminants on innate immune responses of tracheal epithelial cells. Findings may provide a better understanding of effects of contaminants on host immune responses and as a platform for assessing the risk of specific toxicants.

VSP-64: FELINE LEUKEMIA VIRUS IN THE FLORIDA PANTHERS (PUMA CONCOLOR CORYI): EVIDENCE FOR VIRAL PERSISTENCE

Elliott S. Chiu, Simona Kraberger, Mark Cunningham, Dave Onorato, Melody Roelke, Sue VandeWoude

In the early 2000s, endangered Florida panthers (Puma concolor coryi) experienced a feline leukemia virus (FeLV) outbreak, infecting 23 individuals and resulted in the death of at least three animals. Between 2010-2016, the Florida Fish and Wildlife Conservation Commission documented six additional FeLV-positive panthers by commercially available ELISA. We hypothesized that the two outbreaks were due to separate FeLV introductions. To determine the origin of ongoing cases, seven complete FeLV genomes were sequenced from a total of six panthers spanning both outbreaks. They were examined with full domestic cat FeLV genomes from Genbank. Phylogenetic analyses revealed 6 of 7 FeLV genomes form two sister clades together with domestic cat sequences from Genbank belonging to FeLV subgroup A. Two isolates from the contemporary outbreak are highly homologous to isolates from the original outbreak whereas novel FeLV-A's were also detected, indicating a recent FeLV introduction as well as continued circulation of an earlier introduced strain. Additionally, a full length FeLV-B subtype was isolated from one panther co-infected with FeLV-A. Lymphoid tissue proviral loads were greater in Florida panthers (1x10^5 – 3x10^7 copies per 10^6 cells) compared to reported domestic cat FeLV proviral loads. This report documents: (1) potential persistence of FeLV in Florida panthers since a documented outbreak in the early 2000s; (2) at least two FeLV strains circulating in the panther population; and (3) evidence for the first case of a recombinant FeLV (FeLV-B) transmitted horizontally, and (4) evidence of heightened susceptibility of panthers to FeLV infection compared to domestic cats.

VSP-65: Rhodococcus equi Pneumonia in an Adult Cat

Mitchell T. Caudill, Thomas E. Cecere, Stefanie M. DeMonaco, Katie M. Boes

A 9-year-old neutered male domestic long hair cat was referred to the Virginia-Maryland College of Veterinary Medicine Veterinary Teaching Hospital for pleural effusion and a possible thoracic mass. Physical exam revealed decreased compressibility of the cranial thorax and decreased bronchovesicular sounds. Thoracocentesis produced a septic inflammatory exudate (WBC, 46,000/L; TP = 6.2 g/dL) with short chains of coccobacilli within neutrophils and macrophages. Aerobic culture of the fluid isolated a pure culture of Rhodococcus equi. Due to lack of progress on antibiotic therapy, the cat was euthanized and a necropsy performed. Gross evaluation showed a focal, 2-cm umbilicated mass on the right caudal lung lobe, as well as numerous raised, multilobular, coalescing white to tan nodules covering the intercostal, diaphragmatic, visceral and parietal pleura and anterior mediastinum. The left lung lobe was also consolidated with atelectasis in the accessory lobe. Histologically, the focal mass was characterized by a cavitary, necrotic center and epithelioid macrophages, with pyogranulomatous inflammation multifocally found throughout other lung lobes and diffusely in the pleura and pericardium. The macrophages contained Gram-positive coccobacilli. While Rhodococcus equi infections have been reported in cats, primarily as a cause of pyogranulomatous skin lesions, pneumonic infections are extremely uncommon.

VSP-66: THE EFFECTS OF METFORMIN AND RESVERATROL ON CANINE HEMANGIOSARCOMA CELL LINES

Ariel K. Carlson, Karen Herrera, Beshay Zordoky, Marianne Grant, Leslie Sharkey

Canine hemangiosarcoma (HSA) is an aggressive malignant tumor of vascular origin oftentimes with poor patient prognosis. Metformin, a common human anti-diabetic drug, and resveratrol, a natural plant polyphenol, have been shown to possess strong anti-proliferative and/or pro-apoptotic effects in several human cancer cell lines and appear to be non-toxic in dogs. The effects of both metformin and resveratrol are thought to be mediated by activating the AMP-activated protein kinase (AMPK). We hypothesized that metformin and resveratrol have growth inhibitory effects on canine HSA cell lines. HSA cell lines Frog and DD-1 were incubated with increasing concentrations of metformin and resveratrol with or without doxorubicin, a common cytotoxic agent. The growth inhibitory effects of either metformin or resveratrol were assessed by MTT assay. Western blotting was used to assess the activation of AMPK and the expression of several pro-apoptotic proteins. In contrast to human cell line studies, metformin had a minimal inhibitory affect on the growth of Frog or DD-1 cells. Consistent with human cell line studies, resveratrol markedly inhibited both Frog and DD-1 cell growth. In both cell lines, resveratrol induced the cleavage of caspase 3 and increased the expression of phospho-p53, two markers of apoptosis. Resveratrol, but not metformin, increased the growth inhibitory effect of doxorubicin. These findings suggest that resveratrol may have growth inhibitory effects against canine HSA and could be a potential adjunct therapy to doxorubicin use in canine HSA patient treatment.

VSP-67: INTRATHORACIC NEUROBLASTOMA CAUSING COMPRESSIVE PULMONARY ATELECTASIS IN A SUGAR GLIDER (PETAURUS BREVICEPS)

Maggie Buller, Nicholas Crossland, Christine Higbie, Thomas Tully, Leslie Wilson

Background: A 1.5-year-old castrated male sugar glider presented with dyspnea and dysphagia of three days duration, appreciated as a coughing or gagging action during eating. Physical exam revealed the patient to be 7-8% dehydrated with open-mouthed breathing. Diagnostic testing was declined. The animal was hospitalized for four days, treated with oxygen supplementation, subcutaneous fluids, antibiotics, gastroprotectants, and nutritional support via gavage feeding. During hospitalization, the animal appeared interested in food but could not eat. Due to absence of improvement, euthanasia was elected. Method: Postmortem examination with histopathologic and immunohistochemical evaluation. Results: A 4 x 1.8 x 1.5 cm, multilobulated, pale-tan, expansile, firm mass was identified within the cranial mediastinum. The mass caudally displaced the heart and lungs resulting in compressive pulmonary atelectasis, with regional compression of the trachea, esophagus, and great cardiac vessels. Localized infiltration into the thoracic epaxial musculature and diaphragm with pulmonary metastasis were confirmed histologically. Neoplastic cells were characterized by primitive, small, round cells with hyperchromatic nuclei, scant cytoplasm (neuroblastic), arranged in nests and rows supported by stroma resembling nerve fibers. Neuroblastic cells and tumor stroma were S-100 positive. Tumor stroma was strongly positive for vimentin, while neuroblastic cells were negative. Cytokeratin AE1/AE3, GFAP, SMA, desmin, CD3, and CD20 were globally negative. Conclusions: The dyspnea and dysphagia observed in this case are attributed to the intrathoracic space occupying mass, consistent with peripheral neuroblastoma. While not uncommon in pediatric oncology, these tumors are rarely described in veterinary medicine and have not been previously described in sugar gliders.

VSP-69: DEVELOPMENT OF A PCR STRATEGY FOR SENSITIVE DETECTION OF PINWORMS AND FUR MITES IN LABORATORY RODENTS

Kelsey R Brown, Lucio Gama, Julie Watson

Pinworms and fur mites are common in research rodent colonies. Immunocompetent rodents show minimal clinical signs, but subtle changes in immunity and behavior can alter research parameters. Commonly utilized in research, immunocompromised rodents can incur significant effects. One limiting factor in elimination of these parasites is diagnostic methodology. Microscopy of individual cage samples is inconvenient and costly, and sentinels rarely contract fur mites via soiled bedding exposure. This project aims to develop an in-house diagnostic assay for more sensitive detection of pinworms and fur mites in laboratory rodents using quantitative PCR (qPCR) with rack air exhaust dust (AED) as samples. AED acts like a sentinel in that one can sample the entire cage rack with one test. DNA samples from Aspiculuris tetraptera and Syphacia spp. (pinworms) and Myocoptes musculinus and Myobia musculi (fur mites) were isolated and sequenced to develop qPCR TaqMan probes. AED was collected from rodent cages with known parasite infections to validate the isolation protocol and qPCR assay. TaqMan probes were developed for all parasites, but were not successfully standardized, thus future optimization studies are required. Using a SYBR Green protocol instead of probes, we detected pinworms and Myocoptes musculinus in AED from known positive cages at a magnitude of 100 - 1,000 copies per sample. Myobia musculi is not present within the rodent colonies of this institution, thus the qPCR assay could not be validated. We therefore demonstrate that AED assaying by SYBR Green qPCR detects pinworm infection and Myocoptes musculinus infestation in laboratory rodents.

VSP-69: EFFECTS OF 17-α-ESTRADIOL TREATMENT ON NATURALLY OCCURRING STIFLE OSTEOARTHRITIS IN AGED MICE

Channing A. Bancroft, Richard A. Miller, Richard F. Loeser, Cathy S. Carlson

17-α-estradiol (EST), a non-feminizing estrogen closely related to 17-β-estradiol, increased longevity in a natural aging study in male mice. EST has a low affinity for estrogen receptors, but may influence disease processes in multiple tissues. The objective of this study was to determine whether long-term treatment of male mice with EST influenced tissues in diarthrodial joints that are affected by aging. The left hind limbs from 12 control and 12 EST treated male mice (mean age 32 months, range 24-42 months) from a natural aging study were prepared and evaluated by an observer who was blinded to treatment. Measurements of articular cartilage, calcified cartilage, and subchondral bone thickness and area in the medial tibial plateau were taken using the OsteoMeasure bone histomorphometry system. The articular cartilage structure score (0-12) and other recognizable joint lesions, including chondrification and ossification of ligaments, osteophytes, and degenerative changes in patellae and menisci, were recorded. There was a striking range of degenerative changes in these joints, from nearly normal to end-stage osteoarthritis. No effect of treatment on lesion severity was observed and there were no significant relationships between lesion severity and age. Significant treatment effects were identified only for calcified cartilage thickness and width (lower in EST group vs. controls, p=0.03) and subchondral bone and thickness (higher in EST group vs. controls, p=0.04); no other comparisons were significant. Thus, although EST results in increased longevity in male mice, these results indicate that it has does not significantly affect articular cartilage lesions of osteoarthritis.

VSP-70: VACCINE ASSOCIATED OCULAR LESIONS IN A 1 YEAR OLD DACHSHUND

Lawrence C Apgar, Valerie A Schuster, Leandro Teixeira

A 1-year-old male Dachshund presented for bilateral blepharospasm, lateral strabismus, conjunctivitis, and inappetance four days after receiving a commercially available leptospirosis vaccine booster, administered 11 weeks after initial vaccination along with parvovirus and Bordetella vaccines. Despite treatment the patient developed severe periocular pain, acute blindness, bilateral exophthalmos, retinal detachment, panuveitis, scleritis, and corneal ulcers and was euthanized 3 weeks later. Eyes were collected, formalin-fixed and submitted for histopathology. Grossly there was bilateral retinal detachment and tan, firm tissue expanding the episclera. Microscopic lesions were bilateral and symmetrical. The inferior orbital connective tissue and adjacent episclera, sclera, and uvea were markedly expanded and partially replaced by necrosis and hemorrhage. Lymphocytes, macrophages, and neutrophils surrounded blood vessels. Vessel walls were infiltrated by lymphocytes and neutrophils and expanded by eosinophilic material (fibrinoid necrosis) and pyknotic debris. There was a secondary bilateral ulcerative and neutrophilic mycotic keratitis with intrastromal septated hyphae revealed by GMS staining. Immunohistochemistry revealed large numbers of perivascular CD3-positive-T-lymphocytes and IgG staining on endothelium of affected vessels. The chronology of the vaccination booster associated with the clinical signs and ocular lesions are consistent with a vaccine-associated adverse event. The presence of both IgG-immunostaining and predominantly T-lymphocyte infiltration suggests combined type III and IV hypersensitivity reaction likely induced by the second exposure to leptospirosis bacterin. According to the literature, young-adult, small-breed dogs (with an overrepresentation of Dachshunds) receiving multiple vaccines per visit are at higher risk for these reactions. This case highlights an atypical, predominantly ocular presentation, of a post-vaccinal adverse reaction.

VSP-71: DETERMINATION OF IMPORTANT CAUSES OF MORBIDITY AND MORTALITY IN CAPTIVE PSITTACINES SUBMITTED TO THE ONTARIO VETERINARY COLLEGE TEACHING HOSPITAL

Thisuri Eagalle, Nicole M. Nemeth, Hugues Beaufrere, Leonardo Susta

Background: Psittacines are increasingly common in households, aviaries and zoological collections. With the advancement of avian medicine, diagnostic evaluations (i.e., necropsy and biopsy) are increasingly important for early and accurate diagnoses and to predict disease outcomes. However, comprehensive information about disease conditions in birds, including pathological descriptions, is often limited.Objective: We retrospectively reviewed diagnostic data from psittacines submitted to the Ontario Veterinary College (OVC) to assess baseline prevalence and pathological features of the most commonly diagnosed diseases and conditions.Methods: Diagnostic data from psittacine carcasses (necropsy) and samples (biopsy) submitted to the OVC and Animal Health Laboratory from 1995-2015 were reviewed. Data included signalment and morphologic diagnoses, which were categorized and ranked to identify primary cause(s) of death. Results: Eighty-three species (n=1,149) were represented, with the African grey, cockatiel, and green-winged macaw as the most common. Infectious agents were most commonly identified as the primary cause of death (n=409; 35.6%), including 216 viral (52.8%; some suspect), 132 bacterial (32.3%), and 61 fungal (14.9%) infections. The most common respective etiologies were bornavirus (n=154), Mycobacterium spp. (n=18), and Aspergillus spp. (n=18). Neoplasia was diagnosed in 100 cases (8.7%), which most commonly involved the alimentary tract (n=18); squamous cell carcinoma was the most frequently diagnosed neoplasia (n=9).Conclusions: Better recognition and understanding of diseases that impact psittacines will aid clinicians and pathologists to formulate more effective differential diagnoses and targeted diagnostic procedures, ultimately leading to better treatment, management and prophylactic protocols to improve the health of captive psittacines in Ontario and elsewhere.

ACVP Late-Breaking Poster Session

LB-01: PYOMETRA IN AN ADULT MALE GERMAN SHEPHERD DOG

Kristin Vyhnal, Gary Block, S. Christopher C. Ralphs

A 6 year old male German Shepherd Dog was presented for lethargy, vomiting, and inappropriate urination. The dog was reported to be bilaterally cryptorchid. The patient demonstrated abdominal pain. Blood chemistries were normal, and a complete blood count revealed a white blood cell count of 28,000/microliter with a band neutrophilia. Urine sediment contained red and white blood cells and cocci. Ultrasound revealed moderate abdominal effusion and a possible retained right testicle that was mottled and irregular in appearance but well-encapsulated. In the left mid-abdomen, there was a hypoechoic, encapsulated mass caudal to the left kidney, which was contiguous with a fluid-filled structure that appeared to abut/originate from the cranial aspect of the prostate.

At surgery, a septic effusion was identified that appeared to originate from a perforation in a distended uterus. The uterus continued into the pelvic canal so that the caudal most extent could not be located. Grossly, gonads appeared consistent with an enlarged right testicle and atrophic left testicle. Histopathologic findings for both testicles were marked seminiferous tubule atrophy and aspermatogenesis, with a Sertoli cell tumor present within the right testicle. Histopathological findings of the uterus were severe subacute supperative endometritis and pyometra. Culture of the uterine contents was negative, though the patient had been on antibiotics for 24 hours prior to collection of the sample.

Presence of the uterus is consistent with persistent Mullerian duct syndrome. Production of sex steroids by the Sertoli cell tumor may have predisposed the patient to development of pyometra.

LB-02: POTENTIAL PATHOGENS DETECTED IN WILD TURKEYS (MELEAGRIS GALLOPAVO) IN ONTARIO, CANADA.

Amanda MacDonald, Claire Jardine, Jeff Bowman, Evelin Rejman, John Barta, Hugh Cai, Doug Campbell, Nicole Nemeth

Background: Wild and domestic turkeys (*Meleagris gallopavo*) are susceptible to similar pathogen profiles and commonly overlap geographically. Wild turkeys were successfully reintroduced to Ontario, Canada in 1984 after extirpation due to over-hunting, and have since increased in number and geographic range.

Objectives: This study aims to provide information on causes of morbidity and health status of wild turkeys, including surveys for potential pathogens that may circulate between wild and domestic turkeys.

Methods: A retrospective analysis of diagnostic data for wild turkeys over a 20-year period was conducted to assess causes of mortality in wild turkeys in Ontario. In addition, a prospective study surveyed for potential pathogens among hunter-harvested wild turkeys via bacteria culture, fecal float and PCR.

Results: Retrospective data revealed non-infectious mortality causes were common among Ontario wild turkeys (69.6%; n=56); 33.9% were diagnosed with emaciation, followed by trauma (19.6%). Postmortem examination of 152 hunter-harvested turkeys revealed adequate nutritional condition and minimal gross lesions. Samples from most turkeys tested positive for one or more *Mycoplasma* spp. (98.7%), *Eimeria* spp. (76.0%), and lymphoproliferative disease virus (LPDV; 66.4%). Eight (7.9%) LPDV-positive birds had splenomegaly; none had microscopically-evident lymphoid proliferation in liver, spleen, or bone marrow. Two (1.3%) turkeys had poxvirus-positive skin lesions. All turkeys tested negative for avian influenza viruses and *Salmonella*.

Conclusions: These results indicate that Ontario wild turkeys are subject to environmental (emaciation) and human-associated (trauma) causes of death. Further, these birds may serve as apparently healthy carriers of potential avian pathogens, although the source of infection remains unknown.

LB-03: A CRITICAL ROLE FOR CD44-OPN SIGNALING IN GASTRIC CARCINOGENESIS AT THE EPITHELIAL TRANSITIONAL ZONE

Dah-Jiun Fu, Andrea Flesken-Nikitin, Fouad Chouairi, Andrew D Miller, Alexander Y Nikitin

Gastric cancer is the third leading cause of cancer-related deaths worldwide. The incidence of the cancer arising from the epithelial transitional zone (TZ) between the esophagus and the stomach has been increasing over the past few decades. Unfortunately, the pathogenesis of TZ gastric cancer remains poorly understood. Recent studies have shown that many cancers arise from stem cell niches, because the self-renewal ability and longevity of stem cells allow them to accumulate enough genetic mutations required for the malignant transformation. The adult stem cells at the gastric TZ express a member of WNT-signaling pathway, leucine-rich repeat-containing G-protein coupled receptor 5 (LGR5). We have found that conditional knockout of the common tumor suppressor genes *Trp53* and *Rb1* in LGR5-expressing ($LGR5^+$) TZ stem cells leads to metastatic gastric cancer in all mice. At the contrary, only few benign neoplasms develop in the gastric pyloric region, which also contains $LGR5^+$ stem cells. As compared to the pyloric region, the normal gastric TZ, contains significantly larger fraction of highly proliferative $CD44^+Lgr5^-$ immature cells. The CD44 ligand, osteopontin (OPN), significantly facilitates gastric organoid formation by TZ but not pyloric epithelial cells. CRISPR/Cas9-mediated inactivation of the *Cd44* gene abrogates this effect. Taken together, our results suggest that the OPN-CD44 signaling may play a key role in the preferential malignant transformation of the gastric epithelium at transitional zone.

LB-04: BABESIA SPP. INFECTION IN A DOMESTIC CAT FROM BANDEIRANTES CITY, PARANÁ STATE, BRAZIL

Ellen S. Marquez, Luciane Holsback, Ademir Zacarias-Junior, Wanessa Blaschi, Camilla O.R. Alcalá, Angélica L. Sarmento, Stefano Sato, Hugo L. Abate, Odilon Vidotto

Background: A crossbred domestic cat, 5-years-old, castrated female was examined in the Veterinary Hospital from the State University of North Paraná with a one day history of anorexia and apathy. The animal lived in an urban area. When the cat was physically examined on the first time, it showed only apathy. Laboratory exams were solicited.

Methods: Blood samples from the domestic cat were obtained by jugular vein puncture, 3 mL of blood were collected on the 1st day and it was designated to the hemogram and biochemistry analysis. Seven days after the cat has been assisted at the hospital, the hemogram was repeated and the polymerase chain reaction (PCR) was solicited to hemoparasite diagnosis.

Results: The first hemogram revealed microcytic normochromic anemia, leukopenia, jaundiced plasma (2+), increased urea (45 mg/dL) and ALT (96U/L) serum levels, proteinuria (3+), bilirrubinuria (3+), urobilinogen (4+), USG >1050, pH 6,5 and bilirubin crystal 3/lpf. It was recommended treatment for Mycoplasmosis and return after 10 days. On the 7th day the cat got worse and came back to the Hospital. Physical examination revealed jaundice of mucous membrane and skin, weight loss, moderate dehydration, anorexia and apathy. The exams were repeated and demonstrated a normocytic normochromic anemia, jaundiced plasma (2+) and morphological analysis of erythrocytes in blood smear showed piroplasmas that could not be differentiated between *Cytauxzoon* spp. and *Babesia* spp. The PCR confirmed the presence of *Babesia* spp.

Conclusion: To our knowledge, this is the first report of *Babesia* spp. in domestic cat in Paraná State.

LB-05: OBESITY, EXPRESSION OF ADIPOCYTOKINES, AND MACROPHAGE INFILTRATION IN CANINE MAMMARY TUMORS

Ha-young Lim, Hyun-Woo Kim, Jong-il Shin, Byung-joon Seung, Jung-hyang Sur

Background: Obesity influences the development, progression and prognosis of human breast cancer and canine mammary cancer (CMC) but the precise underlying mechanism is not well-documented in the fields of either human or veterinary oncology.

Objective:The objective of this study was to examine the expression patterns of adipocytokines such as aromatase, leptin, leptin receptor (ObR), insulin-like growth factor-1 receptor (IFG-1R) and adiponectin in CNCs on the basis of the body condition score (BCS).

Methods:The expression of major adipocytokines, including leptin, adiponectin, and leptin receptor (ObR) in benign (n=28) and malignant (n=70) canine mammary tumors was investigated by immunohistochemistry and on the basis of the subject's body condition score (BCS). To evaluate the relationship between obesity and chronic inflammation of the mammary gland, macrophages infiltrating within and around tumoral areas were counted.

Results: The mean age of MC development was lower in overweight or obese dogs (9.0±1.8 years) than in lean dogs or optimal bodyweight (10.2±2.9 years), and the evidence of lymphatic invasion of carcinoma cells was found more frequently in overweight or obese group than in lean or optimal groups. Decreased adiponectin expression and increased macrophage numbers in overweight or obese subjects were significantly correlated with factors related to a poor prognosis, such as high histological grade and lymphatic invasion. Leptin expression was correlated with progesterone receptor status, and ObR expression was correlated with estrogen receptor status of MCs, regardless of BCS.

Conclusion:Overweight or obese status may play an important role in tumor progression and metastasis in CMCs.

LB-06: INVESTIGATING THE SUITABILITY OF A LABORATORY MOUSE MODEL TO STUDY THE PATHOGENESIS OF ABORTIFACIENT CAMPYLOBACTER JEJUNI

Victoria Lashley, Michael Yaeger, Orhan Sahin, Zuowei Wu, Ju Ji

Objective: The aim of this study was to assess whether pregnant mice represent a useful model to study the reproductive pathology of *C. jejuni* IA3902.

Methods: BALB/c (inbred) and CD-1 (outbred) mice, at 14 days of gestation, were inoculated orally and intraperitoneally (IP) with 1x10^9 CFU/ml of *C. jejuni* IA3902. Necropsy, microbial culture, histopathology, special stains and immunohistochemistry were performed.

Results: Following challenge, C. jejuni was recovered from the fetoplacental unit in 10 out of 10 (10/10) BALB/c and 10/10 CD-1 IP inoculated pregnant mice, and in 2/7 BALB/c and 3/8 CD-1 orally inoculated pregnant mice. Gross reproductive pathology included necrosuppurative placentitis, fetal resorption, intrauterine fetal death, dead pups and multifocal hepatitis. Histological changes consisted of a locally extensive neutrophilic and necrotizing placentitis with intralesional bacterial colonies of *C. jejuni;* ulcerative endometritis and random multifocal hepatitis with rare cholecystitis. Immunohistochemistry for the major outer membrane protein (MOMP) of *C. jejuni* revealed positive staining within trophoblasts at the periphery of the placental discs and extracellularly with invasion into the placental disc largely via the vascular network. The organism was particularly prevalent in locations in which neutral mucin, iron and L-fucose was detected.

Conclusion: *C jejuni* IA3902 has an affinity for the murine fetoplacental unit where it is capable of producing a necrotizing placentitis with positive microbial recovery after both IP and oral challenge in BALB/c and CD-1 pregnant mice. Similar to findings in infected guinea pigs, the organism appears trophic for neutral mucin, iron and L-fucose within the fetoplacental unit.

LB-07: FELINE LEUKOCYTE ADHESION DEFICIENCY CAUSED BY DELETION IN CD18 BETA-INTEGRIN

Suzanne M Pratt, Thomas R Bauer, Jr., Christina M Palena, Urs Giger

Backround: Leukocyte adhesion deficiency (LAD), an autosomal recessive immunodeficiency, has been described in humans, dogs, and cattle. We report on LAD in a cat, with characterization of the clinical features, leukocyte function abnormalities and molecular defect. The proband, a 5-year-old male domestic longhair cat, had a life-long history of infections, inflammation and granulomas. CBC showed severe leukocytosis (50,000-91,000/ul) due to mature neutrophilia.

Methods: Flow cytometry evaluated leukocyte surface adhesion proteins, CD18 and CD45. In vitro neutrophil adhesion and T-cell proliferation assays examined cell functions. DNA sequencing was used to identify any CD18 variants.

Results: The proband's neutrophils, compared to feline controls, did not express any CD18 on the cell surface as determined by two monoclonal antibodies, but had normal CD45 expression. Cell adhesion experiments demonstrated that adhesion of affected neutrophils was severely impaired with and without PMA activation. Proliferation assays showed that the proband's T-cells responded weakly to 1 pg SEA but was near normal at 100 pg SEA, suggesting a CD18-independent T-cell response. Based upon the study of CD18 beta-integrin cDNA, a deletion in exon 2 of CD18 beta-integrin, leading to a frameshift, alternate ATG, or splicing defect was found in the affected cat but not the healthy control cats and the published feline genome sequence.

Conclusion: Feline LAD due to CD18 deficiency exhibits similar features to LAD in other species. However, feline LAD seems clinically milder allowing (with treatment) for a longer life expectancy, possibly due to T-cell independent leukocyte activities.

Supported by NIH OD 010939 and 1ZIASC010384.

LB-08: EVALUATION OF JOURNAL CITATION INDICES AND BIBLIOMETRICS TO EVALUATE THE IMPACT OF VETERINARY PATHOLOGY

David K Meyerholz, Heather A Flaherty

Background: Journal selection is often based on the "impact" it has in the field. This is typically measured by bibliometrics data, journal impact factor (IF) being the most common metric. While IF data correlates to journal influence, its use is increasingly criticized. We asked whether evaluation of other citation indices or bibliometrics could also show impact by papers published in *Veterinary Pathology*.

Methods: We evaluated *Veterinary Pathology* papers from 2013/14 (n=225, assessed July 2016) and compared the average citation between indices, differences in types of citations, and evaluated different metrics over the past 9 years.

Results: Sage had significantly less and Scopus and Scholar significantly more citations than Web of Science (*P*

Conclusions: Given recent criticisms about IF and changes on the foreseeable horizon, it seems reasonable to consider diversifying evaluation metrics for the journal as well as identifying focused areas of excellence to highlight for potential authors.

LB-10: DEEP SEQUENCING ANALYSIS OF RNA IN THE BONE MARROW FOLLOWING PHARMACOLOGIC INHIBITION OF HISTONE DEACETYLASE 6 IN DISEASED NZB/W F1 MICE

Miranda D Vieson, Song Li, Xin M Luo, Alexander M Gojmerac, Christopher M Reilly

Histone deacetylase (HDAC) enzymes alter protein function by removing acetyl groups from lysine residues. Inhibition of HDAC enzymes decreased disease in various lupus mouse models. The breakdown of tolerance is implicated in the etiopathogenesis of systemic lupus erythematosus (SLE), and is related to altered regulatory checkpoints and proportions of B cells in developmental fractions within the bone marrow. Recently, HDAC6-selective inhibition reversed abnormal proportions of B cells in developmental fractions within the bone marrow of NZB/W mice. We hypothesized that selective HDAC6 inhibition (HDAC6i) will correct aberrant B cell development in the bone marrow. An HDAC6i was administered 5 days/week by intraperitoneal injection to NZB/W female mice in late-stage disease (35 weeks- of-age) for 2 weeks. There were no significant differences in the phenotypic development and differentiation of B cell progenitors in the bone marrow after HDAC6i compared to DMSO-injected controls. However, sequencing and analysis of total mRNA transcripts (RNA-seq) isolated from bone marrow cells showed differential expression of 849 protein coding sequences, 70% of which were down-regulated after HDAC6i treatment. Genes involved in regulation of B cell development and differentiation were among those down-regulated sequences. Additionally, we noted down-regulation of multiple sequences related to interferon (IFN) signaling, a signaling cascade that is often heightened in SLE. Several genes involved in IFN production and regulation are also associated with risk of developing SLE. Therefore, investigations are currently underway to define signaling molecules altered by HDAC6i that regulate B cell development and IFN signaling, and subsequently contribute to decreasing SLE pathogenesis.

LB-11: SUMMARY OF FREE-RANGING SNAKE SUBMISSIONS TO THE SOUTHEASTERN COOPERATIVE WILDLIFE DISEASE STUDY FROM 1977-present

Heather Fenton M. A. Fenton, Sarah Coker, Daniel Mead, Lisa Last, Michael Quist, Hannah Stanford, Jessica Gonynor-McGuire

Background: Snake fungal disease (SFD), a potentially fatal fungal infection caused by the fungus *Ophidiomyces ophiodiicola* (Oo), has been diagnosed throughout the eastern United States and Ontario. The disease has been associated with population declines in several species of rattlesnakes.

Methods: Snake cases (n=54) submitted to SCWDS from 1977 to 2015 were reviewed to investigate causes of morbidity and mortality in free-ranging snakes.

Results: Snake fungal disease was diagnosed in 16/54 (30%) of snake submissions. Diagnosis of SFD was based upon the presence of characteristic microscopic lesions and detection of the fungus either via fungal culture followed by conventional PCR or real-time PCR. In 11 cases, the detection of Oo was determined retrospectively through real-time PCR testing of archived frozen or paraffin-embedded tissue. Samples submitted for detection of Oo included: carcass, biopsy, scale clip, shed, and swabs. Snake submissions to SCWDS and diagnoses of infectious diseases, predominantly SFD, markedly increased since 2013. Cases of SFD were documented from GA, FL, WV, NC, KY, and TN in snake species with varied life histories. Non-SFD diagnoses from snake submissions included bacterial and fungal infections, amebiasis, cryptosporidiosis, trauma, and suspected intoxications.

Conclusion: Further investigation of causes of morbidity and mortality and ongoing population monitoring are needed to understand the impacts of this disease on free-ranging snake populations. It is unclear if the increase SFD cases is true reflection of an increased incidence of the disease in free-ranging snakes and not a bias related to increased awareness.

LB-12: FATAL ENCEPHALITIS CAUSED BY HERPES SIMPLEX VIRUS 1 IN A PREHENSILE-TAILED PORCUPINE (COENDOU PREHENSILIS)

Zachary V Mills, Nancy L Stedman

A 2.5 year old intact male Prehensile tailed porcupine developed progressive inappetance, lethargy, and muzzle pruritus unresponsive to antibiotics, antihistamines, or corticosteroids. Gross necropsy revealed severe ulceration and necrosis of the rostral muzzle and small white plaques in the oropharynx. Histopathology revealed multifocal neuronal necrosis in the frontal and temporal lobes and thalamus and mild perivascular infiltrates of lymphocytes and plasma cells in the neuropil and meninges. Neurons had basophilic to magenta nuclei and peripheral chromatin. In a peripheral ganglion, binucleated necrotic neurons were identified. Lesions were not present in trigeminal ganglia. Muzzle skin had multifocal epidermal necrosis and pustule formation progressing to severe necrosis and ulceration. Necrosis and pustules were also confirmed within oral mucosa. Herpes simplex virus 1 (HSV-1) DNA was amplified from frontal lobe tissue by polymerase chain reaction. Immunohistochemistry confirmed HSV-1 in nuclei and cytoplasm of neurons from the cerebral cortices, thalamus, trigeminal ganglia, and a peripheral ganglion and in intact and necrotic oral mucosal epithelium around pustules, but not within muzzle epidermis. HSV-1 is a reverse zoonosis that causes fatal encephalitis in new world primates and rabbits that have contact with infected humans. Fatal systemic HSV-1 infection also occurred in an African pygmy hedgehog. Natural HSV-1 infection has not been reported in animals of the order Rodentia, although rodents can be experimentally infected. To the authors' knowledge this is the first case of natural fatal HSV-1 encephalitis in a rodent. This porcupine had frequent contact with humans that may have introduced the infection.

LB-13: EXPERIMENTAL PENILE VEIN INJECTION IN MICE: MORPHOLOGIC FINDINGS.

Alessandra Piersigilli, Sara F Santagostino

Background: The dorsal penile vein is an accepted route for intra-venous injection in rodents, including mice, although it is not a commonly used method.

Objective: A series of iatrogenic lesions in the murine male urogenital tract associated with such route is presented.

Methods: NOD.Cg-*Prkdc^{scid}Il2rg^{tm1Wjl}*/SzJ (NSG) mice were experimentally injected into the dorsal penile vein with BV 173 cells and then underwent, following tumor engraftment, humoral and cell-mediated immunotherapy treatment, through a different route. Necropsy was performed to estimate the tumor burden and the efficacy of the anti-neoplastic treatment when the animals survived till the scheduled end points. A total of 4 male NSG mice aged between 3 and 6 months were found moribund or died unexpectedly during the course of the experiment and were submitted for complete necropsy.

Results: On gross examination, 1 mouse presented a partial prolapse of the penis. Moderate to marked necro-inflammatory, thrombotic and fibrotic changes of the corpora cavernosa and other soft tissues of the penis and pelvic region were found in all 4 animals. In 1 mouse, necrosis, vasculitis and neutrophilic inflammation extended to the epididymis and testes. The neoplastic infiltration was minimal to absent in the urogenital tract, but a heavy tumor burden was observed in most of abdominal and thoracic organs.

Conclusions: Due to the size and accessibility of the vessel, the dorsal penile vein is considered an easy injection route approved by the Institutional Animal Care and Use Committee (IACUC); however, challenges and complications must be considered, especially in immunocompromised animals.

LB-14: EVALUATION OF LACTATE DEHYDROGENASE AS TUMOR MARKER IN TRANSMISSIBLE VENERAL TUMOR

Ellen S Marquez, Stefano H. Sato, Mariana P. C. Silva, Celmira Calderón, Mariza F. R. Cruz, André A. S. Rosa, Mariana A Fernandes

Background: The transmissible venereal tumor (TVT) it is a neoplasia that affects canids. This tumor it is most commonly found in the genitalia of dogs. The transmission occurs when there is the social habit of the dogs as sniffing, licking and by coitus. Metastasis, even tough, it's rare, it can happen. The Lactate Dehydrogenase (LDH) converting lactate to pyruvate thereby producing cellular energy. In neoplastic cells, the metabolism is accelerated; therefore, the serum level of the LDH is higher. In cases of lymphoma and mammary tumor, LDH showed to be a good tumor marker.

Objective: The purpose of this study was to evaluate LDH as a tumor biomarker during the chemoterapy treatment for TVT.

Methods: There were evaluated blood sample from 12 animals, during the chemioterapic treatment for TVT. For biochemical analysis was used commercial kit for LDH, it was processed in automated equipment (Cobas®).

Results: After chemoterapy applied on 12 animals, there were serum level's LDH decrease on 6 dogs, along with tumor regression. The other remaining animals had unexpected events, such as concomitant deseases or pregnancy, which caused variances in LDH serum levels, showing that when the tumor returns to growth, this enzyme also increases their levels.

Conclusion: LDH can be used as a good tumor marker considering the decrease in serum levels according to tumor regression as the animals were treated for TVT.

LB-15: LENTICULAR METAPLASIA WITHIN MAMMALIAN GLOBES

Gillian C. Shaw, Felicia D. Duke, Richard R. Dubielzig, Leandro B. C. Teixeira

Lenticular metaplasia is the aberrant expression of a lenticular phenotype in non-lenticular cells. This phenomenon occurs in ocular tissues and has been reported in avian retinas that have sustained trauma; however, it has never been reported in mammalian species. The Comparative Ocular Pathology Laboratory of Wisconsin (COPLOW) database contains 16 cases of mammals with lenticular metaplasia occurring between 1983 and 2016. These include six horses, seven white tail deer, two dogs and a rabbit. Histologically, the lesions consist of one to several large round cells with abundant glassy brightly eosinophilic cytoplasm with a single nucleus, resembling a bladder cell from a cataractous lens. These cells were found within the retina (5/16), ciliary body (4/16), both retina and ciliary body (5/16) or within a tumor (2/16). These lesions accompanied other congenital, inflammatory or neoplastic ocular lesions. The pathogenesis of lenticular metaplasia is unknown; however, the lesions in this series suggest it may represent the aberrant acquisition of a lens epithelial phenotype by reprogrammed, undifferentiated or multipotent precursor cells stimulated by the various ocular lesions observed. This is the first report of spontaneous lenticular metaplasia in mammalian tissues.

LB-16: A Case of Blunt Force Trauma and Admissibility of Forensic Photographic Evidence at Jury Trial

Natalie W Fowlkes, Ken S Kim, Daniel B Paulsen, Joni Buoquoi

A 2-year-old, female pit bull was presented for postmortem examination with suspected blunt force trauma and acute onset of seizures. History also indicated the case was under investigation for abuse. Necropsy findings including open wounds and suffusive subcutaneous hemorrhage with comminuted fractures of the frontal and temporal bones and intracranial hemorrhage were compatible with blunt force trauma to the head. At trial, witness testimony indicated that the defendant had been seen beating the dog with a baseball bat. Necropsy findings, photographs, and interpretation were critical for corroboration of witness testimony at trial; the defendant was found guilty and sentenced to five years in prison with hard labor. In animal cruelty cases, forensic photography can have a major impact on outcome at trial. Special attention must be paid not only to accurate depiction of injuries, but also to producing images that are admissible in court, which is under the subjective discretion of the judge. Factors such as blood in the background or background distractions in general can be enough to cause a photograph to be omitted. Additionally, images may be omitted if the significance is not understood. Placing the most significant images within the text of the report may increase the likelihood that they will be shown to the jury. Herein, we describe a case of blunt force trauma and discuss the value of accurate and admissible forensic photographic evidence, with practical suggestions for the veterinary forensic pathologist in photographic documentation of lesions in abuse cases.

LB-17: SYSTEMIC COCCIDIOSIS CAUSING FULMINANT MORTALITY IN A COLONY OF WILD-CAUGHT EUROPEAN STARLINGS (STURNUS VULGARIS).

Jose G. Vilches-Moure, Stephen A. Felt

Disseminated disease caused by coccidia (systemic isosporosis, also known an atoxoplasmosis) is a common parasitic disease of passerine birds. Twenty-one European starlings (*Sturnus vulgaris*) were caught in California in 2016. The time between arrival and onset of clinical signs ranged from 2 days – 2 months, and signs included anorexia, dropping food, neurologic deficits, conjunctivitis and sudden death.

At the time of death, birds ranged from fledglings to adults (n=21), with most of the birds ranging from 3-6 weeks of age (n=10). Hematology revealed peripheral eosinophilia and lymphocytic intracytoplasmic inclusions in most, but not all, birds. Fecal evaluation revealed cysts measuring 39 x 39 to 39 x 44 um. Necropsy revealed decreased pectoral musculature, blepharedema, prominent keels, enlarged pale livers and spleens and thickened intestinal walls. Histology revealed florid lymphocytic inflammation with intracytoplasmic parasitic inclusions commonly affecting liver, spleen, conjunctiva, connective tissues surrounding the thyroid glands, and bursa of Fabricius. Samples collected for molecular screening of common avian pathogens were negative. Oral ponazuril treatment was initiated, and this improved the clinical symptoms in some, but not all, birds. Because of the variation in clinical manifestations, circulating parasitic load, intestinal parasitic burden and parasitic stages present, we query if these differences represent the true spectrum of lesions observed in infection with a single parasite, or if they represent infection with more than one (or different strains of) coccidian parasites.

LB-18: NEUROLYMPHOMATOSIS WITH MOLECULAR CLONALITY PROFILING IN TWO DOGS

Lauren Himmel, Michael Oglesbee, Rachel Cianciolo

Neurolymphomatosis (NL) describes a rare condition involving infiltration of peripheral nerves by malignant, neurotropic lymphoid cells in the setting of both non-Hodgkin lymphoma and leukemia in humans. In people and animals alike, NL in peripheral nerves, spinal nerve roots, cranial nerves, and neural plexuses has been reported. We describe two cases of NL in aged dogs. One dog presenting with masticatory myopathy was found to have NL of the cranial nerves in addition to disseminated lymphoma, while the second dog presenting with peripheral nervous deficits localized to one forelimb had NL of the brachial plexus nerves alone. A diagnosis of NL was reached based on cytologic features of malignancy assessed by routine H&E staining. In both cases, neoplastic lymphoid cells were characterized immunohistochemically using CD18, CD3, CD20, and Pax5. Subsequently, polymerase chain reaction antigen receptor rearrangement (PARR) assays were performed. Despite our high degree of confidence in the diagnosis of NL by light microscopy, ancillary techniques could not definitively elucidate the immunophenotype of the neoplastic populations. Although compressive and inflammatory neuropathies occur more commonly, CT or MRI findings should raise the differential diagnosis of NL before neuromuscular biopsy to increase the likelihood of achieving an accurate diagnosis. Additionally, NL must be considered as a differential diagnosis along with paraneoplastic polyneuropathy in patients with concurrently diagnosed lymphoma outside the nervous system. Continued reporting of NL cases in veterinary species may help establish diagnostic criteria and elucidate pathogenic mechanisms.

AUTHOR INDEX

AUTHOR INDEX

AUTHOR INDEX

AUTHOR INDEX

AUTHOR INDEX

AUTHOR INDEX

AUTHOR INDEX

AUTHOR INDEX

AUTHOR INDEX

AUTHOR INDEX

AUTHOR INDEX

AUTHOR INDEX

AUTHOR INDEX

AUTHOR INDEX